Sustainable Development and Tropical Agri-chains

Estelle Biénabe • Alain Rival • Denis Loeillet
Editors

Sustainable Development and Tropical Agri-chains

Springer

Editors
Estelle Biénabe
CIRAD – UMR Innovation
Montpellier, France

Alain Rival
CIRAD, DGD-RS
Jakarta, Indonesia

Denis Loeillet
CIRAD – UPR Geco
Montpellier, France

ISBN 978-94-024-1015-0   ISBN 978-94-024-1016-7 (eBook)
DOI 10.1007/978-94-024-1016-7

Library of Congress Control Number: 2017931691

Éditions Quæ, R10, 78026 Versailles cedex, France www.quae.com

Printed on acid-free paper

This Springer imprint is published by Springer Nature
The registered company is Springer Science+Business Media B.V.
The registered company address is: Van Godewijckstraat 30, 3311 GX Dordrecht, The Netherlands

General Foreword

Changes in tropical agricultural value chains profoundly modify, and at the same time reflect, ways of consuming and of producing, as well as the links between them. Be they related to opportunistic behaviours or to long-term strategies and practices, new products, new standards and new organizations affect our daily life. Meanwhile, they feed debates, controversies and polemics. Production and consumption models – as emerged strongly again on the occasion of the Paris climate change agreement – focus attention and are at the heart of concerns about the future of humankind and of the planet.

As part of this movement, the 17 Sustainable Development Goals, negotiated through a broad-based process over several years and approved in New York in September 2015, draw a new global framework for action. As part of a transformative agenda for people and the planet, they emphasize the links between production and consumption. And they stress the urgent need, given the complexity of the processes at stake, to reinforce the resilience of societies and ecosystems and to renew regulation processes.

Scientific books that deal with these links in relation to the high stakes of sustainable development are few and far between. This collection comes at the right time, detailing the multiple and complex configurations of both locally and globally organized agricultural value chains. It shows that these agri-chains are not only technical and economic but also social and political. Through reviewing a diversity of approaches to conceptualizing and understanding these agri-chains in relation to sustainable development and through exploring practical approaches that are taking shape to address current and upcoming challenges, it sheds light on the transformations underway, on the sometimes-implicit political and ideological choices at stake and on the options for action.

From a diversity of perspectives, the various authors draw the reader to the heart of the technical and institutional transformations that take place within agri-chains and confront us with the scientific and political challenges to come. The book provides the international community with a critical contribution by drawing lessons from multiple decades of experience, grounded in the original intellectual investment in the French *filière*[1] approaches. Furthermore, it shows that promising avenues exist, combining private and public actions, technical innovations (ecological intensification, diversity of ways of producing and deriving value, circular economy, etc.) and institutional ones (sustainability standards, payment for environmental services, etc.). In the face of today's high sustainability stakes – natural resource management, environmental and sanitary risks and inequalities – that manifest themselves at different scales, tropical agri-chains that connect the local and the global constitute powerful forms of leverage for inventing new economic and social forms. The analyses also show how the embeddedness of these institutional/economic forms in local territories matters. Local institutional contexts and capacities are critical for the emergence of complementarity between different political spaces – territories, states, international agreements, etc. – without which the impact of acting through agri-chains would be poor.

I highly commend this book, representing as it does an original and substantial contribution to guide the conception and implementation of agri-chain approaches as precursors to more sustainable models of development. Situated at the interface between science, policy and practice, this collaborative work constitutes an important milestone towards a research agenda that tackles some of the most serious global challenges of our time, in ways that remain grounded in the lives and concerns of people in diverse settings around the world.

Institute of Development Studies
University of Sussex, UK

Melissa Leach

[1]*Filière* is a French term, equivalent to some extent to value chain or to supply chain. For more details, see in particular Chap. 1.

Foreword by AFD

Can agricultural markets contribute in a meaningful way to sustainable development in the South?

If we stick solely to the endogenous laws of deregulated markets, the answer will likely have to be an emphatic 'no'. Analytical rigour today obliges us to admit that the arguments in favour of the self-regulation of markets, which have been around for 40 years, are not based on any solid reasoning, as Steve Keen wrote in *Debunking Economics*.[2] Historical experience indicates instead that chronic instability, highlighted by Hyman Minsky in the case of financial markets in the 1970s, has spread to the entire commercial sphere, including to agricultural markets. This instability is detrimental to any effective allocation of resources, and family farms and farmers are its first casualties.

Furthermore, we cannot ignore that some dynamics of growth of agricultural production, driven solely by technical and economic objectives, marginalize the least fortunate, exacerbate disparities and destroy or contaminate ecosystems, sometimes irreversibly. The techno-productivist paradigm can no longer lay claim to agricultural development as its sole preserve.

At the same time, socially, economically and environmentally sustainable trajectories have emerged in a few regions over the last decade and provide some hope for the future of agriculture. To be viable, these initiatives require a common vision, a project shared by all the actors of the agri-chains and territories. In other words, and this is a core idea of this book, it is at the meeting point of the agri-chain and the territory that the future of agriculture will largely be determined.

It is along a *filière* (or value chain) that added-value flows circulate between the many and diverse actors (farmers, processors, traders, consumers) who depend in one way or another on agricultural production. The *filière* also refers to the institutions (organizations, rules, customs, etc.) that private actors set up to deal

[2] Steve Keen, 2001. *Debunking Economics: The Naked Emperor of the Social Sciences*. Zed Books, New York.

and negotiate issues of common interest (quality, volume, costs and prices, etc.) amongst themselves and, whenever necessary, with public authorities (taxation, subsidies, infrastructure, etc.). Equitable and interprofessional governance, in conjunction with regulated market dynamics, allows, ideally, to come close to a fair remuneration for the value chain's actors. But this should in no way be confused with the accounts-centric idea of *fair value* that would supposedly emerge spontaneously from deregulated market interactions and with the political concept of *good governance* popularized in particular by the International Monetary Fund in the 1990s and which has too often been used as a cover for the outright dismantling of a country's public bodies and institutions.

Agreements entered into within a *filière*'s framework also allow partners to implement qualitative goals such as the identity of products, geographical origin, food safety, preservation of natural resources and more environmentally friendly agricultural practices. All these are characteristics that can become competitive factors and that, for some agri-chains, also act as commons. Governance here means dialogue, compromise (in the positive sense) and mutually beneficial cooperation.

However, agriculture cannot be simply reduced to agricultural products. On a given territory, the agricultural system shapes the landscape, allocates the uses of natural capital, underpins social relationships within communities and strongly structures the employment of both men and women.

Thus, the specialization on a product or conversely the diversification into different products, the growth or decline of a particular agri-chain, the expansion of cultivated areas or their intensification and the change in the mode of production such as recourse to irrigation all induce major changes at the territorial and community levels. It is the same with the development of services and local infrastructure which, through the value obtained from production, directly influence the economic and social development of the territories.

Therefore, the governance of rural territories is the second dimension which, together with that of *filière* or agri-chains, provides a framework within which sustainable agricultural development will have to take place, now and in the future. The allocation of the territory, its land and its water resources between agriculture, forests, pastures and habitats and the distribution of the responsibilities of protecting the land and rights to use or develop it between legal right holders and users are only possible by agreements between all local actors, irrespective of the agricultural production that the ecology of the territory concerned and the markets allow. Dialogue, compromise and cooperation are much more effective than the violence of market power relations. And, here too, the structure of the resources as commons – equidistant from the privatization of enclosures and state collectivization – probably provides the secret of human institutions respectful of the unique and fragile planetary ecosystem entrusted to us. This is a secret that humanity has successfully transmitted for millennia but which our contemporary modernity must relearn quickly if we do not want the current systematic destruction of biodiversity to render our planet inhabitable to all who live on it.

Given the present demographic challenges and those of the future, agriculture in the countries of the South has the greatest need for trajectories of sustainable prosperity. We hardly need reminding that, as of now, food security is not ensured for the generation that is about to be born in sub-Saharan Africa. But the heartening news is that such trajectories are already being invented. Their implementation requires the mobilization of economic actors at the intersection of a vertical approach of a *filière* or agri-chain and the horizontal dimension of territorial communities. Shaping the social space and the associated institutions, between agri-chains and territories, is a demanding task. It requires technical, environmental, economic and social innovations, at all scales, from the local to the global, that are made possible by partnerships between the research community and economic and institutional actors and between consumers in the North and farmers in the South. Is it not by learning to dialogue in this way and act with such audacity that humanity will succeed? It is this book's goal to help us all do so.

French Development Agency (AFD),
Paris, France

French National Centre for Scientific
Research (CNRS), Paris, France

Gaël Giraud

Acknowledgements

This book links the issues surrounding tropical agricultural, forestry and agrifood-chains[3] – with which CIRAD and AFD have been involved for many decades – to that of sustainable development, a more recent global concern. To enable readers to assess the relevance and richness of the many interlinked issues involved, we solicited the expertise of CIRAD researchers from across disciplines and of AFD officers to discuss different possible ways of presenting them. This book is the result of their deliberations and suggestions, with its various parts and facets reflecting and consolidating a diversity of contributions and viewpoints.

We thank the authors (see separate list of authors), reviewers and members of the steering committee and the editorial board. We also thank the publishing team for their unfailing support.

Reviewers: Daniel Barthélémy, Dominique Berry, Pascal Bonnet, Estelle Biénabe, Alain Billand, Patrick Caron, François Côte, Benoît Daviron, Pierre Fabre, Etienne Hainzelin, Jean-Louis Noyer, Sylvain Perret, Hervé Saint-Macary and Nadine Zakhia-Rozis.

Members of the steering committee: Didier Baillet, Estelle Biénabe, Dominique Berry, Alain Billand, Patrick Caron, Claire Jourdan-Ruf, Denis Loeillet, Florence Mouton, Sylvain Perret, Hervé Saint-Macary, Alain Rival, Jean-Michel Sourisseau and Emmanuel Torquebiau.

Members of the editorial board: Estelle Biénabe, Patrick Caron, Denis Loeillet, Alain Rival.

Members of Quæ Publishing team: Jean-Marc Barros, Joëlle Delbrayère and Claire Jourdan-Ruf.

Finally, we would also like to extend our thanks to Kim Agrawal for translating this book from the original French.

[3]Collectively referred to as agri-chains in this book.

Contents

List of Authors

Anseeuw Ward – CIRAD – UMR Art-dev – International Land Coalition – Rome, ward.anseeuw@cirad.fr

Bachelier Bruno – CIRAD – UPR Aida – Montpellier, bruno.bachelier@cirad.fr

Baillet Didier – AFD – Paris, bailletd@afd.fr

Bakry Frédéric – CIRAD – UMR Agap – Montpellier, frederic.bakry@cirad.fr

Bessou Cécile – CIRAD – UPR Performance of Tree Crop-Based Systems – Montpellier, cecile.bessou@cirad.fr

Biénabe Estelle – CIRAD – UMR Innovation – Montpellier, estelle.bienabe@cirad.fr

Pascal Bonnet – CIRAD – Environments and Societies Department – Montpellier, pascal.bonnet@cirad.fr

Bosc Pierre-Marie – CIRAD – UMR Moisa – Montpellier, pierre-marie.bosc@cirad.fr

Braconnier Serge – CIRAD – UMR Agap – Montpellier, serge.braconnier@cirad.fr

Briffaz Aurélien – CIRAD – UMR Qualisud, aurelien.briffaz@cirad.fr

Caron Patrick – CIRAD – DG-DRS – Montpellier, patrick.caron@cirad.fr

Cheyns Emmanuelle – CIRAD – Moisa – Montpellier, emmanuelle.cheyns@cirad.fr

Chopart Jean-Louis – CIRAD – UPR Aida – Roujol, Guadeloupe, jean-louis.chopart@cirad.fr

Cruz Jean-François – CIRAD – UMR Qualisud, jean-francois.cruz@cirad.fr

David-Benz Hélène – CIRAD – UMR Moisa – Montpellier, helene.david-benz@cirad.fr

Daviron Benoit – CIRAD – UMR Moisa – Montpellier, benoit.daviron@cirad.fr de Lapeyre de Bellaire Luc – CIRAD – UPR Geco – Montpellier, luc.de_lapeyre_de_bellaire@cirad.fr

Djama Marcel – CIRAD – UMR Moisa – Universiti Putra Malaysia – Kuala Lumpur, Malaysia marcel.djama@cirad.fr

Domaingue Robert – CIRAD – UMR Agap – Montpellier, robert.domaingue@cirad.fr

Dorel Marc – CIRAD – UPR Geco – Neufchâteau, Guadeloupe, marc.dorel@cirad.fr

Drigo Isabel – Nexus Socioambiental – São Paulo, Brazil, isabel.drigo@gmail.com

Ducastel Antoine – CIRAD – UMR Art-dev – Montpellier, antoine.ducastel@cirad.fr

Dury Sandrine – CIRAD – UMR Moisa – Montpellier, sandrine.dury@cirad.fr

Duteurtre Guillaume – CIRAD – UMR Selmet – Montpellier, guillaume.duteurtre@cirad.fr

Dutilly Cécile – CIRAD – UMR Selmet – Montpellier, celine.dutilly@cirad.fr

Ferré Thierry – CIRAD – UMR Innovation – Montpellier, thierry.ferre@cirad.fr

Fouilleux Ève – CIRAD – UMR Moisa – Montpellier, eve.fouilleux@cirad.fr

Fusillier Jean-Louis – CIRAD – UMR G-eau – Montpellier, jean-louis.fusillier@cirad.fr

Gay Frédéric – CIRAD UMR Eco&Sols, frederic.gay@cirad.fr

Gazull Laurent – CIRAD – UPR BSEF – Montpellier, laurent.gazull@cirad.fr

Gérardeaux Edward – CIRAD – UPR Aida – Montpellier, edward.gerardeaux@cirad.fr

Giraud Gaël – AFD – Paris, giraud@afd.fr

Goebel François-Régis – CIRAD – UPR Aida – Montpellier, francois-regis.goebel@cirad.fr

Goli Thicrry – CIRAD – UMR Qualisud, thierry.goli@cirad.fr

Guéneau Stéphane – CIRAD – UMR Moisa – SHIS – Brasília, Brazil, stephane.gueneau@cirad.fr

Guerrin François – CIRAD – UPR Recycling and Risk – Montpellier, francois.guerrin@cirad.fr

Guibert Hervé – CIRAD – UPR Aida – Montpellier, herve.guibert@cirad.fr

Goutard Flavie – CIRAD – UPR AGIRS, flavie.goutard@cirad.fr

Horry Jean-Pierre – CIRAD – UMR Agap – Montpellier, jean-pierre.horry@cirad.fr

Husson Olivier – CIRAD – UPR Aida – Africa Rice – Cotonou, Benin, olivier.husson@cirad.fr

Jagoret Patrick – CIRAD – UMR System – Montpellier, patrick.jagoret@cirad.fr

Jori Ferran – CIRAD – UPR AGIRS, ferran.jori@cirad.fr

Karsenty Alain – CIRAD – UPR BSEF – Montpellier, alain.karsenty@cirad.fr

Kwansirinapa Thanawong – Asian Institute of Technology – Bangkok, Thailand, kwansirinapa.saratis@ait.ac.th

Lançon Frédéric – CIRAD – UMR Art-dev – Montpellier, frederic.lancon@cirad.fr

Leach Melissa – Institute of Development Studies – University of Sussex, United Kingdom, m.leach@ids.ac.uk

Le Bellec Fabrice – CIRAD – UPR HORTSYS, fabrice.le_bellec@cirad.fr

Lecomte Philippe – CIRAD – UMR Selmet – Dakar, Senegal, philippe.lecomte@cirad.fr

Lecoq Jean-François – CIRAD – UMR Art-dev – UNA – Heredia, Costa Rica, jean-francois.le_coq@cirad.fr

Legile Anne – Parc National des Cévennes, Paris, anne.legile@cevennes-parcnational.fr

Lejars Caroline – CIRAD – UMR G-eau, caroline.lejars@cirad.fr

Lemeilleur Sylvaine – CIRAD – UMR Moisa – Montpellier, sylvaine.lemeilleur@cirad.fr

Lewicki Sylvie – CIRAD – UMR Agap – Montpellier, sylvie.lewicki_dhainaut@cirad.fr

Loeillet Denis – CIRAD – UPR Geco – Montpellier, denis.loeillet@cirad.fr

Lortal Sylvie – CIRAD – UMR Qualisud, sylvie.lortal@rennes.inra.fr

Losch Bruno – CIRAD – UMR Art-dev – Montpellier, bruno.losch@cirad.fr

Macombe Catherine – Irstea – Centre de Montpellier, Montpellie, Cedex, catherine.macombe@irstea.fr

Marisa Peyre – CIRAD – UPR Agirs – NIVR Vietnam – Hanoi, Vietnam, marisa.peyre@cirad.fr

Marnotte Pascal – CIRAD – UPR Aida – Montpellier, pascal.marnotte@cirad.fr

Martiné Jean-François – CIRAD – UPR Aida – Montpellier, jean-francois.martine@cirad.fr

Médoc Jean-Michel – CIRAD – UPR Recycling and Risk – LEMSAT – Dakar, Senegal, jean-michel.medoc@cirad.fr

Méot Jean-Michel – CIRAD – UMR Qualisud, jean-michel.meot@cirad.fr

Mestres Christian – CIRAD – UMR Qualisud, christian.mestres@cirad.fr

Moustier Paule – CIRAD – UMR Moisa – Montpellier, paule.moustier@cirad.fr

Mouton Florence – AFD – Paris, mouton@afd.fr

Noël Naomi – AFD – Paris, noeln@afd.fr

Nguyen Mai Huong – RUDEC-IPSARD – Hanoi, Vietnam, maihuong.nguyen@ipsard.gov.vn

Norfaryanti Kamaruddin – Universiti Putra Malaysia, INTROP – Serdang, Malaysia, faryanti@gmail.com

Paillat Jean-Marie – CIRAD – UPR Recycling and Risk – Montpellier, jean-marie.paillat@cirad.fr

Pallet Dominique – CIRAD – UMR Qualisud – Montpellier, dominique.pallet@cirad.fr

Pannier Emmanuel – EFEO – Hanoi, Vietnam, manuelpannier@yahoo.fr

Parrot Laurent – CIRAD – UPR Hortsys – Montpellier, laurent.parrot@cirad.fr

Perret Sylvain – CIRAD – Environments and Societies Department – Montpellier, sylvain.perret@cirad.fr

Piketty Marie-Gabrielle – CIRAD – UPR Green – Montpellier, marie-gabrielle.piketty@cirad.fr

Poccard-Chapuis René – CIRAD – UMR Selmet – EMBRAPA – Belém, Brazil, rene.poccard-chapuis@cirad.fr

Poser Christophe – CIRAD – UPR Aida – La Bretagne – Réunion, christophe.poser@cirad.fr

Pot David – CIRAD – UMR Agap – Montpellier, david.pot@cirad.fr

Queste Jérôme – CIRAD – UPR Green – FOFIFA-DRFP – Antananarivo, Madagascar, jerome.queste@cirad.fr

Rafflegeau Sylvain – CIRAD – UPR Performance of Tree Crop-Based Systems – Montpellier, sylvain.rafflegeau@cirad.fr

Rebuffel Pierre – CIRAD – UMR Innovation, pierre.rebuffel@cirad.fr

Risède Jean-Michel – CIRAD – UPR Geco, jean-michel.risede@cirad.fr

Rival Alain – CIRAD – DGD-RS – Jakarta, Indonesia, alain.rival@cirad.fr

Michel Rivier – CIRAD – UMR Qualisud, michel.rivier@cirad.fr

Roda Jean-Marc – CIRAD – UPR BioWooEB – Universiti Putra Malaisia, Malaysia, jean-marc.roda@cirad.fr

Sainte-Beuve Jérôme – CIRAD – UMR IATE – Montpellier, jerome.sainte-beuve@cirad.fr

Salgado Paulo – CIRAD – UMR Selmet – Antsirabe, Madagascar, paulo.salgado@cirad.fr

Sarraut-Woods Janine – International Consultant, Washington, United States, j.sarraut.woods@gmail.com

Sautier Denis – CIRAD – UMR Innovation – Montpellier, denis.sautier@cirad.fr

Scopel Éric – CIRAD – UPR Aida – Montpellier, eric.scopel@cirad.fr

Taunay Patrice – CIRAD – UMR Qualisud, patrice.thaunay@cirad.fr

Temple Ludovic – CIRAD – UMR Innovation – Montpellier, ludovic.temple@cirad.fr

This Patrice – CIRAD – UMR Agap – Montpellier, patrice.this@cirad.fr

Tillard Emmanuel – CIRAD – UMR Selmet – Ligne-Paradis, Réunion, emmanuel.tillard@cirad.fr

Toillier Aurélie – CIRAD – UMR Innovation, CEDRES, Université de Ouagadougou, Burkina-Faso, aurelie.toillier@cirad.fr

Tonneau Jean-Philippe – CIRAD – UMR Tetis – Montpellier, jean-philippe.tonneau@cirad.fr

Trouche Gilles – CIRAD – UMR Agap – Montpellier, gilles.trouche@cirad.fr

Vayssières Jonathan – CIRAD – UMR Selmet – LEMSAT – Dakar, Senegal, jonathan.vayssieres@cirad.fr

Vigne Mathieu – CIRAD – UMR Selmet, mathieu.vigne@cirad.fr

Wassenaar Tom – CIRAD – UPR Recycling and Risk – Montpellier, tom.wassenaar@cirad.fr

Zakhia-Rozis Nadine – CIRAD – Department Persyst – Montpellier, nadine.zakhia-rozis@cirad.fr

Research Units

Agap	Genetic Improvement and Adaptation of Mediterranean and Tropical Plants
Agirs	Animal and Integrated Risk Management
Aida	Agro-ecology and Sustainable Intensification of Annual Crops
Artdev	Actors, Resources, and Territories in Development
BioWooEB	Biomass, Wood, Energy, Bioproducts
BSEF	Tropical Forest Goods and Ecosystem Services
Eco&Sols	Functional Ecology & Bio-geochemistry of Soils & Agro-ecosystems
Geco	Ecological Functioning and Sustainable Management of Banana and Pineapple Cropping Systems
Green	Management of Renewable Resources and Environment
Hortsys	Agroecological Functioning and Performances of Horticultural Systems
IATE	Agropolymer Engineering and Emerging Technologies
Innovation	Innovation and Development in Agriculture and the Food Sector
MOISA	Markets, Organizations, Institutions, and Stakeholder Strategies
Qualisud	Integrated Approach to Food Quality
Recycling and Risks	Environmental Risk Related to Organic Waste Recycling
Selmet	Mediterranean and Tropical Livestock Systems
System	Tropical and Mediterranean Cropping Systems Functioning and Management
Tetis	Land, Environment, Remote Sensing and Spatial Information

List of Abbreviations

AEF	French Equatorial Africa.
AFD	French Development Agency.
CGIAR	Consultative Group on International Agricultural Research.
Cifor	Center for International Forestry Research.
CIRAD	French Agricultural Research Center for International Development.
CSR	Corporate Social and Environmental Responsibility.
EU	European Union.
FAO	Food and Agriculture Organization of the United Nations.
FSC	Forest Stewardship Council.
Icrisat	International Crops Research Institute for the Semi-Arid Tropics.
IEA	International Energy Agency.
Inra	French National Institute for Agricultural Research.
IPCC	Intergovernmental Panel on Climate Change.
Irad	Institute of Agricultural Research for Development (Cameroon).
Irstea	National Institute for Environmental Science and Research.
OECD	Organisation for Economic Co-operation and Development.
OIE	World Organisation for Animal Health.
RSPO	Roundtable on Sustainable Palm Oil.
RTRS	Roundtable on Responsible Soy.
WTO	World Trade Organization.

List of Text Boxes

Chapter 1
Introduction

Estelle Biénabe, Patrick Caron, Denis Loeillet, and Alain Rival

Agricultural systems across the world are being confronted by the opening up of trade and the intensification of exchanges at unprecedented scales and frequencies (Losch 2004; Robinson and Carson 2015). The past few decades have been marked by profound transformations of production systems brought about by the processes of liberalization and financialization of exchanges, globalization of economies, changes in the role of States, and supranational, regional and international recompositions. These processes pit various agricultural systems against one another in contexts of marked differences in competitiveness (Mazoyer and Roudiart 1997). They are exacerbated and amplified by the phenomena of concentration that the upstream and downstream sectors of globalized agricultural and agri-chains are experiencing. Indeed, these exert an increasing domination over farmers and impose conditions that many of them, in particular small producers, cannot fulfil. In addition, population growth, migrations, and inequalities in the distribution of rights and resources all potentially lead to imbalances that threaten the sustainability of these resources and can engender conflicts between social groups. Agricultural and agrifood dynamics are therefore closely and more and more critically linked to economic, social, and environmental issues. This book proposes to

E. Biénabe (✉)
CIRAD – UMR Innovation, F-34398 Montpellier, France
e-mail: estelle.bienabe@cirad.fr

P. Caron
CIRAD – DG-DRS, Montpellier, France
e-mail: patrick.caron@cirad.fr

D. Loeillet
CIRAD – UPR Geco, F-34398 Montpellier, France
e-mail: denis.loeillet@cirad.fr

A. Rival
CIRAD, DGD-RS, Jakarta, Indonesia
e-mail: alain.rival@cirad.fr

E. Biénabe et al. (eds.), *Sustainable Development and Tropical Agri-chains*,
DOI 10.1007/978-94-024-1016-7_1

address these issues of sustainable development, from the perspective of tropical agri-chains in their wider sense (agriculture, livestock farming, forestry).

1.1 *Filière*, Value Chain and Supply Chain: A Diversity of Concepts and Their Origins

The terms 'supply chain', 'value chain', and their French equivalent *filière* refer in a generic manner to forms of vertical[1] coordination between actors for the production and movement of a product until it is consumed. These forms of coordination encompass much more than purely commercial transactions.[2] As discussed in the first chapter of this book, the notion of *filière* originates from economics. It allows the sub-division of production systems, especially in agriculture, on the basis of vertical linkages between markets (Milhau 1954). Even though the term *filière* was not used until the second half of the twentieth century, it is now widely used in different perspectives. Supply chain, value chain or *filière* analysis, used in particular in the fields of economics and management, deals with the 'special relationships between certain economic agents from a technical viewpoint (input-output relationships), an organizational perspective (integration, specialization) and forms of exchanges' (Temple et al. 2011) with different focuses depending on the concerned discipline and research communities.

From the perspective of targeted-research organizations[3] such as CIRAD, the concept of *filière* corresponds to representations which underpin research in biological, technical, or social sciences undertaken in partnership with socio-economic operators involved in the production of a commodity or end product. Thus, the technical sciences conceive of the *filière* as a succession of material and decision-making processes, 'characterized by inputs (resources) and outputs (results) and by the nature of the flows for the exchange of goods and information' (Fabre et al. 1997). They rely on this concept in particular for the purpose of quality assurance and improvement. From the perspective of the social sciences, the *filière* represents 'sub-sets of agents of an economy linked by the movement of a product through its transformations and who undertake exchanges between themselves' (Griffon et al.

---

[1] The term 'vertical' is used to refer to relationships that exist between actors working at different stages in a product's production, processing, and routing. In contrast, horizontal relationships are those that exist between actors at the same stage and who thus undertake the same type of activities.

[2] At the economic level, the forms of specialization concerning commodities can be interpreted as social processes of division of labour that the expansion of markets make possible and which increase the efficiency of production systems, as Adam Smith (1776) had already noted. But, contrary to his thinking and as argued by institutional economics, the existence of a market usually implies a public action, a collective action, and forms of coordination that go beyond purely market transactions (Daviron and Gibbon 2002).

[3] Targeted research endeavours to produce knowledge that can help solve societal problems. It is thus action oriented.

2001, p. 48). It permits the analysis of actor strategies, the efficiency of resource use and exchanges, and forms of coordination (Fabre et al. 1997). The *filière* has been widely used at CIRAD in the technical and social sciences as a diagnostic tool, in particular in the 1980s, to analyze the performance of agri-chains and improve their competitiveness. *Filière* analysis has been a key tool to facilitate public decision making and consultation between stakeholders as part of interprofessional organizations, at the time also called 'quaternary' institutions (Griffon et al. 2001; Bourgeois and Herrera 2000). This type of analysis was thus used to define agricultural development policies for facilitating the transformation of production systems and the insertion of agriculture into the rest of the economy.

By aiming for a holistic concept of processes and actor systems, *filière* analyses promote interdisciplinary approaches. They are designed as 'a privileged means of dialogue between society and the research community and can thus help scientific activities align with the social demand' (Fabre et al. 1997). These approaches can sometimes refer to partnership-based structures set up by socio-economic operators involved in production at different stages of the agri-chain of which they are part, structures to which the researchers can also contribute. For some researchers, these approaches refer also to the form of partnership formed between them and the agri-chain's operators. Over the course of its involvement in these different *filière* approaches, CIRAD has developed, through privileged long-term partnerships with a wide range of actors of agricultural and agri-chains in several countries, a large methodological corpus for the biotechnical and institutional analysis of agri-chains and for supporting innovation. Enriched by its recent trajectory and the contribution of new disciplinary perspectives,[4] it has acquired an in-depth understanding of biological models, farming systems, and forms of organization of agri-chains in different countries.[5]

More recently, approaches based on global value chains, which is another way of analyzing agri-chains, have put the role of the actors governing the transactions within the value chain, called its 'pilots', at the core of the analysis. They have thus emphasized power relationships to ensure vertical coordination and the conflicting interests between the value chain's actors at the vertical level (distribution of added value) and the horizontal level (competition) (Daviron and Gibbon 2002). These approaches emerged in the 1990s in the context of trade liberalization and the internationalization of agrifood and distribution companies, which, through consolidation, established their domination over most agri-chains. Using these approaches, it is possible to analyze the factors that determine supply strategies and the quality standards that these firms establish, as well as the consequent impact on the insertion of farmers into the markets (incomes, effects of exclusion).

---

[4] Since its creation, CIRAD has been diversifying its research themes and the issues it addresses and has widened its geographic coverage. In the 2000s, it was restructured into theme-based scientific departments: biological systems, performance of tropical production and processing systems, and environments and societies.

[5] This understanding is based on the accumulation of individual and collective know-how, results, publications, and databases. It is a source of expertise recognized and sought after internationally.

1.2 Tropical Agri-Chains and Development

In many tropical regions, in particular in the very poorest countries, a small number of agricultural commodities and agri-chains play a major role in terms of employment, contributions to gross domestic product (GDP), access to monetary resources, land-use planning and organization of services, and food supply and security. As we will discuss below with reference to France's history, these specializations in a few key commodity chains are a legacy of the colonial system and the result of a long history from which tropical agricultural engineering originates. This history has a special meaning for CIRAD for it is in it that are found its own origins.

1.2.1 Agricultural Specialization and International Trade in Countries of the South: A Look Back at the History of Relations with France

Specializations at the country level that are based on the export of agricultural and forest products have their roots in European colonization with the invention of the plantation model. These specializations then continued post decolonization, especially in francophone Africa as part of the French policy of cooperation in rural development, shifting from the objective to ensure uninterrupted supplies to the metropole to that of the national development of the newly independent countries (Sarraut-Woods 1998; Daviron and Sarraut-Woods, this book, Chap. 3). Advances in transportation, methods of processing and preservation, and communication technologies, and the growth of cities marked the expansion of trade over long distances in the nineteenth century. To organize long-distance trading, futures market (or commodity exchanges), which fix reference prices and define standards for the traded commodities (cereals, coffee, etc.),[6] were progressively established in the early twentieth century. The futures markets were the result of collective action by companies, usually trading companies, specializing in specific products. By the same logic, standards were one reason for the establishment of public and professional organizations specializing in a particular product or commodity that also

[6] The historical analysis of the dynamics of specialization associated with long-distance trade presented in this section was offered by Benoit Daviron during this book's preparation. He clarified that 'The definition of agricultural commodities is a social process of which a key component is the definition of standards aimed at specifying the attributes under consideration and the methods of measuring them. Agricultural commodities, understood as goods with clearly defined characteristics, which are known by all participants of a market transaction, do not exist naturally in the wild. The fact that agricultural products are defined with reference to a particular species – in the biological sense – is in itself a situation that is not obvious. Natural rubber was, for a long time, a commercial product extracted from many different species of trees and lianas. *Hevea brasiliensis* prevailed only from the early decades of the 20th century. By ensuring the existence of a reference price for all transactions involving a product, the futures markets play a fundamental role in the very existence of commodities as tradeable goods.'

multiplied in the twentieth century. These organizations, whose history is analyzed by Daviron and Sarraut-Woods (Chap. 3), fulfil a number of functions: stabilization of prices, quality control, supply of inputs, investments in infrastructure, extension, research, etc. They were the main tool of a logic of intervention through public and semi-public projects based on a specialization between products and territories that was widespread in agriculture until the 1980s.

Bonneuil and Kleiche (1993) describe how the value addition process began to be organized at the beginning of the twentieth century on the basis of new research practices coupled with these patterns of exploitation of territories. Research stations specializing in specific plants were set up. This is how tropical agriculture began to be structured, through what these authors call the 'political project of a scientific organization of agriculture', which viewed genetic improvement as a source of progress. Starting in 1936, with IFC for rubber, several institutes were created for many different commodities: IFAC for fruit and citrus in 1942; CTFT, a state-owned corporation, for timber in 1949; IRCT for cotton in 1946; IMVE for veterinary medicine in 1948, which had its origins in the colonial veterinary medicine course created originally in 1921; IRHO for oils and oilseeds; and IFCC for stimulant plants in 1958. Later, at the time of decolonization, new institutions came into being: IRAT for food crops in 1960 and CEEMAT for agricultural machinery in 1962. The consolidation of these French institutes led to the creation of CIRAD (French Agricultural Research Center for International Development) in 1984.[7] At the international level, it was during this same period of the Cold War and of emerging concerns about population growth that the first international research centres in tropical agriculture were set up, which were later attached to the CGIAR (Consultative Group on International Agricultural Research).

1.2.2 *Renewal of* Filière *and Value Chain Approaches and New Public-Private Partnerships*

More generally, agricultural and rural development policies in the South have largely encouraged specializations by product throughout the twentieth century (Chaps. 3 and 4). These strategies of intervention, while they precede the invention of the terms *filière* and value chain, have been described as *filière* or value chain approaches by scientific communities and practitioners of development.[8] We are

[7] For more information on the history of these institutes, see de Padirac (1993), Surre (1993), Tourte et al. (1993), Catinot (1994), Charpentier (1995), and Bichat et al.. (1996).

[8] The aim of the institutes that preceded CIRAD was to organize research and expertise in order to improve production conditions in tropical agri-chains, increase production, and ensure continued supply of commodities and products. They worked primarily by providing support to technical and economic operators of the concerned agri-chains. At the time, this *filière* approach had the goal of identifying and gaining mastery over – or even eliminating – the obstacles to increased production. The scope of study was restricted: it ranged from the field to the initial processing stages and rarely dealt with consumption.

currently witnessing a renewal of these rationales of agri-chain intervention in agriculture by donor agencies and international development organizations in the form of new public-private partnerships. The wide dissemination of approaches based on 'global value chains', which are considered vectors of inclusive development, and the ever-increasing economic and commercial importance of large multinationals are driving this renewal. These interventions are perceived by donor agencies as a means of ensuring impact at scale, by counting on the capacity of these companies to act at different locations in order to exceed local success stories.

1.3 The Putting of Sustainable Development on the Agenda and Its Political Translations

The third millennium has started with the emergence of social criticism, which calls for a recast of the relationships between societies and their agricultural systems through a necessary rebuilding of public development aid and the emergence of new forms of global governance. Summits and international conventions, social forums, and demonstrations by civil society continue to multiply. They are largely marked by the appearance and increasing use of the expression 'sustainable development' in everyday life. The dissemination of sustainable development in the political and economic world and the strong support expressed for it by civil societies reflect a social reaction to changes that have been poorly controlled heretofore and the collective will to solve these pressing problems.

1.3.1 The Origins of Sustainable Development: The Environment as a Global Issue

The concept of sustainable development dates back to the emergence of environmental concerns in the 1960s and the political recognition at the international level of the environment's planetary boundaries. The report by Meadows et al. (1972), written for the Club of Rome, was instrumental in popularizing the idea that growth, uninterrupted since the end of the Second World War, cannot continue indefinitely in an uncontrolled demographic context and raised fears of a depletion of resources and an inability to continue maintaining the biosphere's main equilibriums. The first UN conference on the environment was held in Stockholm in 1972 at the instigation of networks of global experts attached to international institutions and major environmental NGOs. It marked the beginning of political awareness of the environmental crisis and led to a series of reports and conferences that attempted to address the relationship between the environment and development. In 1992, the Earth Summit in Rio de Janeiro, also under the auspices of the UN, formalized the definition of sustainable development proposed by Gro Harlem Brundtland in the report by the World Commission on Environment and Development (1987): 'Development which meets the needs of current generations without compromising the ability of future generations to meet their own needs.' The Earth Summit also led to three environmental conventions: on climate change (United Nations Framework Convention on Climate Change – UNFCCC); biodiversity (Convention on Biological Diversity – CBD); and desertification (UN Convention to Combat Desertification – UNCCD).

Sustainability requires at least four elements (Caron 2011). The first, clearly highlighted by the Brundtland report, concerns the taking into account of time, inter-generational issues, and societal responsibility in view of an uncertain future and the extended timeframe of major biophysical processes (climate, biodiversity, accumulation of pollutants, etc.). The second corresponds to notions of social equity and ethics. The third calls for a rethinking of relationships that societies have with nature, with a scope that goes well beyond the conservationist dimension. The last element is the recognition that public goods can be altered by individual and collective behaviour, especially by productive activities. This represents an implicit emphasis on the need to implement regulatory mechanisms at different organizational levels, from the local to the global. Development can only be deemed sustainable when it is, at the same time, economically efficient, socially equitable and environmentally sustainable, conditions that are reflected in the reference to its three pillars (economic, social and environmental).

1.3.2 International Summits and the Evolution of the Political Agenda of Sustainable Development

At the Johannesburg Summit in 2002, social concerns took centre stage, in a context in which the model of official development assistance that had prevailed since decolonization is being badly shaken. The desire to end extreme poverty was clearly affirmed at this summit, expressed through the Millennium Development Goals.[9] As decided at the United Nations Conference on Sustainable Development held in Rio de Janeiro in June 2012 (Rio+20), these goals were succeeded by the 17 Sustainable Development Goals.[10] Adopted recently, in September 2015, in New York at the United Nations Sustainable Development Summit, these 17 goals build upon those of Johannesburg. At the same time, they reflect a new break from the past. While they reaffirm the link between the tyranny of poverty and the need to

[9] The eight Millennium Development Goals were set in 2000 by the United Nations to reflect the desire to fight poverty and formed, until 2015, the common framework of global development that the UN's Member States committed to strive to achieve by 2015, http://www.un.org/millenniumgoals/.

[10] http://www.un.org/sustainabledevelopment/summit/

take care of the planet and preserve it, they also highlight the interdependence of changes taking place in different regions of the world and blur the differentiations between countries of the South and those of the North. They also reflect concerns about social inequalities that gained prominence after the global economic crisis of 2008 (Genevey et al. 2013) and consider their reduction as an issue separate from that of eradicating extreme poverty. Moreover, these goals were negotiated by government representatives and reflect a framework of overall consistency, which is based on the inseparability of all 17 of them. They are the result of 2 years of consultations and organization of working groups.[11] Compared to the Millennium Development Goals, the Sustainable Development Goals put greater emphasis on equity within and between nations, and on the management of natural resources and ecosystems as a key element of sustainable development. These 17 goals have been broken down into 169 targets with associated indicators. They thus provide a clear framework for measuring progress at a global level or at those of individual countries, but nevertheless run the risk of efforts that are segmented by goal (Voituriez 2013).

The Rio+20 Summit, held against the backdrop of a financial and economic crisis in the North, also promoted a new form of growth, called 'green', which considers the bioeconomy[12] and the circular economy (Chap. 16) as the vectors of a rational management of renewable resources and the environment. The green economy is based on synergies between environmental and economic policies and investments, in particular for energy and the climate, in order to ensure a transition towards sustainable development. Innovations that help an economy become 'low carbon' and eco-efficient (water, energy, natural resources) are considered factors that contribute to job creation and competitiveness. The economic importance of maintaining and restoring natural capital is reaffirmed, especially for poor populations. From this point of view, public intervention (policy and funding reform) is seen as a lever to redirect capital flows towards the green economy and to contribute to the development of innovative financing mechanisms, promoting in particular the 'greening' of financial markets (UNEP 2011).

[11] 'More than 5000 civil society organizations from 120 countries and the heads of hundreds of private companies were involved in the initial consultations', http://www.letemps.ch/opinions/2015/09/24/apres-objectifs-millenaire-nouveaux-objectifs-developpement-durable, retrieved 26 March 2016.

[12] The bioeconomy is defined by the OECD as a 'set of economic activities relating to the invention, development, production and use of biological products and processes', http://www.oecd.org/futures/bioeconomy/2030, retrieved 26 March 2016. The bioeconomy strives for an efficient and innovative use of biological resources and the substitution of fossil fuels by these resources. See also: http://ec.europa.eu/research/bioeconomy/pdf/201202_innovating_sustainable_growth_en.pdf, retrieved 26 March 2016.

1.4 Sustainable Development: Opportunities for Action at All Levels

Paradoxically, despite the risk, widely noted in the literature, of quickly becoming meaningless, the term 'sustainable development' is an indispensable reference for the political, economic, and scientific worlds and for civil society. As a societal project, it embraces the ideals of both social and environmental justice for present and future generations and of deliberation and participation by all in decision-making processes (Pestre 2011). Sustainable development formalizes a political and social awareness of a number of problems that have to be solved and represents a normative category, which alters the concept of development itself, long equated with economic growth. The challenge at the heart of sustainable development is that of the relationships between economic processes and the transformation of the biosphere. Mediation of these relationships takes place, on the one hand, through technology and, on the other, through a concern for social equity in conjunction with the environmental issue and the new shortages and limitations that this question forces us to think about (Godard and Hubert 2002).

The inclusion of sustainable development in the policy agenda at all levels, from the global to the local, and its adoption by different types of economic and social actors (civil society, large companies, local authorities), as well as the desire to establish decision-making processes that are more participatory, are grounds for hope for dialogue and improved control over ecological and social transformations. Thus, the regulatory requirements linked to globalization processes enshrine civil society and the various groups that represent it (NGOs, indigenous peoples, professional organizations, etc.) as actors and spokespersons of a new supra-national system that has to be constructed in partnership with intergovernmental organizations, through a transnational civil society that is gradually taking shape. The principles of corporate social responsibility (CSR) and the activities it encompasses have been widely disseminated.[13] Many codes of conduct and voluntary commitments to responsible behaviour have been put in place; they generate standards and communication mechanisms for data and reporting. 'New forms of coalitions bringing together States, multinationals, NGOs and scientists in specific programmes in which participants contribute legitimacy, financial resources, skills, guarantees, and local knowledge' are also emerging (Godard and Hommel 2005, p. 108–109). As noted in the next section, the agri-chains participate in a substantial way in these dynamics. A recent expression of this is the introduction of sustainability standards that promote good social and environmental practices and the roundtables associated with them, such as the Roundtable on Sustainable Palm Oil (RSPO).

[13] Corporate social responsibility programmes are considered by the European Commission as contributions by firms to sustainable development and form part of its strategy (http://eur-lex.europa.eu/legal-content/EN/TXT/?uri=URISERV:n26034, retrieved 26 March 2016).

The necessary rehabilitation of public action is also accompanied by a rediscovery of institutional spaces and forms required for its promotion, not only at the global or supra scale, but also at the national or infra levels. The notion of territory assumes here its full significance: it becomes a space to stimulate initiatives, to ensure the provision or preservation of public goods, and to address emerging issues (Caron 2011).

1.5 Agriculture, Tropical Agri-Chains, and Sustainable Development

Agriculture is the planetary activity that most impacts natural resources. It is also a major source of livelihood for the poorest people. As such, on the one hand, it is perceived as a threat to the environment (loss of biodiversity, pollution, resource depletion) and, on the other, it is at the very core of the issues of poverty reduction, job creation, food security, and the green economy, especially in Africa. In view of these issues, questions have resurfaced on the viability of technical agricultural systems and of production choices and on the evaluation and indicators of performance. Reflections are currently taking place on agroecology and ecological intensification in the South to help provide solutions. It is in this context that CIRAD undertakes activities for the design and evaluation of cropping and livestock systems and for the mobilization of biological cycles and a new ecological engineering.

As far as food and its supply are concerned, perceptions regarding quality and health are evolving, not only in the North, but also, and increasingly so, in the South. The growing importance being accorded to environmental and social concerns by consumers, which has led NGOs and the agrifood industry to come up with new labels and certifications (organic, fair trade, sustainability standards), has given birth to the term 'consumactors'. Environmental and development NGOs have taken advantage of this new movement to apply economic pressure at the level of the agri-chain. The agrifood firms see this as an opportunity to differentiate themselves and to respond to this social pressure, thereby protecting their reputations and promoting their brands. With the success of these strategies, deviations have emerged. Opportunistic behaviours are being reported, both at the social level and in the form of greenwashing.[14] These new quality standards are the most recent manifestation of the turn to quality taken by agri-chains in the 1990s. They are clearly part of the impetus towards sustainable development at the global level. They contribute significantly to the growing influence of private and public governance of the environment on agriculture. Debates on the use of payments for

[14] Greenwashing refers to business practices or organizational communications that claim to embrace or champion environmental arguments that are, in reality, deceptive or exaggerated in order to construct an environmentally responsible image.

environmental services for regulating ecosystem services related to agricultural activities are another expression of this movement. The growing number of instruments and standards aimed at reducing the environmental and social footprints of the agri-chains is being accompanied by increasingly sophisticated monitoring and evaluation systems.

Furthermore, new ways of deriving value from resources are leading to many innovations in agriculture and agro-industry. Examples include the recycling of agro-industrial and urban residues and waste. Policies for and investments in energy- and climate-related endeavours are contributing in a big way to large-scale non-food uses of agricultural biomass, especially in the form of biofuels. These new uses and the growing interest of the financial sector in agriculture are thus creating new markets and opportunities, but also exerting further pressure on resources, especially land and water.

1.6 The Goals of This Book and Its Structure

The new debates around sustainable development – and the hopes they arouse – not only call for a change in attitudes and practices, but also require us to reconsider the very elements and conceptual categories on which we base our actions. Research plays an important role here. Sustainable development is discussed in this book as both a framework of transformation of public and private practices and regulations that are associated with agri-chains, and as a framework for assessing the performances of these agri-chains and shaping transformations therein. In addition, planning a transition towards sustainable development, both globally and at various territorial levels, requires us to question the relevance and limitations of various forms of vertical coordination between the actors grouped together under the terms *filière*, 'value chain' or 'supply chain', and referred to generically in this book as agri-chain. This question contributes to a broader understanding of how taking sustainable development issues into account modifies agricultural and rural development models and how, in turn, changes in forms of organization of agri-chains impact sustainable development.

To understand these links between tropical agri-chain transformations and sustainable development, this book's different parts delve deeper into the topics briefly presented in this introduction: the agri-chains as vectors of development, as spaces of innovation, as objects of evaluation, and as arenas of regulation. The first part views agri-chains as vectors of development at various levels: those of the individual actors and the collective actions that these actors undertake, those of territories and public policy. It aims to draw useful lessons in a perspective of sustainable development. Continuing from where this introduction lets off, it discusses the long history of agri-chains from the perspective of the social sciences and development policies. The second and third parts present agri-chains as spaces of innovation for sustainable development. The second part discusses the transformation of practices and of research and partnership mechanisms in pursuit of innovation, as well as the evolution of research themes. It relies more specifically on the experience acquired by CIRAD in undertaking research in partnership. Varietal improvement, crop systems, and artisanal and industrial processing are topics that are explored in depth. In a perspective of the bioeconomy and energy transitions, the third part focuses on the increasingly new ways of using biomass and their impacts on agri-chains and research. The fourth part discusses agri-chains as spaces of evaluation and investigates evaluation methods that link agri-chains and sustainable development. It is based on life cycle assessment (LCA) methods. Finally, the fifth part considers agri-chains as spaces of regulation and discusses the scope and limitations of various public and private instruments for governing the sustainability of agri-chains and agriculture. These include sustainability standards and other environmental commitments by agro-industry (labels, zero deforestation approaches), corporate social responsibility, and payments for environmental services. The final chapter offers a prospective reflection on the financialization of agriculture, currently being manifested through new investments and new actor strategies, and its impacts on coordination within agri-chains and on sustainable development.

In addition, short chapters (Chaps. 7, 13, 17, and 21) at the ends of the first four parts discuss research agendas or present an original perspective on transversal issues: food security, the notion of waste, and evaluation methods.

The chapters that make up this book are the fruits of the experience and research in the biological, social, and technical sciences of nearly 80 researchers from CIRAD and its partners, and officials of the French Development Agency (AFD). The wide diversity of contributions collected here confirm the timeliness and usefulness of further exploration of the links between transformations of tropical agri-chains and sustainable development.

References

Bichat H, Chèze B, Gaury C, Herblot G, Le Moigne M (1996) Le Centre d'études et d'expérimentation du machinisme agricole tropical. 1962–1992. Cirad, Montpellier, 123 p

Bonneuil C, Kleiche M (1993) Du jardin d'essais colonial à la station expérimentale 1880–1930: éléments pour une histoire du Cirad. Cirad, coll. Autrefois l'agronomie, Montpellier, 107 p

Bourgeois R, Herrera D (2000) Actor-led change for efficient agrifood systems: handbook of the participatory actor-based CADIAC approach. CGPRT Centre-Cirad, Bogor, 135 p

Caron P (2011) Réconcilier agricultures et sociétés: dévoiler les territoires et repenser les limites, 3 volumes, habilitation à diriger des recherches. Université Paris Ouest Nanterre La Défense, Nanterre, 39 p., 254 p., 242 p

Catinot R (1994) Le Centre technique forestier tropical. 1916–1984. Cirad, coll. Autrefois l'agronomie, Montpellier, 175 p

Charpentier JM (1995) L'Institut de recherches sur les fruits et agrumes 1942–1984: de l'Ifac à l'Irfa. Cirad, coll. Autrefois l'agronomie, Paris, 117 p

Daviron B, Gibbon P (2002) Global commodity chains and African export agriculture. J Agrar Chang 22(2):137–161

de Padirac R (1993) L'Institut de recherches sur le caoutchouc. 1936–1984. Cirad, coll. Autrefois l'agronomie, Paris, 79 p

Fabre P, Bonnet P, Despréaux D, Freud C, Lassoudière A, Raoult-Wack AL (1997) Le concept de filière: un outil pour la recherche. Notes et documents, Cirad, 24. Cirad, Montpellier, 28 p
Genevey R, Pachaury RK, Tubiana L (dir.) (2013) Reducing inequalities: a sustainable development challenge. Teri Press, coll. A planet for life – Sustainable development in action, New Delhi, 271 p
Godard O, Hommel T (2005) Les multinationales, un enjeu stratégique pour l'environnement et le développement durable ? Revue internationale et stratégique 60(4):101–112
Godard O, Hubert B (2002) Le développement durable et la recherche scientifique à l'Inra. Rapport intermédiaire de mission pour la directrice générale de l'Inra, décembre. Inra, Versailles, 58 p
Griffon M, Boutonnet JP, Daviron B, Deybe D, Hanak-Freud E, Losch B, Moustier P, Ribier V, Bastianelli D, Ducrot R, Duteurtre G, Leverrier B, Ruf F (2001) Filières agroalimentaires en Afrique: comment rendre le marché plus efficace? Ministry of Foreign Affairs, Paris, 314 p
Losch B (2004) Debating the multifunctionality of agriculture: from trade negotiations to development policies by the South. J Agrar Chang 4(3):336–360
Mazoyer M, Roudiart L (1997) Histoire des agricultures du monde. Seuil, Paris, 546 p
Meadows D, Meadows D, Randers J, Behrens WW (1972) The limits to growth. Universe Books, New York
Milhau J (1954) Traité d'économie rurale. Presses universitaires de France, Paris
Pestre D (2011) Développement durable: anatomie d'une notion. Nat Sci Soc 19:31–39
Robinson GM, Carson DA (2015) Handbook on the globalisation of agriculture. Edward Elgar Publishing, Cheltenham/Northampton, 528 p
Sarraut-Woods J (1998) La politique française de coopération en matière de développement rural. Des certitudes à la dérive. In: Observatoire permanent de la coopération française, Rapport 1998 (OPCF). Karthala, Paris, pp 47–94
Smith A (1776) An inquiry into the nature and causes of the wealth of nations. Londres, Methuen & Co
Surre C (1993) L'Institut de recherches pour les huiles et oléagineux. 1942–1984. Cirad, coll. Autrefois l'agronomie, Paris, 100 p
Temple L, Lançon F, Palpacuer F, Paché G (2011) Actualisation du concept de filière dans l'agriculture et l'agroalimentaire. Écon Soc 33:1785–1797
Tourte R, Borget M, Celton J, Tufféry R (1993) L'Institut de recherches agronomiques tropicales et des cultures vivrières 1960–1984. Cirad, coll. Autrefois l'agronomie, Montpellier, tome 1, 230 p., tome 2, 217 p
UNEP (2011) Towards a green economy: pathways to sustainable development and poverty eradication. A synthesis for policymakers. UNEP, Nairobi. Retrieved 28 March 2016. http://www.unep.org/greeneconomy/Portals/88/documents/ger/GER_synthesis_en.pdf
Voituriez T (2013) À quoi servent les objectifs du développement durable. Working Papers 13/13. Iddri, Paris, 20 p
World Commission on Environment and Development (WCED) (1987) Our common future (Brundtland report). Oxford University Press, Oxford/New York, 383 p

Part I
The Agri-Chain as a Vector of Development?

Chapter 2
The Concept of *Filière* or Value Chain: An Analytical Framework for Development Policies and Strategies

Frédéric Lançon, Ludovic Temple, and Estelle Biénabe

Widely used from the 1960s to the 1980s in formulating industrial policies, the concept of *filière* was also called upon to accompany the agricultural sector's transformations in the North and to orient development policies in the South. The *filière* was thus constituted as an analytical framework for the vertical structuring of production systems, the production of intermediate and final goods, and forms of coordination between economic agents. It has been promoted historically by governments to define their intervention methods. If the use of the term *filière* in the francophone academic world was at the core of social science research programmes, in the English- and Spanish-speaking worlds the concept is manifested through variant terms (supply chain, value chain) that are relevant to different user communities. These terms are, in turn, increasingly finding use in French without always a clear explanation of whether this semantic shift corresponds to new analytical frameworks and new issues, linked to new forms of public and private actions.

We therefore define here these terms, their uses, and their analytical bases so that we can better analyze how agri-chains are transforming in the face of the challenges of sustainable development and examine how the changes brought about address these issues. To this end, we offer an historical exploration of the emergence of this concept in the literature and the diversification of contexts of its use. We analyze their effects on the evolution of the concept of *filière* not only into an analytical

F. Lançon (✉)
CIRAD – UMR Art-dev, Montpellier, France
e-mail: frederic.lancon@cirad.fr

L. Temple • E. Biénabe
CIRAD – UMR Innovation, Montpellier, France
e-mail: ludovic.temple@cirad.fr; estelle.bienabe@cirad.fr

E. Biénabe et al. (eds.), *Sustainable Development and Tropical Agri-chains*,
DOI 10.1007/978-94-024-1016-7_2

framework, but also into various operational tools that are now part of public and private modes of intervention.

2.1 The Concept of *Filière*

2.1.1 A Bridge Between Micro- and Macro-economics: A Disaggregation of the Economic System

Paradoxically, although the French term *filière* for value chain or supply chain is connected to the francophone academic world, it was used for the first time in 1947 at a convention of the Econometrica journal in the United States when B. Chait presented a theory of the relationships between agents to explain the flows connecting them (the output of an agent corresponding to an input for his client) and inventory levels needed to regulate the system. The initial objective was therefore to provide an analytical framework that overcame one of the main limits of the pure and perfect competition model by introducing inventories, and thus time, in the design of a general model to explain economic coordination. This initial conceptualization of a *filière* on the basis of an input/output system is based on separating out the economic system's components (agents). The objective is to define subsets of agents in which the internal relationships (interdependences) are stronger than this subset's relationships with the rest of the economy. Implications for the formulation of industrial policy or development strategies were not discussed at the time.

In France, the term was gradually defined in a very close sense through studies of industrial policies that used the tools of national accounts (Supply and Use tables; Aujac 1960). The concept of *filière* was then developed further to overcome the limits of an input/output analysis, whose division into sectors and branches was not sufficient for defining strategies to support economic activities. The *filière* was thought of here as a tool for subdividing the economic system on the basis of the relationships shown in the table of inter-industry trade, by highlighting the backward linkages effects that an agent has on the next. This concept was then applied to the implementation of development and industrialization policies in developing countries as a tool to orient the strategy of 'industrializing' industries, as argued by Perroux and Destanne de Bernis in particular. The *filière* approach also allows the creation of added value, and its distribution between agents or between production locations, to be analyzed.[1] It played an important role in the development of the 'effects method', a method for the economic evaluation of investment projects,

[1] Added value is conventionally understood as the wealth created by the difference between the value of goods and the cost of intermediate resources. It can be calculated at the level of each agent and then consolidated for an entire value chain. This value, or created value, then pays for the labour factor and 'finances' investment.

which was a key tool of French cooperation for 30 years, from the 1970s to the late 1990s (Chervel and Le Gall 1976).[2]

In addition to helping identify the relationships that structure an economic system, the *filière* was also a framework for representing the sequence of techniques that lead to the development of ever more complex products (for example, the graphite-gas *filière* or supply chain). The concept of *filière* gradually became more structured as analyses of power relationships and technological changes were incorporated into it. The *Revue d'économie industrielle* (Industrial Economics Journal) founded in 1977 was the main medium of this enrichment, with the *filière* becoming an analytical framework shared by a growing community of economists working in different fields of application, such as those of the agricultural and rural economy and of development economics (Hugon 1988).

2.1.2 The Production of Food Goods: A Privileged Topic of Study and Conceptualization

Authors such as Davis and Goldberg (1957) in the US have used the term 'agribusiness' to conceptualize the growing role of processes of transforming agricultural products into food and of their distribution, as well as the industrialization of agricultural inputs with the rise of the agrifood and agrochemical sectors after the Second World War. Their approach has highlighted the role of the agricultural sector as a supplier of raw materials and, therefore, the necessity of analyzing the links between agricultural production and upstream and downstream functions.[3] In France, the work of Milhau (1954) on the vertical links within agricultural markets (similar to the work of Chait) also led to the recognition of vertical economic coordination. This logic of organization into *filières* has been central to the development of agricultural policies that have structured the processes of specialization by product. A similar approach was proposed in the 1970s by the Food Research Institute at Stanford University. It advanced the concept of the commodity system to help understand the processes of transformation of agriculture in the South, which had started facing increasing competition from international markets in a rapidly globalizing agrifood system.

The analysis of connections between agricultural production and other sectors of the economy led to an increasing reliance on this concept. Thus, in France, a specific scientific programme in the agricultural and food sector was set up within

[2] The effects method aggregates the various added values, direct and indirect, and permits the assessment of the economic impact at a national level and not for the investor alone. Bridier and Michailof's (1995) practical guide for project analysis, which gives prominence to the effects method, was republished five times between 1980 and 1995.

[3] The term 'agribusiness' is used less and less to designate the *filière* or value chain. It refers today solely to firms investing in agrifood production and distribution.

INRA and national higher agricultural engineering schools. In the South, *filière* analyses played a major role in the formulation of national agricultural policies such as those for dealing with challenges of international competition (policy analysis matrix). In francophone Africa, they formed the basis for the organization of the agricultural sector.

The major agri-chains (livestock meat, milk, grapes and wine, etc.) have thus been characterized within the framework of policies of agricultural transformation under the impetus of the common agricultural policy. The *filière* or value chain, defined as the linked sequence of technical, logistical, and commercial operations necessary to produce and distribute a food or agro-industrial product, from production to consumption, thus became a useful reference for the research community as well as for policymakers. *Filière* and value chain economists then more explicitly mobilized systemic reference bases to account for intended consistencies of the proposed analytical division. *Filière* or value chain analysis thus seemed to have been accepted as a new concept able to take into account intermediate categories (mesoeconomics) that structure the coordination of economic agents between the micro- and macro-economic dimensions, thus breaking with the restrictive assumptions of neo-classical economics.

The food system was then defined in agriculture, in complement to the concept of *filière* or value chain, as the set of all food chains that link producers to consumers (Malassis 1979). In developing countries, these two concepts were used to analyze the performances of different export crops. They were also subsequently used in studies on food supply channels to cities (Baris and Couty 1981). Due to the way CIRAD used to be organized – by broad commodity chains – this work was further advanced in this context by many of its researchers and those of its partners (Griffon 1989; Fabre et al. 1997) for analyzing tropical export chains (coffee, cocoa, oilseeds, cotton, banana, etc.) and food supply to urban markets (rice, maize, plantain, vegetables, milk, meat, etc.). During the severe economic crisis in the African countries of the CFA franc zone in the early 1990s, several studies on the competitiveness of agri-chains were carried out at the behest of or by the French government (for example, Jouve and de Milly 1990; Daviron and Fousse 1993; Ministry of Cooperation 1994). French public aid contributed to the dissemination of studies of agri-chains at the FAO and the European Commission, and remained focused predominantly on the 'efficiency' of agri-chains (Griffon et al. 2001). Although starting in the 2000s, tools for analyzing the place of agricultural systems of the South in social and economic dynamics became widely diversified, CIRAD continued to use the *filière* as a reference (Temple and Lançon 2008; Temple et al. 2011).

It is not by chance that the concept of *filière* or value chain acquired a predominant role in analyzing and defining public policies for the integration of agricultural and livestock production into the economy. Historically, the processing of agricultural products into food is based on a small number of stages. The processes involved consist primarily of extraction and sorting. Very rarely are combination and assembly also part of the processing; these steps are usually left to the consumer or the cook at the time of preparing the dishes – though in recent decades the agrifood industry has also started producing ready-to-eat products. Even if the use of this concept is not exclusive to the food sector and extends to some industrial sectors (textiles at the beginning of the century and then energy, etc.), agrifood processes are especially well-suited to a value chain or supply chain analysis since the initial agricultural product used serves as a frame at every stage. In contrast, manufacturing of industrial products by assembling components is less conducive to this method of analysis. The value chain is also an important instrument for analysis and a method of intervention in the South because it can be used in contexts of limited or incomplete information. It also gained legitimacy through the strong polarization and dependence of the dynamics of development on a few agro-industrial chains in which the processes of accumulation concentrate (Hugon 1985).

This value chain approach based on a combination of the various functions necessary to produce the final product has been gradually revised with new analytical references that reflect the growing importance of the globalization of production systems, a process in which companies, initially part of industrial groups and now increasingly of financial ones, occupy a prominent place. Thus, in the work of Bandt, and its extension into the contexts of the countries of the South (Hugon 1994), the concept of structuring an economy in terms of value chains has been advanced, with the theory of regulation and the changes in modes of accumulation in countries of the North and the South as frames of reference. The economics of conventions is also mobilized, in particular to clarify the role of quality standards in the control exerted over the value chains by the companies that define them (Ponte and Gibbon 2005).

2.2 The Value Chain: An Instrument for Analyzing Business Strategies in a Context of Globalized Competition

Major technological advances in long-distance transportation, the liberalization and financialization of trade, and the increasing mobility of non-tangible goods such as information intensified competition in the 1980s in the industrialized economies and the expansion of international markets to new products. Value chain analyses were frequently called upon at the time to determine and assess the determinants of competitiveness of industrial as well as agro-industrial companies. The performances of these companies were no longer determined solely by the cost and productivity of factors of production as argued by the standard theory of comparative advantage. They also depended on the ability of these companies to leverage a set of external resources, to create a framework for institutionalized coordination between economic agents, notably through conventions of quality, and to position themselves at strategic points in production systems (Porter 1985).

The need to consider the configuration of value chains not solely as a process of division and specialization of productive tasks, but also as arising out of the strategies of companies to manage their position in these systems had already been addressed in the methodologies for analyzing *filières*. Nevertheless, it was not until the 1990s that companies began to be widely considered as entry points for analysis.[4] This development arose from two different strands of thought. The first belongs primarily to the management sciences and shows how a company's competitiveness is determined by its position and its relationships with other actors of the same production system. The second strives to understand how economic globalization leads to a reorganization of production systems at the planetary scale. Both unite in confirming the importance of the strategies of agro-industrial companies in the governance of globalized value chains.

2.2.1 Value and Supply Chain: A Strategic Framework for the Competitiveness of Companies

If sociologists of development questioned very early on the global logic of organization of production chains as a way to capture value, they did so with the aim of explaining the inequalities of development. Other authors, however, conceptualized value chains as a tool for optimizing the strategy of firms in a rationale of supporting decision-making processes. The value chain then designates the boundary within which the firm optimizes its relationships, not only with upstream and downstream actors but with all the services it mobilizes to become and remain competitive (research and innovation, maintenance, transportation, training, etc.). The concept of the value chain was thus initially conceived of as a management tool. The notion of value is here understood in a wider sense than that of added value of *filière* analysis; it is thus often associated, in an *ex ante* situation, to the difference between the price the consumer is willing to pay and the producer's reserve price.[5]

The management sciences also offer a tool similar to that of value chain but more operational in nature: the supply chain. The supply chain, formalized as far back as the 1950s (Forrester 1958) and which then gave rise to the supply chain management approach (Harland 1996), focuses on the efficient flow of goods and services between a company, its suppliers, and customers in order to coordinate production chains that involve several industrial entities. This analytic framework is an application of the concerns first advanced by Chait (1949) to conceptualize inventories and time in a framework of pure and perfect competition. It uses the tools of information and of network theories, with its applications multiplying as new information and communication technologies develop (Omta et al. 2008). It is the counterpart of just-in-time production methods, i.e., business-to-business computer

[4] A distinction is proposed between nation-based value or supply chain and company-based value or supply chain (Jacquemin and Rainelli 1984).

[5] The reserve price is the floor price below which the producer will not sell.

applications aimed at reducing inventories and ensuring efficient coordination of multiple production-related decision making.

2.2.2 From National Production Systems to Transnational Production Systems: The Global Value Chains

As mentioned above, another approach was proposed for an improved understanding of the internationalization of business activities, which accelerated in the 1980s with the advent of neoliberal policies that promoted free trade and the financialization of the economy and with the emergence of new countries and the associated restructuring of the international division of labour. This change in ambit within which the companies defined their strategy became visible in the 1970s when large industrial groups began reorganizing themselves. Their international departments were eliminated and replaced by the grouping of their various business activities by major world regions according to available resources (raw material, technology, other production factors) or end-product demands (Michalet 1976). This transformation was theorized by Gereffi and Korzeniwicz (1994), first under the concept of 'global commodity chain', then as the 'global value chain' (Gereffi et al. 2005). The concept of the global value chain refers to the same subdivision of production systems as in *filière* analysis but it highlights the role of 'driver' in the chain who have the ability (in the form of technology, financial capital, network) to govern this process. Global value chains differentiate themselves depending on the link in which the governance capacity resides, upstream or downstream, according to the types of more or less sophisticated products manufactured therein. As far as public policies are concerned, if globalization calls national industrial policies into question, the conceptual framework provided by the global value chain allows governments to target their actions (on the type of value chain or level of intervention within the value chain) in order to anchor global value chains in national territories and induce backward linkages on their development (Palpacuer et al. 2005).

Agriculture occupies a paradoxical position in this globalization process. Colonial trade in tropical products (coffee, tea, cocoa, sugar, cotton, banana, etc.), which combined production locations in the South with those of processing and value addition in the North, can be considered the first forms of global value chains. However, their rationale of accumulation was limited to the ambit of the colonial empires. The current phenomenon of globalization of agri-chains is more recent than that of the industrial sector. Globalization has really come into its own since the 2000s; it corresponds to a process of dissemination of food consumption patterns around the world (processed cereals, chicken, vegetable fats, dairy products, chocolate) and of domination by internationally renowned brands, whether in the supply of upstream inputs (seeds) or the downstream distribution of finished products (brands, ready meals, fast food restaurant chains). 'The supermarket revolution', as T. Reardon called it, is one manifestation of this dynamic, especially in emerging countries (Reardon et al. 2003). It is resulting in a restructuring of domestic forms of distribution, based in varying degrees on the development of specialized wholesalers, the consolidation of logistics and distribution centres, contracts with preferred agricultural suppliers, and private standards (Reardon and Berdegué 2002). By modifying the relationships between producers and the downstream parts of the value chain, this dynamic has generated substantial effects of exclusion for small producers (Weatherspoon and Reardon 2003). Another manifestation is the emergence of globalized companies in the South, for example in the production of instant rice noodles which, though they originated in Southeast Asia, are now being widely distributed in Africa.

2.2.3 An Integrated Conceptual Framework of the Value Chain, a Predominant Reference for the Development of Rural Areas

Since the early years of this century, the scientific community and international institutions (for example, the World Bank) have most frequently relied on the concept of the value chain for analyzing relationships between actors in a production system. The business strategies approach dominates and is enriched and articulated by transaction cost theory and its qualification of modes of coordination (market, contract, and hierarchy) (Gereffi et al. 2005).

The value chain serves as a reference for strategies of development in the South, especially those pertaining to agriculture, formulated a decade ago by international institutions (Werner et al. 2014). This reference base renews the initial concept of *filière* by introducing the study of governance mechanisms, while retaining the methodological fundamentals. Value chain developments offering stable outlets for farmers, particularly the smaller ones amongst them, are considered a preferred vehicle for inclusive growth, resulting in a permanent reduction of rural poverty (Vermeulen et al. 2008). These strategies have been formulated in response to the limited successes in alleviating poverty, if not the counterproductive consequences, of the liberalization policies of the 1980s (Casadella et al. 2015). These policies were based on the strong assumption that a spontaneous development of all-inclusive markets would connect rural producers to the globalizing economy, an assumption belied by 20 years of increased differentiation.

The value chain is understood as a framework for constructing markets that allows the formulation of integration strategies consistent with a market economy, and hence as a framework for defining interventions in pursuit of development. The intervention of the State is acknowledged in producing an institutional framework for market functioning, but it remains limited. This context explains why value chain R&D programmes assign significant importance to forms of coordination such as contracts, considered a controlled form of trade, which overcomes the disadvantages of trade outside a formal framework (asymmetrical position, risk, uncertainty, etc.). By the same logic, studying the effects of standardization and codification of exchanges and trade is necessary in order to determine the steps to take so that rural producers can meet the demands of the dominant actors of the agrifood systems (agro-industries, distribution centres, consumers, etc.) (see for example, Dolan and Humphrey 2000). The same causes underpin the emphasis on the development of market information systems (bids, availability, etc.).

The institutionalization of value chains as a development tool is widely promoted as a way to link the macro-economic level, in the form of investments and international trade, with development issues at the micro-economic level, in the form of improved livelihoods. But it is also denounced as a way for dominant actors to push and perpetuate the neoliberal development agenda (Neilson 2014) and, in the agricultural domain specifically, as a new technology of economic and ecological power for the appropriation of value produced by small farmers (MacMichael 2013).

2.3 Conclusion: Incorporating Sustainable Development Issues in the Concepts of *Filière* and Value Chain

While the various different conceptualizations of the *filière* show a close similarity in the way they represent it – a system of interdependent actors fulfilling complementary functions for the manufacture of a product or a group of technologically homogeneous products –, each relies on a different perspective to analyze these interdependencies. Some lay emphasis on functional interdependencies in terms of inputs, outputs, and management of flows, others on forms of coordination between the actors. These distinctions reflect the issues that these analytical frameworks are meant to address, the scale and the scope of the analysis, and, finally, the roles assigned to public policies, and forms given to them, for addressing these issues.

Since *filière* and value chain approaches have been used to structure the productive dynamics of agricultural specialization (territorial or sectoral), whose environmental and social impacts continue to be controversial, they have come in for criticism for their impacts on the sustainability of development. Indeed, as Bolwig et al. (2010) emphasize, analyses of value chains only partially address the impacts of these value chains' activities on poverty, inequalities (including gender issues), food security, and the environment. As far as social impacts are concerned, these analyses are generally limited to opportunities of generation of income without a consideration of the exposure of poor populations to various risks. Furthermore, they take the environment into account only through the role of quality standards.

Indeed, the issue of sustainable development is historically quite alien to these analytical frameworks. This explains the use, since the 1970s, of life cycle assessment (Boustead 1996) to develop a method to evaluate environmental externalities and then the material benefits of recycling materials. This method has also been used to construct indicators for the consumption of material and energy in order to develop environmental standards or to meet them. In methodological terms, life cycle assessment is very similar to *filière* analysis. It favours an approach by technical functions, with an evaluation system in physical units on one side and in monetary values on the other. Nevertheless, life cycle assessments remain very focused on the technical goal of measuring environmental effects and developing environmental standards. The difficulties of identifying the social impacts and dynamics that govern the coordination of actors and technological choices make life cycle assessments useful but imperfect tools for determining the sustainability of production systems and for examining the relationships between a value chain and sustainable development.

Without getting into a discussion here on the definition of 'sustainable development', we can note that attempts to analyze sustainability are confronted by the challenges of scale and the feasibility of the methods of analysis. In its most comprehensive and rigorous sense, the sustainability of a human activity can be really appreciated only by taking into account its interaction with the entire biophysical system at the planetary scale. It is a matter of analyzing to what extent the production of a final good and the entire value chain associated with it increases entropy and reduces non-renewable resources. This would require, once the value chain's system has been clearly defined, the qualification of relationships in terms of inputs and outputs with the system 'Earth' in its entirety, and this in a dynamic perspective.

Another issue is that of mapping forms of physical sustainability to forms of governance and of mapping relationships between actors. In this regard, an approach solely based on standards, which are the products of a social construction, cannot suffice to analyze the effects of different forms of organization and of the structuring of economic activities on sustainable development. From the agribusiness to the *filière* to the global value chain, the various analytical frameworks all aim to characterize these processes of structuring and organization of production systems and to evaluate their performances according to criteria that encompass only partially the imperatives of sustainable development.

An analytical framework that allows all three dimensions of sustainable development to be articulated – preferably one that uses a dynamic approach (modelling) – remains yet to be developed. Approaches in terms of sustainable food systems, in which the concept of the *filière* or value chain can characterize supply chains and their shortcomings constitutes possible directions for future research.

References

Aujac H (1960) La hiérarchie des industries dans un tableau des échanges interindustriels. Revue économique 11(2):169–238

Baris P, Couty P (1981) Prix, marchés et circuit commerciaux africains. Amira, Paris, 52 p

Bolwig S, Ponte S, du Toit A, Riisgaard L, Halberg N (2010) Integrating poverty and environmental concerns into value-chain analysis: a conceptual framework. Dev Policy Rev 28(2):173–194

Boustead I (1996) LCA – How it came about. Int J Life Cycle Assess 1(3):147–150
Bridier M, Michailof S (1995) Guide pratique d'analyse de projets. Évaluation et choix des projets d'investissement. Economica, Paris
Casadella V, Liu Z, Uzunidis D (2015) Développement économique et capacités d'innovation dans la mondialisation. ISTE, coll. Innovation, entrepreneuriat et gestion, London, 116 p
Chait B (1949) La loi de divergence. Econometrica, 17, 117–123. Supplement: Report of the Washington meeting.
Chervel M, Le Gall M (1976) Manuel d'évaluation économique des projets. La méthode des effets. Ministry of Cooperation, série Méthodologie de la planification, Paris, 204 p
Daviron B, Fousse W (1993) La compétitivité des cafés africains. Ministry of Cooperation, Paris, 252 p
Davis HJ, Goldberg RA (1957) A concept of agribusiness. Harvard University, Boston, 136 p
Dolan C, Humphrey J (2000) Governance and trade in fresh vegetables: the impact of UK supermarkets on the African horticultural industry. J Dev Stud 37(2):147–176
Fabre P, Bonnet P, Despréaux D, Freud C, Lassoudière A, Raoult-Wack AL (1997) Le concept de filière: un outil pour la recherche, Notes et documents, vol 24. Cirad, Montpellier
Forrester JW (1958) Industrial dynamics: a major breakthrough for decision makers. Harv Bus Rev 36(4):37–66
Gereffi G, Korzeniwicz M (1994) Commodity chains and global capitalism. Praeger, Westport, 334 p
Gereffi G, Humphrey J, Sturgeon T (2005) The governance of global value chains. Rev Int Polit Econ 12(1):78–104
Griffon M (1989) Économie des filières en régions chaudes: formation des prix et échanges agricoles. Actes du 10e séminaire d'économie et de sociologie. Cirad, Montpellier, pp 11–15
Griffon M, Boutonnet JP, Daviron B, Deybe D, Hanak-Freud E, Losch B, Moustier P, Ribier V, Bastianelli D, Ducrot R, Duteurtre G, Leverrier B, Ruf F (2001) Filières agroalimentaires en Afrique: comment rendre le marché plus efficace? Ministry of Foreign Affairs, Paris, 314 p
Harland C (1996) Supply chain management: relationships, chains and networks. Br J Manag 7:63–80
Hugon P (1985) Dépendance alimentaire et urbanisation en Afrique: un essai d'analyse mésodynamique en termes de filières. In: Bricas N, Courade G, Hugon P, Muchnik J (eds) Nourrir les villes en Afrique subsaharienne. L'Harmattan, Paris, pp 23–46
Hugon P (1988) L'industrie agro-alimentaire. Analyse en termes de filières. Revue Tiers Monde 29(115):665–693
Hugon P (1994) Instabilité et organisation des filières coton en Afrique. Économie rurale 224 (1):39–44
Jacquemin A, Rainelli M (1984) Filières de la nation et filières de l'entreprise. Revue économique 35(2):379–392
Jouve P, de Milly H (dir.) (1990) Compétitivité du cacao africain. Analyse du marché mondial et des principaux producteurs. Ministry of Cooperation and Development, Paris, 279 p
MacMichael P (2013) Value-chain agriculture and debt relations: contradictory outcomes. Third World Q 34(4):671–690
Malassis L (1979) Économie agro-alimentaire. Tome I: Économie de la consommation et de la production agro-alimentaire. Cujas, Paris, 402 p
Michalet CA (1976) Le capitalisme mondial, vol 9. Presses universitaires de France, Paris, 368 p
Milhau J (1954) Traité d'économie rurale. Presses universitaires de France, Paris, 264 p
Ministry of Cooperation (1994) Cafés. Études de cas sur la compétitivité des principaux pays producteurs. Ministry of Cooperation, Paris, 367 p
Neilson J (2014) Value chains, neoliberalism and development practice: the Indonesian experience. Rev Int Polit Econ 21(1):38–69
Omta S, Trienekens J, Beers G (2008) Chain and network: a research framework. J Chain Netw Sci 1(1):1–6

Palpacuer F, Gibbon P, Thompsen L (2005) New challenges for developing country suppliers in global clothing chains: a comparative European perspective. World Dev 33(3):409–430

Ponte S, Gibbon P (2005) Quality standards, conventions and the governance of global value chains. Econ Soc 34(1):1–31

Porter ME (1985) Competitive advantage. Creating and sustaining superior performance. The Free Press, New York, 557 p

Reardon T, Berdegué J (2002) The rapid rise of supermarkets in Latin America: challenges and opportunities for development. Dev Policy Rev 20(4):317–334

Reardon T, Timmer P, Barrett C, Berdegué J (2003) The rise of supermarkets in Africa, Asia, and Latin America. Am J Agric Econ 85(5):1140–1146

Temple L, Lançon F (2008) Atelier Concepts et méthodes en économie des filières. Montpellier, 16–19 October 2007, Montpellier, Cirad, 1 CD-Rom

Temple L, Lançon F, Palpacuer F, Paché G (2011) Actualisation du concept de filière dans l'agriculture et l'agroalimentaire. Économies et sociétés 33:1785–1797

Vermeulen S, Woodhill J, Proctor FJ, Delnoye R (2008) Chain-wide learning for inclusive agrifood market development: a guide to multistakeholder processes for linking small-scale producers to modern markets. IIED/WUR, London/Wageningen, 112 p

Weatherspoon DD, Reardon T (2003) The rise of supermarkets in Africa: implications for agrifood systems and the rural poor. Dev Pol Rev 21(3):1–17

Werner M, Bair J, Fernández VR (2014) Linking up to development. Global value chains and the making of a post-washington consensus. Dev Chang 45(6):1219–1247

Chapter 3
History of Public Organizations and Associations Specializing in a Single Agricultural Commodity and Related to Francophone Africa

Benoit Daviron and Janine Sarraut-Woods

The agricultural sector was characterized in the twentieth century by the proliferation of public organizations and associations whose identities were tied to a particular agricultural commodity (wheat, coffee, etc.) or a group of commodities (cereals, fats, tropical timber, etc.).[1] For the countries of francophone Africa, this process started with colonization and continued after independence, without any great disruption, until the 1980s. This chapter first covers this history before briefly describing the far-reaching reconfiguration of agriculture in the wake of liberalization policies implemented over the last 30 years.

3.1 In Pursuit of Imperial Autarky

In the French colonial empire, organizations and collective entities specializing in a given commodity were first created in the early twentieth century, even before the end of the pacification of the newly acquired colonies. But these initiatives were subjected to new dynamics with the two world wars and the crisis of the 1930s. Indeed, just like other European powers, France too eschewed globalization and opted for an imperial strategy of autarky, i.e., a strategy to obtain most, if not all, of the products it was not self-sufficient in from its colonies. This strategy continued

[1] We avoid the use of the terms 'supply chain' or 'value chain', or their equivalent in the French language (*filière*), in this chapter since they appeared only in the second half of the 20th century.

B. Daviron (✉)
CIRAD UMR-MOISA, Montpellier, France
e-mail: benoit.daviron@cirad.fr

J. Sarraut-Woods
International Consultant, Washington, DC, USA
e-mail: j.sarraut.woods@gmail.com

E. Biénabe et al. (eds.), *Sustainable Development and Tropical Agri-chains*,
DOI 10.1007/978-94-024-1016-7_3

until the late 1960s in view of the continuing problems of balance of payments due to the country's reconstruction. Measures favouring colonial production over 'foreign' products were adopted. The policy adopted for the trade in coffee is an apt example (Box 3.1).

Box 3.1 French Policy for Regulating the Coffee Trade

Benoit Daviron and Janine Sarraut-Woods

There were four distinct phases in French policy for regulating the trade in coffee:

- from 1929 to 1939, the policy explicitly strove to promote colonial production, which it sheltered from falling prices and competition from foreign coffees. The customs tariff, from which colonial coffees had been exempt since 1913, was increased. A special import tax (10 French francs/100 kg in 1931, increased to 20 in 1936, and to 40 in 1938) was imposed to fund aid to colonial coffee cultivation (direct subsidies to growers and exporters). Finally, the import of foreign coffees was subjected to a quota (November 1932) and an additional tax was imposed on licences to import them. Through these measures, the colonial robusta coffees enjoyed an import price well above that of competing coffees: in 1933, 708 French francs/100 kg against just 392 for Indonesian robusta; in 1935, 578 against 254;
- from 1939 to 1948, in the context of the war and the immediate post-war period, the priority became to manage shortages. Colonial coffee was prohibited from being exported outside the empire. Indeed, almost no other coffee was available on the French market at the time. Consumption was rationed, and deliveries of coffee, and of its imitations (barley, chicory, etc.), to roasters were subject to quotas. The few supplies of foreign coffee in the immediate post-war period were the result of government-to-government negotiations;
- the period between 1948 and 1954 were marked by a return to a certain freedom of trade. The import tax was abolished and foreign coffees were freely allowed to enter France. The export of colonial coffees outside the empire was also permitted. These market-opening measures were made possible largely by higher coffee prices which made colonial protection less necessary;
- finally, from 1955 to 1958, the protection returned. A 20 % tax on foreign coffees was used to fund aid to colonial coffee cultivation. Colonial coffee enjoyed export subsidies for export to the dollar zone. In 1955, a national price stabilization fund for colonial products was established and minimum export ratios to foreign destinations were fixed for the colonies[2]. In 1956, stabilization funds were created for Côte d'Ivoire, Cameroon,

[2] Côte d'Ivoire: 1 tonne for 4 to France; Madagascar: 1 for 5; Cameroon and Guinea: 1 for 6.

Box 3.1 (continued)

Guinea, and French Equatorial Africa. Finally, in 1958, import quotas for coffee from the French franc zone were established at a guaranteed price 30 % higher than international prices.

This policy of supporting the development of colonial production can be considered a success in terms of the objective of imperial autarky. Between 1913 and 1938, the share of the colonial empire in food imported into mainland France went from 29 to 71 %, a level that was maintained until 1958 (Table 3.1).

In this context of progressive orientation towards imperial autarky, it is possible to differentiate between three different types of initiatives that formed the bases of the organizations specializing in a product:

- initiatives of French industry keen to secure supplies for their factories and thus to promote the expansion of a particular production in the empire. The most representative product in this category is cotton. The Colonial Cotton Association (*Association cotonnière coloniale*, ACC) was founded in 1903, modelled on the British Cotton Growing Association, to promote the growth of this crop in the French colonies with a view to overcoming the dependence on US cotton. The first president of the ACC was Robert Esnault-Pelterie, president of the general trade union of the French cotton industry;
- initiatives by public authorities wishing to ensure the supply of products to France for reasons of national security and thus, here too, to boost production in the empire. An example is tropical timber that was needed to build warplanes. Research was thus undertaken as early as the time of the First World War to test different timber species;
- the initiatives of French Planters undertaking their activities within the empire and anxious to protect their activities from competition from other production areas. Rubber is an apt example here. A French Institute of Rubber (*Institut français du caoutchouc*, IFC) was founded in 1936,[3] under the ambit of the Union of Indochinese Rubber Planters (*Union des planteurs de caoutchouc d'Indochine*, UPCI). Its first president was Philippe Langlois-Berthelot, also president of the plantation companies belonging to the Rivaud group from Indochina, then the largest group in this field. The institute was party to the supplementary scheme to the international agreement signed in 1934 to regulate the production and trade of natural rubber in order to support prices. In addition to the French Institute of Rubber, the *Rubber Stichting* in the Netherlands and the British Rubber Development Board in Britain were also set up. The activities of

[3] IFC began its research activities at the Collège de France, and in 1939 bought a building at 42 rue Scheffer, which is where the headquarters of CIRAD's general directorate are still located.

Table 3.1 Share of the colonial empire in French imports of agricultural products

	1913 (%)	1938 (%)	1958 (%)
Wine	57	97	71
Cereals	12	80	78
Dessert fruits	17	49	72
Coffee	2	43	76
Cocoa	2	88	85
Oilseeds	25	54	78
Sugar	100	78	94
Total agricultural products	29	71	71

Source: based on Marseille (1984)

these three organizations were coordinated by the International Rubber Research Board[4] and they were all funded by export taxes. In addition, the International Rubber Development Committee was created to promote the use of natural rubber.

These different initiatives converged during the Second World War when the Vichy government promoted corporatist policies. That period was marked by the creation of several institutions that continued to exist after the liberation of France. Indeed, the movement was given further impetus by the new government in the post-war years. One man in particular, Robert Michaux, former director of the Rubber Finance Company (Socfin) in Malaysia and later its administrator, as well as of the Terres Rouges plantation company, also linked to the Rivaud group, played a prominent role in these efforts. He worked for the French Institute of Rubber and was involved in the creation, in 1940, of its offshoot, the Indochina Rubber Research Institute. He became president of the Confederation of Overseas Agricultural Producers after the capitulation of France, in June 1940, in the Second World War. A mission to establish a programme of development for agricultural production in French overseas territories was entrusted to him by the Secretary of State for the Colonies. Following the mission, which he undertook in Africa, two institutes were set up: the Institute for Research in Oils and Oilseeds (*Institut de recherche sur les huiles et oléagineux*, IRHO) in late 1941 (with himself as its chairman) and the Institute for Rubber Research in Africa (*Institut de recherche sur le caoutchouc en Afrique*, IRCA) in 1942. The creation, in the same year, of what later became the Institute of Colonial Fruits and Citrus (*Institut des fruits et agrumes coloniaux*, IRFA) rounded off the support available to farmers in the colonies. A grouping under the umbrella of the Colonial Agricultural Institutes brought together IRHO, IRCA and IFC. Later on, they were joined by the Cotton

[4] The *International Rubber Research Board* and the *International Rubber Development Committee* merged in 1960 to form the *International Rubber Research and Development Board* (IRRDB) which exists to this day.

Union of the French empire (*Union cotonnière de l'empire français*, UCEF), which had succeeded the Colonial Cotton Association.

This approach to colonial agricultural products, driven by the State's self-centred economic interest, became even stronger in the post-Second World War period, especially since significant public budgets could then be mobilized (various plans and the Investment Fund for Economic and Social Development of the colonies) for the creation of infrastructure, facilities, and increased operational research intended – also – for indigenous producers, always in conjunction with the agricultural companies. The segmentation of supplies to mainland France according to a logic of territory-product specialization (cotton in French Sudan, coffee in Côte d'Ivoire and Madagascar, groundnuts in Senegal, banana in Guinea, rubber in Indochina, copra in French territories of Oceania, etc.) strengthened the connection with applied agricultural research, which continued its structuring by plant – with a separate research institute for each major crop type – and thus consolidated a vertical approach to interventions, more so since there existed specialized technical operators that ensured the approach's coherence and success (Sarraut-Woods 1998). The promotion of cotton cultivation, which, along with coffee and cocoa cultivation, was undertaken by African producers, was reorganized. Two separate but closely linked entities were created to replace UCEF: the Cotton and Exotic Textile Research Institute (*Institut de recherche du coton et des textiles exotiques*, IRCT) in 1946 and the French Textile Company (*Compagnie française des textiles*) in 1949. Finally, in 1958, the French Institute of Coffee and Cocoa was founded, meant to promote these crops in the colonies.

3.2 European Integration and Independence: Ruptures and Continuities

The signing of the Treaty of Rome by France in 1957 marked the abandonment of its imperial strategy of autarky. The creation of the European Economic Community led to a customs union, and therefore a harmonization of tariff policies. This harmonization necessitated the dismantling of the protection enjoyed until then by the former French colonies vis-à-vis their 'foreign' competitors. It therefore immediately increased competition from other agricultural exporting countries.

Changes in composition and geographical sources of imports of vegetable oils into France during this period probably constitute the best illustration of the new situation that confronted export agriculture in francophone African countries (Table 3.2). The withdrawal of the protection enjoyed by groundnut cultivation in Senegal against other vegetable oils in the French market forced it into competition with US soya beans, cheaper and benefiting from strong expansion due to its growing use in animal feed.

The loss of the protection afforded to former colonies in the French market was partly offset by the implementation of the European policy of development aid within the framework of the Yaoundé Convention, signed in that city in 1963,

Table 3.2 Imports of groundnuts and soya into France (oil and grain in equivalent oil) in the 1961–1963 and the 1981–1983 periods

	1961–1963 (thousand tonnes)	1981–1983 (thousand tonnes)
Groundnuts	325	218
Soya	18	209

Source: Daviron (2014)

between the European Economic Community and former colonies in Africa and complemented in 1975 by the creation of Stabex. However, these efforts could not prevent the complete marginalization of Africa in international trade. 'Aid not trade' was, at the time, the implicit mantra of Europe.

In francophone African countries, the project of European construction was also accompanied by increasing specialization in tropical drinks, mainly coffee and cocoa. The share of this product group, which already accounted for 46 % of agricultural exports from francophone Africa in the early 1960s, reached 62 % in the early 1980s. On the coffee market, the former colonies benefited from the quota system created in 1960 within the framework of the International Agreement.[5] On the cocoa market, francophone Africa, led by the Côte d'Ivoire, profited from the challenges facing Ghanaian and Nigerian cocoa farming in the form of high taxes imposed on it by these newly independent countries.

Strangely enough, independence did not create a major disruption in either the goals or in the methods of intervention vis-à-vis export crops.[6] There was a smooth transition from the extraction of value from the overseas territories in pursuit of imperial autarky to aid for the development of the new States, in part due to the characteristics of the French system of cooperation and of its political particularism. A series of bilateral and regional agreements made it possible to maintain and even strengthen budgetary support and teams of technicians (mass technical assistance being a French speciality), and consequently efforts to create standards and reference bases and to develop methods of intervention.

Continuity was also ensured with regard to sectoral policies, as the main objective remained unchanged: to produce more. Technicians adapted easily to the new political situation: instead of working for balancing the French balance of payments, they were, from then on, acting for national development by increasing exports of agricultural products whose taxation played a vital role in financing partner States and allies of France. In line with the prescriptions of the model of economic development in vogue in the 1950s, plans for economic and social development depended on interventions using a preferred tool: the project. To

[5] This agreement was specifically perceived by the French government as a substitute for the protection previously accorded as part of the empire.

[6] However, we must qualify our remarks on the absence of disruptions when decolonization took place by noting that the French military defeat in Vietnam, followed by the break with Guinea, obviously did have serious consequences. They led to a redeployment of some of the French plantations present in these two countries to the Côte d'Ivoire and, of a few, to Cameroon. The break with Guinea also resulted in the loss of IFAC's main research station.

overcome the shortcomings of the local environment – paucity of capital and entrepreneurs, services, and infrastructure – the State strengthened or created structures specialized by product responsible for developing production, marketing, and processing. Specialized research helped reinforce a productivist perspective in which the agronomic optimum formed the reference, in an environment (relatively) protected by price-stabilization mechanisms and, for products for the domestic market, by tariff barriers.

It was in Côte d'Ivoire that this model found its fullest expression. Indeed, the path of agricultural development in the country and the extension system implemented by the Ivorian government were directly linked to French tropical agricultural research.[7] The various specialized research institutes were maintained without any changes after independence, with their facilities remaining under their original ownership.[8] They then became the organic interface of an agricultural framework that adopted a sectoral approach through the creation of 'development companies' (*sociétés de développement* or *Sodé*) by major product: Sodepalm (1963) for oil palm and coconut with the support of IRHO; Sodefor (1966) for forests with CTFT; Sodefel (1968) for fruit and vegetables with IFAC; Sodepra (1970) for animal production with IMVT; CIDT (1974) for cotton with IRCT; Sodesucre (1971) for sugar; and Soderiz (1970) for rice, with the support of IRAT.[9]

This Ivorian model provided former French institutes specialized by product a basis to continue their activities, from then on within the framework of international cooperation.[10] Public development corporations specializing in a particular product were set up in many countries, such as Sodecao for cocoa in Cameroon. This model was designed to promote more intensive farming techniques to farmers through advice provided to them, dissemination of improved varieties, and distribution of fertilizers and phytosanitary products. The implementation of this model was facilitated by high prices on international markets – and therefore abundant export earnings – and public investment funded by cheap international credit. Public action sometimes took the form of direct investment in production with the creation

[7] As A. Sawadogo, Minister of Agriculture from 1966 to 1977, noted: '(...) the new agricultural development policy, characterized by a withdrawal of the public administration, the creation of State corporations and the establishment of specific channels for transfer of scientific knowledge from the research institute to the farmer (...)' (Sawadogo 1977, p. 97). '(...) The choice has been made in favour of an agricultural development supported by the results of agricultural research, (it) consists of taking stock of the opportunities offered by institutes in the country' (*Ibid.*, p. 227).

[8] This ownership was maintained until 1984 and then transferred to the Ivorian government. The institutes, however, retained their names and CIRAD was entrusted with their management until 1991.

[9] The Institute of Tropical Agricultural Research (*Institut de recherches agronomiques tropicales*, IRAT) was established after independence in 1960 when it inherited the research stations of colonial agricultural services. Like them, it was focused mainly on 'food' crops.

[10] The length of tenure of a number of personalities within these institutions reflects this remarkable continuity. René Carrière de Belgarric retained his position as Director of the IRHO from 1941 to 1972 before becoming its president until 1975. Edouard Senn, president of UCEF since 1940, remained the President of IRCT until 1974.

of a number of agro-industrial companies such as Sodeblé for wheat in Cameroon, the Boumango Industrial Company for Agriculture and Livestock Farming (*Société industrielle d'agriculture et d'élevage de Boumango*) for poultry in Gabon, and the Comoé Sugar Corporation (*Société sucrière de la Comoé*) in Burkina Faso.[11]

3.3 The Great Dismantling

While the 1970s were characterized by high prices on international commodity markets, a period of low prices began in the 1980s that would last almost 20 years. Several organizations specializing in products exported by francophone African countries did not survive this period. In the short term, falling prices[12] led to bankruptcies or paralysis of various public organizations in charge of stabilizing domestic prices. In fact, few of them had the financial reserves to compensate the difference between the export price and guaranteed producer price, which took time to be adjusted downwards. Therefore, the States appealed to bilateral and multilateral funding entities, which provided support on the condition of reforming these organizations or even eliminating them altogether. Indeed, while many of these commodity-centred organizations were meant to address lacunae in the markets (credit, insurance, supplies, marketing) and to provide some public services (basic infrastructure, technical advice), their design had also generated many inefficiencies, which were sources of additional costs and a low productivity of factors (problems of allocation in a context of monopolies, of governance, of rent-seeking behaviour, inadequate technical decisions with little control over project implementation, etc.). These diagnoses were used as a basis for the liberalization policies that were subsequently adopted.

Table 3.3 reports on part of the liberalization process. It shows the extent of institutional changes that occurred for coffee, cocoa, and cotton. These are all crops that were originally characterized by the omnipresence of public bodies in the management of exports, domestic marketing, and setting of prices. For these products, two decades of 'reforms' have led to a situation in which local and producer prices fluctuate in tandem with international prices and in which private operators dominate, especially as powerful transnational companies have invested

[11] This proliferation of new projects in Africa was not without impacts on the French system. One of many examples was the creation, in the early 1970s, of SOMDIAA (*Société d'Organisation de Management et de Développement des Industries Alimentaires et Agricoles*), the multinational corporation for the development of food and agricultural industries, set up with the objective of coordinating the interventions in Africa of a major French miller, Grand Moulins de Paris, and a major bank, BNP Parisbas. SOMDIAA also played a role in the management of a number of development companies mentioned above.

[12] The fall in prices was felt only with some delay in francophone African countries due to the appreciation of the US dollar in the early 1980s. Thus, it was only after 1986 that the export earnings of the countries using the French franc began to fall.

Table 3.3 Reforms in structures of export and domestic trade for coffee, cocoa, and cotton

Product	Country	Structure of export trade		Structure of domestic trade	
		Before the reform	After the reform	Before the reform	After the reform
Cocoa	Cameroon	Limited number of export licences	Liberalized, private	Prices and margins fixed by the government	Liberalized
	Congo	State monopoly on exports		Public monopsony	
	Côte d'Ivoire	Limited number of export licences		Prices and margins fixed by the government	
	Togo	State monopoly on exports		Prices and margins fixed by the government	
Coffee	Cameroon	State monopoly on exports	Liberalized, private	Prices and margins fixed by the government	Liberalized
	Central African Republic	State monopoly on exports			
	Congo	State monopoly on exports			
	Côte d'Ivoire	Limited number of export licences			
	Gabon	State monopoly on exports			
	Guinea	State monopoly on exports			
	Madagascar	Limited number of export licences			
	Togo	State monopoly on exports			
Cotton	Benin	State monopoly on exports	Regional private monopolies	Public monopsony	Regional monopsonies; interprofessional bodies
	Burkina Faso	State monopoly on exports	Regional monopolies		Regional monopsonies; interprofessional bodies

(continued)

Table 3.3 (continued)

Product	Country	Structure of export trade		Structure of domestic trade	
		Before the reform	After the reform	Before the reform	After the reform
	Cameroon	State monopoly on exports	State monopoly		Public monopsony, producer organizations
	Côte d'Ivoire	State monopoly on exports	Regional private monopolies		Regional monopsonies; interprofessional bodies
	Central African Republic	State monopoly on exports	State monopoly		Regional private monopsonies
	Mali	State monopoly on exports	State monopoly		Public monopsony and interprofessional organizations
	Senegal	State monopoly on exports	Private/PO monopoly		Private monopsony, producer organizations
	Chad	State monopoly on exports	State monopoly		Public monopsony
	Togo	State monopoly on exports	State/PO monopoly		Private monopsony, producer organizations

Sources: Akiyama et al. (2003) and Fok (2010) for cotton

in production, processing, and marketing of agricultural and agrifood products. The public organizations specializing in particular products for the domestic market have not fared any better. As most of them had enjoyed strong protection from imports ever since they were created, they were unable to survive the opening up of markets and the reduction in public subsidies imposed by structural adjustment policies.

However, in view of these policies' goal – the creation of competitive markets that ensure efficiency – the earlier technical choices and the organizational structures proved very restrictive. Based on organizations specialized by commodity, which internalized access to the various inputs and services, they held back, if not prevented outright, the development of markets for credit, inputs, insurance, and services.

In very many cases, the earlier public and parastatal monopolies were replaced by private oligopolies, usually under the control of large foreign firms, which took over the supply of these inputs and services. The recurrent problem of the organization and development of markets of factors of production thus remained unresolved. Nevertheless, as is the case with Côte d'Ivoire, there has been a rethinking in recent years of integrated approaches to investment and agricultural development specializing in products (not only coffee, cocoa, rubber, oil palm, cotton, banana, but also poultry, aquaculture, etc.). Built around interprofessional bodies encompassing major commodity value chains, this type of development recasts the earlier public-private partnerships and revives the association between economic operators and the research community. This was manifested in the creation, in 2002, of an Interprofessional Fund for Agricultural Research and Advisory Services (*Fonds interprofessionnel pour la recherche et le conseil agricoles*, FIRCA), funded by professional and private contributions, as well as by governments and funding entities. FIRCA finances research programmes, technical advice, support to its members, and training (Losch and Michaud 2015).

3.4 Conclusion

In francophone Africa, the colonial period gave rise to a large number of institutions specializing in particular agricultural products, especially for products exported to mainland France. These institutions carried out many functions that, as was believed at that time, could not be undertaken by private operators or left to the markets alone: stabilization of prices, quality control, supply of inputs, extension, research, etc. When these former colonies gained independence, this structuring by product of institutions was left unchanged. Indeed, these new States relied on it for the implementation of their development policies.

It was only with the structural adjustment policies of the 1990s that this continuity was broken. In an almost opposite perspective to the one that prevailed during the colonial era and the three subsequent decades, private operators and the markets were seen as the only way to guarantee an optimal allocation of resources. Within a span of a few years, public institutions were dismantled in order to allow free competition and encourage investment.

However, over the past few years, this perspective has shown its limitations and it seems that the need for new institutions, rules, and organizational structures is imposing itself, as attested by the current proliferation of public-private partnerships. They have resulted, for example, in the assigning of regional monopsonies[13] to private companies, not only in charge of collecting certain agricultural products but also of providing inputs to producers or their organizations. It is probably too early to assess these new mechanisms – and the opportunities they may be able to offer in terms of sustainable development –, but, as some studies (Willoughby 2014) have already pointed out, they are not without risk. Monitoring and analyzing them will represent a major challenge in the coming years.

[13] A monopsony is a market structure in which only one buyer interacts with many would-be sellers of a particular product.

References

Akiyama T, Baffes J, Larson DF, Varangis P (2003) Commodity market reform in Africa: some recent experience. Econ Sys 27(1):83–115
Daviron B (2014) Turning organic economies into mineral economies: two centuries of oils and fats use and supply in France. In: European social science history conference, Vienna, 36
Fok M (2010) Facteurs d'efficacité des arrangements institutionnels en politique cotonnière africaine. Cahiers Agricultures 19(1):68–74
Losch B, Michaud A (2015) Evolution of rural development strategies and policies: lessons from Côte d'Ivoire. Report for the OECD development centre project on "Rural development policies in perspective. From challenges to action". Cirad, Montpellier
Marseille J (1984) Empire colonial et capitalisme français: histoire d'un divorce. Albin Michel, Paris, 464 p
Sarraut-Woods J (1998) La politique française de coopération en matière de développement rural. Des certitudes à la dérive. In: Observatoire permanent de la coopération française, Rapport 1998 (OPCF). Karthala, Paris, pp 47–94
Sawadogo A (1977) L'agriculture en Côte d'Ivoire. PUF, Paris 368 p
Willoughby R (2014) Moral hazard? "Mega" public-private partnerships in African agriculture. Oxfam International, Oxford 40 p

Chapter 4
Evolution of Donor Intervention Modalities on Agri-Chains: The AFD Experience

Anne Legile, Naomi Noël, Florence Mouton, and Didier Baillet

The French Development Agency (*Agence française de développement*, AFD) has used the term *filière agricole*[1] for the past 50 years to describe all the operations and stakeholders involved in an agricultural product, from its production and up to its processing and marketing on national, sub-regional, or international markets. This concept takes in consideration the production of value by the different stakeholders, also called links, of the value chain. The term *filière,*[2] seems equivalent to 'value chain' in this regard. However, the *filière* is not just a useful concept for analyzing economic flows. It also refers to the institutions (organizations and rules) that private actors (farmers, traders, processors, exporters) create in order to deal with and negotiate issues of common interest (quality, volumes, costs and prices, innovation, promotion) among themselves and with public authorities when required (taxation, public support, infrastructure, etc.). The *filière* is therefore a space for interprofessional and joint governance of a set of agricultural products that can be grouped together for technical or commercial reasons. This importance of governance led agri-chains to be first organized (or structured) for processed and/or exported products for which formal enterprises and explicit public policies play an important role (rice in Asia; cocoa, cotton, rubber, palm oil in sub-Saharan Africa). Some products for domestic markets (poultry, milk) have also given rise to

[1] In English, this term has different translations, most notably agricultural supply chain, agricultural value chain (see Chap. 2 for more detailed information on the concept).
[2] Translated in English as agri-chain in this chapter.

A. Legile (✉)
Parc National des Cévennes, Paris, France
e-mail: anne.legile@cevennes-parcnational.fr

N. Noël • F. Mouton • D. Baillet
AFD, Paris, France
e-mail: noeln@afd.fr; mouton@afd.fr; bailletd@afd.fr

E. Biénabe et al. (eds.), *Sustainable Development and Tropical Agri-chains*,
DOI 10.1007/978-94-024-1016-7_4

specific chain organizations. The term *filière* or agri-chain can be used for any agricultural product, even for those in which industrial and commercial stakeholders are not involved. Thus, the production, exchange, and processing of tubers, roots, and rhizomes in Africa should be analyzed as *filière*, even though these activities are still not very structured.

The boundaries or scope of a *filière*, in its meaning of a common project necessary for economic and social stakeholders, must be clearly – and flexibly – specified depending on their interest in acting together. It is easy to define a rubber chain. In contrast, there is no single all-inclusive animal chain, but many separate ones: meat cattle chain, milk cattle chain, pork chain, broiler chain, layer chain, etc. The 'cotton' chain can be structured around the fibre market, but since cotton seeds are crushed, this cotton chain is connected to those of peanuts, soya beans, or sunflowers, and those of oilseeds for human food (edible oil) and for animal feed.

Of course, one must be careful not to consider the *filière* as an exclusive analytical concept or framework for action. On the one hand, since the vast majority of farms in countries in which the AFD intervenes are diversified (crops and livestock, food and fibre, etc.) and sell on different markets (local, domestic, and export), they are participants in different agri-chains, at any one instant and historically. On the other hand, these farms have to deal with issues that are common to all their activities (land, training and advice, access to inputs, credit, etc.) even if some of them benefit from being part of a given agri-chain (inputs, credit, and advice in the cotton chains, as we shall see).

In this chapter, we offer a quick look at the evolution of AFD's *filière*-based interventions in recent decades, its concrete illustration for some agri-chains, and a reflection on the necessary enrichment of the *filière* or agri-chain approach and on its limitations.

4.1 A Brief History of French Support to the Structuring of Agri-Chains

4.1.1 From Independence to Structural Adjustment

Following the independence of its former colonies, French cooperation supported agriculture in its countries of intervention through two types of project: (1) integrated rural development projects, investing in rural facilities, not targeted at specific products and benefiting all farmers and (2) the development of key agricultural export chains, in a continuity of the investments made during the colonial period to supply the industries of mainland France. In both cases, public development companies, whose capital was wholly owned by the State (the 'Sodé'), were created for an agri-chain or territory.

The public development companies worked for the comprehensive development (from production to export) of one or two crops each in the context of a national monopoly. They also contributed to the funding of some services and infrastructure in rural areas: not only for inputs (planting material, fertilizers, processing) and agricultural extension required for the targeted production, but also for general public utility services whose responsibility the States delegated to them, such as schools, clinics, and roads.

In addition to the industrial units for primary processing (ginning, crushing), the public development companies often owned, in the case of perennial crops (rubber, oil palm), large-scale industrial plantations (blocks ranging in size from a few hundred to several thousand hectares, awarded to them as part of the State's domaniality over untitled land that prevailed at the time) that constituted centres of production ensuring a significant part of supply to factories. Supplies for the factories were also obtained through contractual relationships with smallholders or with peasants.

This development model was based on the following simple principles:

- the public development company enjoyed a monopoly over the purchase of agricultural production in the country and its export; it had access to concessional external short- and long-term funding, which was guaranteed by the State; its experts established the formulas for calculating procurement prices paid to farmers;
- the public development companies provided farmers with technical advice,[3] pre-financed inputs on an *ad hoc* basis, and additional financing for equipment (animal traction); the farmers had an obligation to deliver their production only to this company; they had the guarantee of being able to sell their entire production at a price announced before planting (for annual crops such as cotton) or before the harvest (negotiated in advance);
- the newly independent States fixed the purchase prices paid to farmers (based on information provided by the public development companies and on political objectives, such as maintaining peace among rural populations). They enforced financial levies on exports either for fiscal purposes (budgetary support) or to build up price stabilization funds that could be drawn upon to pay producers in case world prices would drop.

In the late 1980s, the fall in prices of agricultural commodities on international markets, combined with shortcomings in the governance by States (disproportionate standards of living, excessive debt, financial mismanagement, etc.), led to a generalized debt crisis. Structural adjustment was then requested by the international financial institutions (led by those of Bretton Woods). This resulted in

[3] Which were often developed in partnership with research institutes specializing in tropical products (IRCA, IRHO, IRCC, etc.). These separate institutes would later be merged into a single entity: CIRAD.

particular in the collapse of a large number of public development banks and agricultural products price stabilization funds. The public development companies, which States could no longer support financially, found themselves in difficult circumstances. The Bretton Woods institutions then insisted on the privatization of these companies' industrial assets, the ending of national monopolies through their breaking up into different regional entities, the abandoning of vertical integration policies of the agri-chain (for example, separation of the oil mills from cotton ginning factories), and, in general, the withdrawal of States from commercial activities and their refocusing on their sovereign, normative, fiscal, and regulatory functions. At the same time, the economic crisis in West and Central Africa (and of the agri-chains in particular), due partly to their lack of competitiveness because of the overvaluation of the CFA franc, led to a devaluation in 1994. This contributed to renewed profitability of agri-chains.

During this period, in West and Central Africa, AFD funded public development companies for an expansion of their industrial assets (plantations and factories), usually in the form of loans. This was the case for the rubber and palm oil chains (SAPH in Côte d'Ivoire, Hevecam and Socapalm in Cameroon, Agrogabon and Hevegab in Gabon, Grel in Ghana, etc.). In addition, AFD provided loans to States for the expansion of small private plantations (as opposed to industrial production centres). This work was entrusted to public development companies of the concerned agri-chain and territory since they had the expertise, ensured an outlet, and were expanding their collection base. For cotton, all the national companies (Mali, Burkina Faso, Senegal, Côte d'Ivoire, Benin, Togo, Chad, Central African Republic) that had taken over the activities of the French Company for the Development of Textiles (*Compagnie française pour le développement des fibres textiles*, CFDT) after independence were financed.

The evolution of food chains (livestock as well as crops) for domestic markets was not ignored. Hydro-agricultural schemes (irrigated or flood-recession agriculture), whether linked or not to processing units, were funded along the Sahelian rivers (SAED in Senegal, Office du Niger in Mali, Soderiz in Côte d'Ivoire, Semry in Cameroon, etc.) while short animal chains (poultry and pork in southern Côte d'Ivoire) and extensive pastoral chains (Sodepra-Nord in Côte d'Ivoire, Pastoralism in Tchad) were supported.

4.1.2 From Structural Adjustment to the Surge in Prices in 2008

During this period, the reductions in State investments in agriculture and rural development went hand in hand with those of donors. This disengagement was driven by the need to prune public debt and the hope of a spontaneous commitment by the private sector. However, rural agriculture funding by donor agencies was not replaced by banks and microfinance institutions, who focused instead on less risky activities in urban areas.

The financing of family farms, in particular the expansion of smallholder plantations of perennial crops, thus came to a standstill for a time in Africa because of the investments required, the lack of enforceable guarantees and risks, due to the end of monopolies and non-delivery of products to companies that had provided pre-financing (side selling or pilfering).

During this period of liberalization, with more or less privatized agri-chains, AFD considered it appropriate to focus its resources on two areas:

- support for interprofessional organizations as institutions capable of prioritization and regulation according to a market and not to a political logic. In this perspective, AFD supported the creation of a price smoothing fund (*fonds de lissage*) for the cotton sector under the responsibility of the interprofessional organizations;
- professionalization of producers by strengthening their organizations so that the latter could fulfil the functions previously handled by public development companies and public services (literacy, training, technical and economic advice). Producer organizations in the Senegal River valley, the Office du Niger in Mali, and the cotton-growing zones of West and Central Africa became therefore among the continent's better structured, capable of defending their interests at the political level, and to contribute to the supply of inputs to their members and the marketing of their production.

At the same time, in countries where public agricultural development banks and the capacities of intervention of the State had not undergone structural adjustment, AFD continued to fund perennial crop plantations: village plantations for rubber[4] and oil palm[5] in Ghana, rubber in Vietnam[6] and Cambodia,[7] and tea[8] in Vietnam. The intervention methods were similar. A loan to the government was partially on-lent to a state development bank, which granted loans to farmers for buying planting material and inputs. The rest of the loan to the State was allocated, as a grant, for funding advisory services for farmers and, if necessary, for research and capacity building of the organizations of agricultural producers. Part of the funding was also sometimes used to open or rehabilitate access roads. In all cases, contractual relationships between the company concerned, the bank, the State, and farmers were closely scrutinized, in particular regarding the sharing of value between the various economic stakeholders and the genuineness of prices of inputs and exported products.

[4] Rubber Outgrower Projects 1–5.

[5] Oil Palm Smallholder Project.

[6] Agricultural diversification project, cofinanced by the World Bank.

[7] Smallholder rubber development project.

[8] Project for the development of tea production in Phu Tho province.

In the domestic food chains, the large number of stakeholders, especially downstream of production, makes it difficult to build sufficiently secure contractual relationships. Nevertheless, in Guinea, progress made in mangrove rice cultivation (Maritime Guinea) and vegetable gardening, promoted by the Fouta Djallon Farmers' Federation, shows that it is possible to act on a large scale. This federation had over 20,000 members in 2008, 70 % of whom were women. During the 1990s, it benefited from several financial assistance grants for the structuring of three agri-chains (potatoes, onions, and tomatoes). It is now capable of exporting potatoes to neighbouring countries. Its profitability and certified accounts allowed it to be financed by national banks, which have mobilized a partial guarantee from AFD.

In 2008, the sudden rise of commodity prices on world markets, including of some agricultural products, triggered awareness of governments and donor agencies at two levels on the pitfalls of a hands-off approach to agriculture. On the one hand, a deteriorating trade deficit in essential food crops such as rice, wheat, or animal products could, in case of soaring prices on the world markets – increasingly driven by highly speculative pressures –, jeopardize poor and low-income urban households that spend a significant part of their budget on food, and thus lead to the destabilization of incumbent political powers. Even though the term is a misnomer as rising fuel prices also bore part of the responsibility, the 'food riots' came as a political shock. On the other hand, the rise in prices of most traded agricultural products (largely abated since), perceived as a reversal of the downward trend of the past decades, emerged as an opportunity for growth over the long term, due to the increasing global demand driven by demographics, rising living standards, and new non-food uses of agricultural products.

4.1.3 From 2008 to the Present

All in all, the events of 2008 triggered a renewed interest of States in agriculture, especially food production, and a new commitment of institutions financing development for food security. Furthermore, often encouraged by States and some financial institutions, there appeared a wave of private investment in agricultural production, including by foreign investors hoping for large-scale and double-digit returns, regardless of the historic rights of indigenous peoples. In many instances, NGOs denounced these agreements between States and companies as abusive. Even though family farms have proven their ability to be productive and profitable as part of different agri-chains when they are given the opportunity to grow and capitalize, we are witnessing a resurgence of the dichotomous colonial vision, which views small subsistence farmers as having no future and commercial industrial plantations as the panacea for world hunger.

In this context, AFD continues to support the evolution of its partners through the structuring of agri-chains that are based on family farms (and their associations) for production and, upstream and downstream of these farms, on private companies of varying sizes depending on markets and products, i.e., ranging from small and medium enterprises (including cooperatives) to large agro-industries. Since it is virtually impossible for downstream companies to finance and undertake activities for the public good, as public development companies used to do, it is essential to involve domestic financial institutions in financing farms for the short and medium term. It is also important to strengthen systems to help farmers secure land tenure and to provide agro-economic advice to them.

4.2 Lessons Learnt From the Experience of Some Agri-Chains Supported by AFD

4.2.1 The Rubber Chain in Ghana

In Ghana, from 1995 to 2012, AFD supported three successive phases of a programme to support village rubber cultivation through the provision of loans to the State, which assumed responsibility for the currency risk. These funds were allocated to a technical operator for the implementation of the project (the programme included access roads, research, and support for professional farming organizations) and to a financial operator to provide investment credit to farmers (securing land tenure, inputs, and investment labour[9]). Ghana Rubber Estate Limited (GREL), privatized in 1996, fulfilled the function of the project's technical operator and industrial partner. Two public banks, the Agricultural Development Bank and subsequently the National Investment Bank, assumed the commercial risk of loan repayment as soon as the second phase. Village rubber plantations were developed through a tripartite contractual relationship between a private industrial company (GREL), a financial institution, and smallholders. This scheme and favourable rubber prices made this programme a success for both farmers and industry. From 2010, a fourth and a fifth phase (in progress) have benefited from two non-sovereign loans from the AFD to the Agricultural Development Bank. This bank grants loans to outgrowers to cover investment costs (securing land tenure, technical assistance, inputs, and investment labour). In this arrangement, GREL continues to fulfil the functions of technical operator and industrial partner. The Agricultural Development Bank assumes the currency risk and the commercial risk of loan recovery. At the end of this fifth phase, AFD will have financed 9500 farms cultivating 29,000 ha of rubber.

[9] Which represents all the labour used in the rubber plantation before the rubber tree starts production (i.e., for the first 5 or 6 years).

4.2.2 The Cotton Chain in Burkina Faso

In the entire CFA franc zone, cotton development public companies intervened in an integrated manner at the different agri-chain levels, with a high level of coherence. Programmes funded by French aid thus involved monitoring and evaluation systems,[10] as well as industrial ginning facilities, research and development, and structuring of farmer organizations. The results, in terms of production and yields (more than one million tonnes of fibre in the mid-2000s) as well as of the abilities of farmer organizations, demonstrate the effectiveness of this integration. Cotton farmers' organizations from this African region manage supply and credit, provide the primary selling of seed-cotton and have made their voices heard up to the World Trade Organization (complaints against US subsidies, in Cancun in 2003).

However, in the 2000s, with the world cotton market becoming very volatile, the traditional intervention mechanisms to support prices paid to farmers, sometimes poorly managed, found themselves overwhelmed. In this context, after consultation with its European and African partners, AFD funded extensive studies in 2004 and 2005 that led to the creation of a new system to limit too sudden variations in prices paid to producers without, however, opposing market trends or jeopardizing the financial sustainability of the cotton companies.

The smoothing fund set up in Burkina Faso incorporates the principles of previous price support funds, with a guaranteed minimum price paid to the producer, guaranteed from the beginning of the season, and a fund on which the three cotton companies could draw upon when their selling price on the international market was not high enough for them to buy cotton from producers at the guaranteed price. However two significant changes were incorporated into the smoothing fund:

- the price paid to producers for cotton is fixed on the basis of world prices and not determined by national negotiations, which were in the past often disconnected from the market. At the beginning of the cropping season, a fibre trend price is determined from reliable and accepted public sources of world cotton prices and exchange rates. The initial buying price paid to producers is fixed on the basis of this trend price, through a distribution of the fibre value between producers and ginners agreed to within the interprofessional organization, with a safety margin of 5 %;
- the contribution to the smoothing fund, in case the sale price of fibre is higher than forecast, and, in the opposite situation, the determination of drawing rights by each company are automatic since they are determined by rules and formulas approved by the interprofessional organization.

An association for the management of the smoothing fund was created, including representatives of cotton companies and producers. A depository and fund managing bank was recruited. At its inception, the Burkinabe government

[10] Allowing reliable information about production systems and POs to be available for these areas.

Table 4.1 Annual production of seed cotton from the 2006/2007 season to the 2014/2015 season

Production	Annual season								
	2006/ 2007	2007/ 2008	2008/ 2009	2009/ 2010	2010/ 2011	2011/ 2012	2012/ 2013	2013/ 2014	2014/ 2015
Seed cotton production (thousand tonnes)	649.0	366.5	447.2	361.1	335.4	417.4	601.0	650.0	708.0[a]

Source: AICB, June 2015
[a]Excluding production of organic cotton.

Table 4.2 Market conditions and margins of seed-cotton producers from the 2006/2007 season to the 2014/2015 season

Economic data	Annual season								
	2006/ 2007	2007/ 2008	2008/ 2009	2009/ 2010	2010/ 2011	2011/ 2012	2012/ 2013	2013/ 2014	2014/ 2015
Market price of seed cotton (FCFA/kg)	263	294	267	268	386	478	393	395	344
Producer bottom price (start of the season)	165	145	165	160	182	245	245	228	225
Additional price during the season	0	10 (+3)	0	8	28	29 (−5)	8	17	0
Producers gross revenue (Billion FCFA)	107.0	57.9	73.7	60.6	70.4	114.4	152	159.2	159.3
Estimated margin[a] (thousand FCFA/ha)	84	68	78	81	113	159	156	140	est. 120

[a]Margin: margin of producers after repayment of cotton inputs calculated for a fixed yield of 1000 kg/ha of cotton seed

contributed to this fund with funding provided by AFD, as a grant (for the share of the National Union of Burkina Cotton Producers – UNPCB) and through a loan retroceded to the association managing the smoothing fund.

The mechanism has been operating from the 2006/2007 season to the present, in very different market situations (contributions or drawings, large variations), which has led to the amendment of some its rules. Its financial balance remains precarious because contributions are insufficient as compared to the volume of cotton produced in Burkina Faso (Tables 4.1 and 4.2). AFD is the only funding entity that has contributed to the creation of this mechanism.

A similar mechanism is operational in Cameroon. Feasibility studies have been commissioned in Côte d'Ivoire and Zambia but have not arrived at any conclusions so far.

4.2.3 The Rice Chain in Senegal

The construction of the Diama dam[11] on the Senegal River and the Manantali dam on one of its tributaries, under the responsibility of the Organization for the Development of the Senegal River Valley (*Organisation pour la mise en valeur de la vallée du Sénégal*, OMVS) which brings together the three riparian countries (Mali, Senegal, and Mauritania), allowed the National Corporation for the Development and Use of the Senegal Delta and the Senegal and Faleme Valleys (*Société Nationale d'Aménagement et d'Exploitation des Terres du Delta du fleuve Sénégal et des vallées du fleuve Sénégal et de la Falémé*, SAED) to develop 95,000 ha of public and private irrigated perimeters on the banks of the Senegal river. In addition to infrastructure development, SAED was also mandated with extension, management and maintenance of facilities, supply of water and inputs, and the collection and processing of paddy. Loans for the purchase of inputs, originally provided through SAED, subsequently became the responsibility of the Senegalese National Fund for Agricultural Credit (*Caisse nationale de crédit agricole du Sénégal*, CNCAS). In 1984, forced to make changes, the Senegalese government transferred management of the irrigation schemes, until then the responsibility of SAED, to producer organizations. They are now in charge of operating, maintaining, and refurbishing facilities and associated equipment. The devaluation of the CFA franc in 1994 did not sufficiently improve the competitiveness of Senegal's rice sector and infrastructure started deteriorating due to lack of maintenance. In 1997, European donor agencies – the World Bank had lost interest in this sector – proposed a moratorium on investments for new perimeters. A fresh strategic reflection was undertaken to create a new institutional framework for irrigated agriculture. These developments led AFD to support four key reforms:

- in 2002, the three riparian States signed the Senegal River Charter to avoid conflicts on water sharing;
- a policy for maintaining infrastructure was adopted. It relied on an irrigation infrastructure maintenance fund, financed by contributions from the State and users of major infrastructure;
- land use and land allocation plans were established to secure land for irrigation schemes for rural communities. The Charter specifies the rights and obligations of the land beneficiaries towards rural communities, and those of the beneficiaries of the water service towards SAED;
- modernization of the agri-chains was ensured through the creation of interprofessional organizations (industrial tomato, rice, onion) and the creation of management and rural economy centres in the delta (Podor, Matam). These organizations ensure transparency in contractual relationships and in financial

[11] On the saline lands of the delta, private irrigation areas were made possible by the construction of the Diama dam and dyking of banks.

flows (maintenance funds, accounts of water unions, CNCAS credit accounts of producers' economic interest groups, accounts of service providers and intermediary actors, etc.).

In 2008, since Senegalese cities were being supplied with rice mainly through imports, the increase in rice prices triggered protests. The government then launched the Great Agricultural Offensive for Food and Abundance (*Grande offensive agricole pour la nourriture et l'abondance*, Goana) and the national programme for rice self-sufficiency (*Programme national d'autosuffisance en riz*, PNAR) to organize and plan the 'march of Senegal towards self-sufficiency in this commodity for 2012,[12] with a production target of 1.5 million tonnes of paddy, equivalent to one million tonnes of white rice.' In 2009, when a study on the rice chain concluded that Senegalese rice production[13] was competitive on the world market, AFD resumed financing irrigation schemes, first in the delta, and then in the middle and upper valley of the Senegal River, covering a total of 12,500 ha. These new programmes included the construction of infrastructure to create viable farms (ranging from a minimum size of 5 ha to maximum of 50 ha), institutional support to the rice chain's actors (producers, millers, traders), and management of land of the rural communes[14] involved.

The development of new irrigable areas is only justified if, at the same time, the tenure of the irrigated land is secured, facilities are maintained, credit is made available to the rice chain's actors, especially the farmers, and tariff and tax policies are fine tuned so as to maintain attractive prices for the valley's farmers.

4.2.4 The Mangrove Rice Chain in Guinea

In Guinea, a country outside the CFA franc zone, AFD has provided, since the 1990s, support to the rice chain which is competitive compared to Asian imports. This support initially focused on mangrove rice (in Maritime Guinea). In mangrove rice cultivation, sea water is allowed to enter rice fields during the dry season to limit diseases and weeds, and bring in silt. During the rainy season, the rains wash the soils clean and hydraulic arrangements prevent ingress of saltwater.

AFD first funded a traditional project for hydro-agricultural infrastructure, with a project team organized around a French operator. The organization of water management and maintenance was entrusted to producer committees. This rice chain also benefited from support provided to downstream operators in the sector (drying, processing, and marketing of rice) and from the presence of two

[12] This production target was later postponed to 2018.

[13] This study was updated in October 2012.

[14] In Senegal, a commune is a fourth-level administrative division.

microfinance institutions, *Crédit rural de Guinée* and *Crédit mutuel de Guinée* (also supported by AFD), that funded the operators downstream of the rice chain. In general, all stakeholders, public (research, extension, rural engineering) as well as private (NGOs, banks, traders), have coordinated their activities with those of the farmers to support mangrove rice production.

AFD focused in particular on building local capacity. The maintenance of roads and tracks (the opening up of remote areas was considered a key factor for the rice chain's success) and construction of irrigation infrastructure were offered to local businesses through tenders. Local consultation companies were relied upon to undertake development studies. A producer organization now handles the marketing.

In the early 2000s, the preservation of the mangrove, a particularly fragile ecosystem, became a cause for concern for the authorities. It was not possible to declare a coastal area with high population dynamics a protected area. The mangrove rice production system had to preserve mangrove areas from destruction while developing the rice chain. An observatory for the Maritime Guinea region was thus set up and later extended to encompass the entire country under the name National Observatory of the Republic of Guinea. It undertakes scientific monitoring of the balance between the environment and agricultural production, and of territorial dynamics and populations.

Moreover, the sale of this rice in the capital, Conakry, revealed that some urban consumers were willing to pay more for mangrove rice than for imported rice. After additional market research and taste tests, the creation of a collective trademark and/or a protected geographical indication for Guinean mangrove rice under the name Bora Malé will be considered.

4.3 Conclusion: A Progressive Enrichment That Has to Continue

The examples presented above illustrate the progressive enrichment of the support provided by AFD to the agri-chains. This is reflected, in particular, in the establishment of participatory and shared governance of these chains, instead of an exclusive public control.

The situation has gone from one of export chains that were built around a public company enjoying a national monopoly and which provided the upstream supply, credit, processing, and marketing to one with institutional and contractual arrangements, bringing together several enterprises (private or public), banks, and producer organizations which fulfil economic, advisory, and commercial functions of varying importance. Interprofessional bodies with representatives from downstream businesses and producers and who are answerable to an increasingly disengaged, but not disinterested, State ensure the regulation of the agri-chain, in the sense of implementing a common development strategy and ensuring a fair distribution of added value between the chain's actors. Two goals were foremost in the construction of this new form of governance. First, it was necessary to ensure that family farms are the ultimate beneficiaries of the concerned agri-chain's development, despite the social and economic asymmetries inherent in their position vis-à-vis agro-industries. This concern is addressed by strengthening producer organizations and unions, involving them in the economic functions, and helping them become shareholders of agro-industries. Second, it is necessary to base decisions (prices, investments) primarily on medium-term market dynamics, irrespective of the short-term variability. The Burkina Faso smoothing fund is an apt illustration of this mechanism.

The issues of short- and medium-term farm funding and of covering risks (climate, pest) inherent in any agricultural production, remain to be addressed. Since the downstream companies, even those belonging to the most robust export chains (rubber, palm oil), can no longer pre-finance plantations, the only option left is to make banking services accessible to farms and their organizations. Contracts and agreements have to change from a bipartite format (farmer/industry) to a tripartite one (farmer/bank/industrial buyer). But the partial guarantee that the buyer can provide to the lender is insufficient. It is thus necessary to design a farm financing ecosystem based on tailored financial products (in terms of maturities, rates, guarantees, insurance), financial institutions conversant with the agri-chains' economics, independent accounting firms, provision of technical and economic advice adapted to farmers, and judicious State intervention in the form of risk coverage and targeted aid that does not distort market dynamics.

The development of food chains for regional, national, and local markets requires specific approaches that take all the downstream links into consideration: marketing starting from the farm gate, local processing, storage, logistics, urban commercial structures (wholesale and semi-wholesale markets), and out-of-home catering and marketing (brands, labels). In Guinea, the rice and potatoes chains are successful illustrations of these comprehensive approaches that address aspects of national, regional, and international trade policies. Undoubtedly, more support (technical, financial, commercial) should be provided to all downstream economic actors of national food chains (collectors, traders, processors) who form the link between producers and urban consumers, with the latter having rapidly evolving qualitative requirements (health aspects, ease of preparation, shelf life, taste).

But a product-centric or *filière* approach also has its limitations. It does not address land issues and those pertaining to the diversification of production systems, which is becoming increasingly necessary to combat climate change and biodiversity loss.

At present, broader issues that concern rural communities and their relationships with cities must also be considered in the development of agriculture. These include territorial planning, usage of spaces, and restoration of ecosystem services. We cannot choose between territorial approaches and *filière* approaches. They must be combined. Of course, a territorial project is primarily a matter of engagement of all local actors, regardless of the agri-chains they are part of.

Nevertheless, due to the considerable ecological, social, and economic impact that a transformational or major production could have in a given territory, the partners of an agri-chain have to make choices, not only in terms of competitiveness, productivity, volume, and sustainability of their agri-chain, but also as regards the sustainability of the very territories in which the agri-chain is located and which it structures through its activities.

This implies a further enrichment of the *filière* approach, beyond reflections on contracting, sharing of added value, and financing, based on its contribution to the production of public goods, i.e., to ecological transitions (the technical management and bearing of the cost of reducing the negative environmental and health related externalities associated with the improper use of fertilizers and pesticides, extension of cultivated areas at the expense of areas of high biodiversity conservation value) and to societal impacts of their development (land inequality and insecurity, social inequality, child labour). These new obligations require agronomic innovations (how to produce more, more sustainably?), social innovations (how to conclude and implement agreements on the adoption of technical innovations at large-enough scales, within the agri-chain and across the territory?), and commercial innovations (how to trace, raise awareness of, promote, and convince the market to pay for the new ecological and social quality of the products?).

Nevertheless, given the scale of the needs in terms of food and nutrition security (for individuals), food sovereignty and balance of trade (for States), and employment and decent incomes (for young entrants to the job market), the *filière* or agri-chain remains an effective framework for mobilizing private investment and regulating relationships between its links, especially in the countries in which AFD operates. Any limitations such an approach may have should be considered with respect to the numerous shortcomings and failures that often characterize markets (credit, inputs, information, products, etc.).

Chapter 5
Alliances Between Agri-Chain Actors for a Sustainable Development of Territories in Vietnam

Guillaume Duteurtre, Denis Sautier, Manuel Pannier, and Nguyen Mai Huong

In the countries of the South, the shift towards liberal thinking in the late 1980s favoured the progressive rise to prominence of the market in agricultural systems. Within a short period of 25 years, this major movement of commoditization of peasant economies had permanently changed rural areas. Farmers have themselves striven towards increased commercial exchanges leading to an unprecedented increase in food production. But at the same time, agricultural intensification, industrialization of economies, and urbanization have put growing pressure on natural resources and have upset social balances in the countryside. Agriculture has thus become one of the sectors most concerned by doubts currently surrounding the development model of this liberal period. And the future of rural areas in developing countries now finds itself at the core of the debates on sustainable development.

This chapter aims to explain the mechanisms through which agri-chains develop territories and to assess to what extent this development is sustainable. To this end, one has to understand territorial agrifood dynamics in a global context (Réquier-Desjardins et al. 2004; Biénabe et al. 2004; Muchnik et al. 2007). We will rely on three case studies conducted in Vietnam, which reflect contrasting ecological and

G. Duteurtre (✉)
CIRAD – UMR Selmet, Montpellier, France
e-mail: guillaume.duteurtre@cirad.fr

D. Sautier
CIRAD – UMR Innovation, Montpellier, France
e-mail: denis.sautier@cirad.fr

M. Pannier
EFEO, Hanoi, Vietnam
e-mail: manuelpannier@yahoo.fr

N. Mai Huong
RUDEC-IPSARD, Hanoi, Vietnam
e-mail: maihuong.nguyen@ipsard.gov.vn

E. Biénabe et al. (eds.), *Sustainable Development and Tropical Agri-chains*,
DOI 10.1007/978-94-024-1016-7_5

economic situations. Vietnam was chosen because of the rapid changes that have taken place in that country. In 25 years, Vietnamese agriculture has gone from a regime characterized by a collectivist economy, marked by chronic food shortages, to one based on an export market economy. These rapid changes have led to local disruptions that underline the need to regulate this advent of the 'market' and to progressively take new social and environmental issues of development into account.

5.1 Methodological Elements

Our approach implicitly recognizes the role of several types of factors in the functioning of markets: price and cost adjustments, of course, but also collective organization, modes of access to resources, spatial organization, and technical rationales. To take these factors into account, we propose to combine a value chain approach, useful for analyzing vertical relationships (Kaplinsky and Morris 2001), with a territorial approach which is more suitable for examining horizontal relationships (Pecqueur 2005).

We define territory here as a space appropriated by a social group with a clear sense of co-ownership (Brunet et al. 1992). In the Vietnamese context, the district represents the most relevant scale for this concept. The territory's actors include farmers, socio-professional representatives, actors upstream and downstream of the agri-chains, residents, local authorities, officials of supporting services, and policymakers. The territory is considered, at the same time, a framework for regulation, within which and with which changes take place, an organization of space for production, a place of compromise and adjustments between social demands and production, and a tool for intervention by the public actor (Caron 2005). It is thus thought of as a material, identity, and organizational framework (Angeon et al. 2006). In other words, the territory is 'a privileged space for coordination between market relationships, State power, and participation of civil society' (Lardon et al. 2008).

Territorial development reflects the capacity of local actors to control and monitor developments that affect them and to anticipate the hazards and pitfalls that are likely to confront them (Deffontaines et al. 2001). Sustainable territorial development is thus a 'social construct' aiming at a particular trajectory that 'incorporates in particular the co-evolution of economic, social, and ecological systems'. In other words, it 'is defined as the construction and articulation by local actors of economic, ecological, and ethical norms' (Angeon et al. 2006). Each case study is thus analyzed in terms of the capacity of local communities to collectively take charge of sustainable territorial development, and thus to articulate those norms. We compare these situations by using a common analysis grid, which considers in particular the critical points defined with respect to local contexts for each dimension of sustainable development (economic, environmental, and social) (López-Ridaura et al. 2002). We then identify the risk of contradictions between these three dimensions and necessary trade-offs (Callois 2006). Finally, we include in the analysis the levels of cohesion and identity that reflect the collective appropriation by the communities of the challenges of sustainable development (Callois 2006). We identify in particular local sources of cohesion and collective identity related to commercial exchanges, as well as the degree of advancement of a project likely to unite the agri-chain's actors around a common goal of sustainable development.

5.2 The Vietnamese Agri-Chains That Are Transforming Territories

The three case studies presented here reflect contrasting situations from the perspective of agricultural systems, remoteness of the market, and the social context. These studies concern the milk shed in Ba Vì, the production of healthy vegetables at Mộc Châu, and maize production in Mai Sơn.

5.2.1 The Ba Vì Milk Shed, Between City and the Countryside

The inclusion of dairy products in Vietnam's urban food culture has led to the recent emergence of several milk production areas in the country. The growth of the Ba Vì milk shed is emblematic of these transformations (Duteurtre et al. 2015).

Ba Vì district is located in the northwest part of the administrative province of Hanoi, on the low hills surrounding the Ba Vì mountain. Dairy cows were introduced to the area by French colonists in the 1920s. Starting in the 1960s and during the collectivist period, these farms were nationalized. Milk production was then concentrated in a single State farm. In the early 1990s, the sector was gradually liberalized, and the animals and the land of the former State farm were entrusted to some hundred households which embarked on dairy farming through sharecropping agreements. Today, this very small-scale family production has gradually spread into the neighbouring rural communes. Some of these private farms also grow rice, tea, fruit trees, or raise pigs. In 2014, about 1500 households raising a total of 9000 dairy cows between them were involved in dairy farming in the district.

Most of the farms are very small, with an average of six dairy cows per farm raised in stalls and with less than one hectare of cultivated land. Animal feed is based on the intensive cultivation of elephant grass (*Pennisetum pupureum*) throughout the year and on maize cultivation in the rice fields during the winter. The milk is collected by several dairies through a network of private collectors. Between them, two firms collect 93 % of the milk produced in the district on contract basis: the IDP company and the Ba Vì dairy company (Mai Huong et al. 2014).

Dairy farming is one of the most profitable agricultural activities in the district, especially for large farms and mixed farms (Duteurtre et al. 2015). The long experience of Ba Vì in milk production has endowed local farms with recognized expertise. Ba Vì dairy farmers also benefit from the experience of the old State-owned farm which has been converted into a research centre for the development of locally suitable fodder and livestock. The certification of a 'Ba Vì milk' mark, registered in 2009, has brought recognition to the district in the eyes of consumers in the capital.

However, the gradual expansion of family farms and the recent appearance of a very large industrial farm have generated new problems in the territory. Biogas production, the installation of effluent treatment units, and marketing of local effluents constitute options to be further explored to ensure a smooth growth of dairy farming in this relatively touristic area subject to high peri-urban pressures. Moreover, the recent fluctuations in milk prices have been vigorously condemned by farmers and have highlighted the strong dependence of farming communities on the strategies of dairy firms that collect their milk. In the absence of powerful and dynamic farmer organizations, local authorities play a crucial role in negotiations between dairy farmers and these firms.

5.2.2 *Healthy Vegetables in Mộc Châu, an Innovative Agri-Chain*

To feed rapidly growing cities, vegetable production has to expand beyond peri-urban spaces into new production areas, where it soon alters the territory. This is what the example of Mộc Châu shows (Sautier and Nguyen 2015).

Located in a mountainous area at an altitude of 400–1500 m above sea level, the Mộc Châu region (Sơn La province) is experiencing a boom in vegetable production. This area, traditionally populated by highland communities (H'Mông, Thái, and Mường), has been receiving, since 1950, a large number of migrants (Kinh) from the Red River Delta. Resettled densely into small plots and without mastery over local technical systems, these newcomers have logically continued their activities of rice farming and vegetable cultivation. Cabbage, onions, tomatoes, and beans were produced, initially only for the family and the local market, later for Hanoi's urban market.

These vegetable-growing neo-highlanders of Mộc Châu have experimented with vegetable production and adapted it to this region's cooler climate, growing in what is the off-season for vegetable growers in the delta. Lacking the support of technicians, they sought advice from their contacts back in their home villages, and obtained suitable seeds and other agricultural inputs from them. In doing so, migrants have introduced innovations in highland agriculture, especially for ridge farming. But their neighbours, the H'Mông, Thaí, and Mường, have not so far adopted commercial vegetable farming.

In the 2000s, with growing urban demand, the sale of Mộc Châu vegetables in the delta has increased, and the reputation of this area of plentiful land and water has grown. Wholesalers have encouraged the production of vegetables and even sometimes entered into contracts with farmer groups. The market at Kilometre 20, originally the main location for retail sales, has gradually evolved into a transhipment point for loading trucks with vegetables bound for Hanoi.

Under the impetus of a project to develop off-season vegetable farming, experiments were conducted in Mộc Châu to produce healthy vegetables (*rau an toàn*), i.e., growing vegetables in conformance with reasoned farming standards in force in Vietnam. Village producer groups have been created to facilitate the implementation of environmental and marketing specifications. Vegetable growers now interact regularly with local authorities to fine tune these specifications in association with the traders involved. To ensure continued sustainability and emphasize the territorial linkage of this agri-chain, a certification mark bearing the name of the territory is being launched (Sautier and Nguyen 2015). It will join the existing 'Mộc Châu milk' trademark and the 'Mộc Châu tea' protected geographical indication.

5.2.3 Maize Cultivation at Mai Sơn, Between Firms and the Peasantry

Maize production has grown spectacularly in Vietnam since the mid 1990s. But since it is practiced on recently cleared slopes, it creates risks of soil erosion. This is especially true in Mai Sơn district (Thao 2005; Hauswirth et al. 2012).

Mai Sơn district is one of the most famous highland areas for hybrid maize production in the country. Yields are as high as 4.66 T/ha and production can reach 90,000 T/year (SLSO 2012). Hybrid maize arrived in the region due to a government initiative. After a few years of experiments, a national maize seed production centre opened in 1996, and the district's agricultural extension centre disseminated the seeds and techniques to farmers. A few independent farmers adopted the innovation first, and once the incomes from it became apparent, others followed suit. From the 2000s, Vietnamese and foreign private companies installed themselves upstream and downstream of the agri-chain. The private sector gradually replaced the State in organizing the trade in this crop.

Hybrid maize seeds are made available to farmers by large national and international companies (An Giang, Syngenta, Monsanto, Pioneer, CP, etc.) through an extensive network of wholesalers and retailers. These retailers also sell other inputs used in growing maize (fertilizers, pesticides, insecticides). The majority of inputs are sold to producers on credit, creating informal debt relationships. Seed companies also provide local technical training and undertake field trials.

The farmers sell most of their maize to collector-processors with large warehouses, trucks, drying facilities, and maize husking machines. The presence and efficiency of these facilities has led to an improvement in post-harvest processing and made Mai Sơn maize more competitive as well as more attractive for traders. The advent of these collector-processors has strongly contributed to an increase in local production. The dried maize kernels are sold to independent brokers or commercial agricultural product companies who then resell the production to animal feed companies (Dabaco, Proconco, East Hope, Cargill, CP, etc.) (Bocquillet 2013).

The entire maize chain at Mai Sơn is thus dominated by large private companies that intervene locally through a network of retailers of inputs and maize collector-driers. This network allows farmers to practice remunerative agriculture and have a guaranteed outlet for their products, but, in return, they have to conform to technical conditions and accept prices imposed on them. This situation leads to a partial loss of control of the domestic economy; high levels of indebtedness; increased dependence on suppliers, markets, and intermediaries; and a vulnerability of producers to market risks (Culas and Pannier 2015). In this context where the State is reducing its active involvement in the market, the absence of collective organizations of producers contrasts with the domination of large firms.

5.3 Summary of the Observed Transformations

These three case studies highlight the fact that marketing channels are not only radically upsetting local equilibriums, but they are doing so in several different ways.

5.3.1 The Emergence of Markets Impels an Agricultural Transition Which Allies the State, Producers, and Private Companies

The trade in agricultural goods induces a transformation of agricultural systems: they diversify or specialize by integrating new activities. Examples are the emergence of dairy farming in Ba Vì, new types of vegetables at Mộc Châu, and maize at Mai Sơn. The commoditization of production systems is a process highly dependent on the relationships that form between farmers, businesses, and traders. Upstream industries play an important role (seed maize, vegetables) as do downstream ones (milk, maize) and State services. Traders play an important role in all the three agri-chains studied. These commercial relationships are based, depending on the cases, on formal contracts, personal networks, credit relationships, or the exchange of information and expertise, all of which engender confidence and make trade more secure.

5.3.2 *The Growth of Agri-Chains Leads to an Intensification of Technical Systems and a Higher Economic Risk for Farmers*

The analysis of these three agri-chains highlights the impact of the upstream and downstream actors on the emergence of new, more efficient production methods. We see the appearance of new crops or new varieties and the introduction of new breeds or even new types of animals. Production and processing methods become more intensive in capital and land, which lead to productivity gains and additional value creation. New processing techniques also emerge, such as maize drying and milk processing in the cases we studied, but there are also associated risks for producers who agree to go into debt. Price fluctuations can then lead to acute social crises, as was the case for the Ba Vì dairy sector in 2015. The management of these risks is at the heart of relationships between the rise of commercial agri-chains and sustainable territorial development.

5.3.3 *From the Derivation of Economic Value from Local Natural Resources by the Market to Their Tegulation by Local Organizations*

The agri-chain allows the management and derivation of value from natural resources. As illustrated by the soil erosion problems in the maize chain and by the risk of soil contamination in the milk and vegetables chains, the impact of an agri-chain on local natural resources is one of the essential data for determining the sustainability of commercial agricultural development. Indeed, it seems important to put in place procedures for defining and managing limits, or tolerance thresholds, in consultation with all actors involved in production, trade, and environmental management, i.e., the territorial organizations tasked with managing these resources. This observation is illustrated in another context in Box 5.1.

Box 5.1 Reconciling the Development of Agri-Chains and the Sustainable Management of Groundwater Resources: Examples From France and Morocco

Jean-Louis Fusillier and Caroline Lejars

Globally, one-third of irrigated land for agriculture is based on the exploitation of groundwater resources. The growth in agricultural use of available resources, and their consequent depletion, can lead to competition and conflicts of use. While farmers dependent on irrigation and already in competition for access to water are undoubtedly affected, so are their downstream partners in the agri-chains, those involved in collection or initial processing.

Water Sharing Between Agri-Chains in Beauce

The Beauce plains, a major breadbasket of France, have seen significant diversification of crops since the 1960s, thanks to the extensive development of irrigation from groundwater. Rotations on the farms have become more complex under the combined effect of the introduction of crops with higher added value (potato, onion, vegetables for canning, seeds), development of specific contracts that include irrigation arrangements, and qualitative segmentation of markets in all the agri-chains. In the 1990s, a decline in the water table led to the imposition of regulated access to it, through the establishment of extraction quotas per farm. The annual setting of quotas, based on the winter groundwater recharge, induces risks of severe restrictions for farmers dependent on irrigation.

Significant water restrictions would make many of the current production systems unviable, in particular by impacting the production of vegetable and seed crops for which irrigation is essential (Lejars et al. 2012). While changes in crop rotation at the farm scale are *a priori* possible, the significant decrease in production volumes at the territorial scale would generate competition between irrigated agri-chains and would not allow all specialized downstream operators to survive, leading to issues of broken contracts and lost investments.

Impact of Massive Drawing of Groundwater on Agri-Chains in Saïss, Morocco

Historically, agriculture in Morocco's Saïss region has been rainfed, based on cereals and sheep farming. Following the droughts of the 1980s, the relaxation of restrictions on digging wells, and the development of drip irrigation, the region experienced a significant increase in vegetable cultivation areas and yields. Saïss has increased its production tenfold in 15 years and now accounts for 50 % of national onion production. As irrigated areas have expanded, water withdrawals have increased. At certain locations in Saïss, the groundwater levels have fallen by 60 m in 20 years.

(continued)

Box 5.1 (continued)

The exploitation of groundwater in Saïss has had a significant economic impact on the development of the vegetable chains and all the downstream actors (Lejars and Courilleau 2015). 80 % of production is sold through informal marketing networks and only 20 % via the wholesale markets. There are 2000 intermediaries for 4000 producers. These intermediaries are involved in the fields of marketing, brokering, and delivery of minor services. Although the margins are very unevenly distributed between them along the agri-chain, these intermediaries are highly dependent on vegetable production.

In the medium and long term, increases in production costs – coupled with lower sales prices due to market saturation – call into question the very viability of these agri-chains and the future of these actors. This is in addition to the issues of access to water resources for direct users. The opportunism and multifunctionality of the actors of this informal economy could be a key factor of adaptation in the case of a water or market crisis. However, the region's actors are struggling to adapt to market changes and are unprepared for the decline in water resources, in particular due to the speed of these changes.

Conclusion

In practice, the actors of irrigated agri-chains who are located downstream of production are rarely taken into account in the water management processes, despite their key role in the development of production and in enhancing the value it generates. It seems essential to analyze their relationships, and to include in the analysis a *filière* or value chain approach, in economic and financial terms, as well in terms of organization and contractual modalities, in order to assess the economic impacts of reduced availability of water. Including these actors in water management processes would also lead to an improved understanding of the knock-on effects of water restrictions.

5.3.4 Agricultural Value Chains Based on the Building Up and Leveraging of Local Expertise and Know-How

The commercial dynamics at play leverage the skills and know-how of the actors in agriculture and agrifood processing. In the cases studied, these skills are in a state of upheaval. While local populations do have the ability to engage in new activities (vegetables, milk, maize), they also suffer from very large disparities in accessing these skills. New opportunities are offered by trade channels opening up for these new products. The emergence of distinctive quality marks specific to these products allows more value to be assigned to them, and in a more sustainable manner, on the markets: trademarks, certification marks, and geographical indications. The distribution of the quality rent is then subject to negotiation between farmers, industry, and local authorities, with industry extracting the greatest advantage when producers are poorly organized. This observation is illustrated in another context in Box 5.2.

Box 5.2 Fences, Animals and People: What Sustainable Alternatives for the Development for Bovine Meat Chains in Southern Africa?

Pascal Bonnet and Ferran Jori

In Namibia and Botswana, beef production far outstrips domestic demand. The priority accorded to exports has led to an organization of territories based on the establishment of zoo-sanitary zones separated by immense fences (Veterinary Cordon Fencing). Existing policies are based on the separation of the domestic and export beef chains and complex zoning of livestock farming territories. These territories are classified according to the standards of the OIE (World Organisation for Animal Health) into infected, buffer, vaccinated, surveillance, and free-to-export areas. This zoning is associated with the establishment of zoo-sanitary barriers and veterinary control fencing to separate livestock cattle from wildlife. The fences are veritable borders, internal to States and between them; they run for about 7000 km in Namibia and 3700 km in Botswana (Bonnet et al. 2010).

In addition to their beneficial effect on beef exports to the EU, these areas segregated by animal health levels have a major socio-economic impact. Indeed, they separate the export zones, where most of the slaughter services and facilities are located, from production areas for the local market from where exports are forbidden and where services and infrastructure are lacking. The socio-economic segregation of Namibia between the North, more densely populated with poor rural households, and the South, protected by the zoo-sanitary fencing and essentially occupied by large commercial farms, seems clearly to be the result of this geographical configuration based on animal health. Furthermore, these zoo-sanitary fences have a negative impact on wildlife management as they prevent natural migrations across these zones (Jori et al. 2009).

Under these conditions, the interprofessional body representing the association of small livestock farmers (Namibia National Farmer's Union) and of commercial livestock farmers (Namibia Agricultural Union) has embarked on an extensive development programme, with State support, to enable people of the North to sell their beef production on the export market. Complementary approaches, based on specifications and a quality label (Farm Assured Namibian Meat), have been proposed, in addition to the OIE zoning certification. Other models of economic and social development

(continued)

Box 5.2 (continued)

of rural areas have also been proposed as alternatives, such as national quality supply chains (Nature Reserves Beef for free-range beef in Namibia) and the promotion of export industries under standards other than those of zoning schemes (Commodity-Based Trade). In this way, the populations that have hitherto been neglected and marginalized could sell their beef on the regional market.

Finally, new forms of development based on revenues from the exploitation of wildlife have also existed in Namibia and Botswana for a decade (wildlife ecotourism, sport hunting, extensive wildlife farming). The current debate aims to determine the activity (cattle farming or exploitation of wildlife) that provides the most income, employment, added value, and environmental benefits, in order to inform future choices.

The model of zoo-sanitary segregated areas has therefore been called into question in the quest for more equitable and sustainable forms of development.

5.3.5 *The Growth of Agri-Chains Modifies the Governance of Territorial Development*

In conclusion, these case studies show the role played by agri-chains in collective organization and in local regulatory bodies. Partnerships between farmers and commercial and processing actors lead to the emergence of agricultural professionals' groups by agri-chain. We are also witnessing the emergence of a process of partnership between local authorities and private companies. In the case of Ba Vì and Mộc Châu, local authorities have managed to convince private companies to participate in a collective project for the sustainable development of the territory as a whole. In contrast, in the case of Mai Sơn, the role of public authorities is more limited and there is a growing influence of private firms in the trajectories of territorial development.

5.4 Applying a Comparative Approach to the Links Between Agri-Chains and Sustainable Development

Table 5.1 lists the critical points of the three case studies for each dimension of sustainable development and the risks of contradictions associated with the expansion of each of the three agri-chains. From this table, it appears that the maize boom in Mai Sơn is generating major risks for the district's development, while these risks are lower in the other two case studies.

Table 5.1 Risks arising from the development of the three agri-chains studied for the sustainable development of territories

Dimensions	Critical points	Ba Vì milk	Mộc Châu vegetables	Mai Sơn maize
Economic	Dependence on inputs	+ Feed concentrate and access to land	+ Seeds, fertilizers, and pesticides	+++ Seeds, fertilizers, and pesticides
	Impact of price volatility of products	++ Price of milk	– Diversified production	+++ Price of maize and risk of indebtedness
Environmental	Impact of agricultural practices on land and water quality	+ Management of livestock farming effluents	+ Use of phytosanitary products	+++ Soil erosion and use of phytosanitary products
Social	Capacity of producers to organize themselves	+ Some small and relatively weak groupings	+ A few producer organizations	+++ No professional organization
	Strengthening links between actors of the chains	– Tripartite contracts between industry, collectors, and dairy famers	+ Agreements between organizations and wholesalers	+ Trade links and credit for inputs

–: little or no proven risk; +: existence of a risk; +++: existence of a major risk

These risk levels pertain to differences in the levels of collective appropriation of the issues of sustainable development. Table 5.2 presents an analysis of the sources of cohesion and collective identity, and of levels of advancement of a collective project that could unite the agri-chain's actors around a common goal of sustainable development. This table shows that in the case of Mai Sơn, the acute competition that currently exists between firms in the upstream and downstream markets makes it difficult to build a sustainable local community project.

5.5 Conclusion: Sustainable Development and Governance of Agri-Chains

This chapter sheds light on the mechanisms through which agri-chains contribute to territorial development. It also presents a method to assess how far this development is sustainable. This method relies on identifying critical points and contradictions between the three dimensions of sustainable development. The analysis of forms of organization of the agri-chains' actors and the regulatory institutions overseeing them are key in this evaluation method.

In the case of 'healthy vegetables' from Mộc Châu, the role of traders in the establishment of a collective approach helps solve issues pertaining to the use of chemical products. At Ba Vì, the sustainable development of the dairy chain seems to be guaranteed through close coordination and consultation between local authorities and locally established agro-industries, although risks arising from the

Table 5.2 Levels of collective appropriation of issues of sustainable development by local entities in the three agri-chains studied

Indicators of collective appropriation of issues of sustainable development	Ba Vì milk	Mộc Châu vegetables	Mai Sơn maize
Source of cohesion within the agri-chain	Producer-collector links	Producer-trader links	Producer-trader links
Topics discussed by the actors	Role of groups – Price of milk – Certification mark – Management of effluents	'Healthy vegetable' certification initiative and technical model	Negotiations on maize prices – Access to credit – Soil erosion
Collective identity generated by the agri-chain	Strong (mark)	Strong (certification)	Weak (opportunism)
Level of advancement of a collective project to deal with the challenges of the territory's sustainable development	Project for obtaining protected geographic indication – Local authorities 2020	Project being constructed around certification	No long term project concerning the maize chain
Conclusion: level of appropriation of the issues	Strong	Strong	Weak

Sources Culas and Pannier (2015), Sautier and Nguyen (2015), Duteurtre et al. (2015), and Mai Huong et al. (2014)

fluctuations in milk prices still persist. In the Mai Sơn maize chain, private seed companies and the animal feed industry dominate, which poses risks for the environment and, more generally, for the sustainability of territorial development.

These three cases illustrate the significance of the local presence of commercial and agro-industrial actors, in addition to the appropriation by farmer communities themselves of issues of these agri-chains' development. Thus, the sustainability of development largely depends on how agro-industries organize their local presence. Alliances between firms, local authorities, and farmers are at the core of the collective mechanisms of appropriation of issues of sustainable development. For territorial actors to curb the risks of a rapid erosion of resources due to practices driven by opportunistic actors, the State and local authorities have to support historically established local institutions so that they can have a say in the trade-offs between goals of economic, environmental, and social development. In addition, agri-chain professional organizations that emerge must increasingly participate in this territorial regulation.

References

Angeon V, Caron P, Lardon S (2006) Des liens sociaux à la construction d'un développement territorial durable: quel rôle de la proximité dans ce processus. Dév Durable Territ 7. doi:10.4000/developpementdurable.2851 Retrieved 15 April 2016

Biénabe E, Boucher F, Cerdan C, Fournier S (2004) Durabilité des systèmes agro-alimentaires localisés dans les pays en développement: le rôle des ressources spécifiques et de l'action collective. In: Gervais JP, Larue P, Rastoin JP (eds) Actes du colloque développement durable et globalisation dans l'agriculture AIEA, Quebec, 23–24 August 2004, p 1301–1317

Bocquillet X (2013) Animal feed industry in Vietnam. Revalter working paper. Ciradm, Hanoi, 56 p. Accessible at http://www.futurelivestock.net/index.php?language=1&module=document&cateId=1290. Retrieved 15 April 2016

Bonnet P, Bedane B, Bheenick KJ, Juanes X, Girardot B, Coste C, Gourment C, Wanda G, Madzima W, Oosterwijk G, Erwin T (2010) The LIMS community and its collaborative livestock information management system for managing livestock statistics and sharing information in the SADC region (Southern African Development Community). In: Information scientifique et technique et développement rural: éclairages sur des pratiques innovantes. IAALD XIIIth World Congress, Montpellier, France, 26–29 April

Brunet R, Ferras R, Théry H (1992) Les mots de la géographie, dictionnaire critique. Reclus/La documentation française, Montpellier/Paris 334 p

Callois JM (2006) Les relations sociales, frein ou moteur de la durabilité. Développement durable et territoires, 8/2006, made available online on 7 January 2013, http://developpementdurable.revues.org/3284. Retrieved 15 April 2016

Caron P (2005) À quels territoires s'intéressent les agronomes? Le point de vue d'un géographe tropicaliste. Nat Sci Soc 2(13):145–153

Culas C, Pannier M (2015) Dynamiques du territoire à travers le prisme des filières maïs et café dans le district de Mai Sơn: relations entre acteurs, gouvernance et impact sur le secteur élevage, Rapport de terrain. Projet Revalter, Hanoi, CNRS/Cirad, 41 p

Deffontaines JP, Marcelpoil E, Moquay P (2001) Le développement territorial: une diversité d'interprétations. In: Lardon S, Maurel P, Piveteau V (eds) Représentations spatiales et développement territorial. Hermès, Paris, pp 39–56

Duteurtre G, Pham DK, Cesaro JD (2015) Bassin laitier d'élevage façonné par les politiques publiques, entre logiques industrielles et soutien à la paysannerie. In: Napoléone M, Corniaux C, Leclerc B (eds) Voies lactées : dynamique des bassins laitiers entre globalisation et territorialisation. Cardère éditeur, Paris, pp 67–87

Hauswirth D, Pham Thi Sen, Nicetic O, Le Quoc Doanh, Kirchof G, Boulakia S, Chabierski S, Husson O, Chabanne A, Boyer J, Auffray P, Lienhard P, Legoupil JC., Stevens ML (eds) (2012) Proceedings of the 3rd international conference on conservation agriculture in Southeast Asia "Conservation agriculture and sustainable upland livelihoods. innovations for, with and by farmers to adapt to local and global changes", 10–15 December, Montpellier, Cirad, p 246–249

Jori F, Vosloo W, Du Plessis B, Bengis R, Brahmbatt D, Gummow B, Thomson GR (2009) A qualitative risk assessment of factors contributing to foot and mouth disease outbreaks on the Western Boundary of Kruger National Park. Rev Sci Tech 28(3):917–933

Kaplinsky R, Morris M (2001) A handbook for value chain research. Institute of Development Studies, University of Sussex, Brighton 133 p

Lardon S, Tonneau JP, Raymond R, Chia E, Caron P (2008) Dispositifs de gouvernance territoriale durable en agriculture: analyse de trois situations en France et au Brésil. Norois 209:17–36

Lejars C, Courilleau S (2015) Impact du développement de l'accès à l'eau souterraine sur la dynamique d'une filière irriguée. Le cas de l'oignon d'été dans le Saïs au Maroc. Cah Agric 24 (1):1–10. doi:10.1684/agr.2014.0729 Retrieved 15 April 2016

Lejars C, Fusillier JL, Bouarfa S, Coutant C, Brunel L, Rucheton G (2012) Limitation of agricultural groundwater uses in Beauce (France): what are the impacts on farms and on the food-processing sector? Irrig Drain 61:54–64

López-Ridaura S, Masera O, Astier M (2002) Evaluating the sustainability of complex socio-environmental systems: the MESMIS framework. Ecol Indic 2(1–2):135–148

Mai Huong N, Duteurtre G, Moustier P (2014) Governance and economic results of dairy value chain in Ba Vi (Vietnam). 8e journées de recherches en sciences sociales, Grenoble, 24 p

Muchnik J, Requier-Desjardins D, Sautier D, Touzard JM (2007) Systèmes agroalimentaires localisés. Économies et sociétés 29:1465–1484

Pecqueur B (2005) Le développement territorial: une nouvelle approche des processus de développement pour les pays du Sud. In: Antheaume B, Giraut F (eds) Le territoire est mort, Vive les territoires! IRD éditions, Paris, pp 295–316

Réquier-Desjardins D, Boucher F, Cerdan C (2004) Globalization, competitive advantage and the evolution of production systems: rural food processing and localized agri-food systems in Latin American countries. Enterp Reg Dev Rev 15(1):49–67

Sautier D, Nguyen Thi Thanh Loc (2015) Strengthening an emergent horticulture cluster in Vietnam: Interest group and certification trademark. In: Proceedings of 29th International Congress of Horticulture, 17–22 August 2014, Brisbane (Australia). Acta Horticulturae, forthcoming.

SLSO (2012) Son La Statistical Yearbook. Son La Statistics Office (SLSO), June

Thao TD (2005) Maize production and selling maize in Son La. In: Vien TD, Tung NV, Calkins P, Allbritton A, Hiesinger MA, Thao VT, Huong PM, Anh TN, Phuong LTT, Quang DV (eds) Marketing and agroforestry development in Vietnam's uplands. Hanoi, Agricultural Publishing House, pp 212–227

Chapter 6
Collective Action in Agri-Chains

Pierre-Marie Bosc, Sylvain Rafflegeau, Hélène David-Benz, Sylvaine Lemeilleur, Paule Moustier, and Marisa Peyre

The linkage between agri-chains and sustainable development is not an abstract issue. It is rooted in highly diverse realities that we will examine, with an emphasis on the possible forms of collective action by different actors operating within the value chains. Maître d'Hôtel et al. (2011) underline the importance of collective action for agri-chains in developing countries in order to respond to various constraints: limited production and productivity, barriers to entry to some value chains, high production and marketing risks, high transaction costs, lack of economies of scale, weak bargaining power, low human and social capital, and increased health risks.

The question addressed here is: under what conditions can collective action induce changes in agri-chains that could be considered more sustainable? We use real-world examples to illustrate the diversity of collective actions in the South and their impact on the various aspects of sustainability. By complementing this overview with a North/South perspective, we conclude by noting the institutional conditions that link collective action and the sustainability of the evolution undergone by agri-chains.

P.-M. Bosc (✉) • H. David-Benz • S. Lemeilleur • P. Moustier
CIRAD – UMR Moisa, Montpellier, France
e-mail: pierre-marie.bosc@cirad.fr; helene.david-benz@cirad.fr; sylvaine.lemeilleur@cirad.fr; paule.moustier@cirad.fr

S. Rafflegeau
CIRAD – UPR Performance of Tree Crop-Based Systems, Montpellier, France
e-mail: sylvaine.lemeilleur@cirad.fr

M. Peyre
CIRAD – UPR Agirs – NIVR Vietnam, Hanoi, Vietnam
e-mail: marisa.peyre@cirad.fr

E. Biénabe et al. (eds.), *Sustainable Development and Tropical Agri-chains*,
DOI 10.1007/978-94-024-1016-7_6

6.1 A Diversity of Forms of Collective Action

The oil palm chain illustrates the diversity of collective action for categories of organized actors, situated in contrasting socio-economic and political contexts.

In Indonesia, the development model that dominates is one based on industrial plantations associated with individual plots that are more or less autonomous with respect to agro-industry. The State has insisted on the creation of cooperatives to serve as intermediaries between the oil palm mills and farmers with individual plots. These pseudo-cooperatives have no real freedom of action; the firms control and orient, without any consultation, the technical and economic aspects of cultivation, following a model of inclusion that could be best described as paternalistic (Barral 2015; Ramirez et al. 2014).

In Côte d'Ivoire, schemes to develop oil palm smallholdings – planted by smallholders, on their own land, with credit for financing the plantation – did not survive liberalization due to the collapse of development banks (Cheyns et al. 2000). Here too, the State imposed the creation of cooperatives to serve as an interface with the privatized oil palm mills, originally part of public development companies. Founded on a paternalistic inclusion model, the cooperatives can increase their flexibility in undertaking their own projects only if they are able to rely on and increase their human and social capital.

In Cameroon, a national smallholders' union was created by urban investors who hired managerial farmers specialized in oil palm cultivation. However, small producers – who the national union claims to represent – do not trust these leaders, who have failed to broaden their base to defend their interests. Collective action remains ineffective in this model of reluctant inclusion.

In contrast, in Ecuador, the national association of oil palm producers (Ancupa) brings together, on a voluntary basis, all producers, ranging from small family farms to large industrial plantations. Funded by a special tax on products, Ancupa undertakes several roles: dissemination of technical and economic information, provision of technical on-field support, forum for negotiation between the State and the oil palm chain's actors (defining the method of calculating the purchase price to be paid by oil palm mills for bunches), geolocation-based censuses of oil palm plantations, and interaction with research centres on behalf of its members. The dual principle of voluntary membership and financing through a tax on production ensures dialogue, representation of and responsiveness to all the members, and increased transparency in a model of participatory inclusion.

We thus propose to distinguish schematically the dynamics of collective action initiated by actors operating in one or more agri-chains from those induced or brought about by interventions external to these actors, via the State or development aid agencies. Very often, agricultural producers[1] are at the origin of collective dynamics that enable them to solve problems beyond the reach of individual action,

[1] This chapter deals mainly with collective action at the producer level. However, it also discusses how other actors of the value chain (traders, processors, industries, exporters, or importers) affect

in particular to benefit from economies of scale and stronger power of negotiation with upstream and downstream value chain's actors. Collective action is also a tool to improve productivity through innovation or to influence the direction of public policy. Changes in economic policies since the phase of liberalization of the 1990s have led States and donor agencies to reconsider their position on intervention for restructuring agri-chains, thus leading to policies that favour a strengthening of organizations, but sometimes with an interventionism not conducive to autonomy and self-reliance. This has resulted in hybrid forms of governance in which collective action can potentially play a strategic role. However, collective action can be truly meaningful only if institutional and political conditions allow it to be.

These two types of collective dynamics can coexist and combine, both when created and as they evolve over time, according to a myriad of institutional configurations resulting from histories unique to each context. These configurations can lead to collective dynamics that bring together various categories of actors in 'interprofessional' consultation, cooperation, or negotiating platforms. The example of Guinea's national coffee growers' federation illustrates how collective action initiated and supported by bilateral aid, in this case French cooperation, is mobilized by the producers – once the initial investments in skills training have borne fruit – to chart their own future through the projects they initiate. We are thus witness to a transition from a specialized sectoral organization dedicated to the expansion of the coffee chain[2] (Rafflegeau et al. 1999) to an organization that provides support to the diversification of family farms by facilitating access to seedlings of fruit trees and selected oil palm seeds (bulk purchases from Côte d'Ivoire, steps to import the first small-scale mills, etc.).

6.2 Understanding Collective Action and Sustainability Through Some Illustrations

We illustrate here how collective action within an agri-chain can be effective, on the one hand, in inducing economic and social inclusion and, on the other, in improving the environmental and health dimensions of sustainability. It is a matter mainly of examining how collective action can help actors draw benefits from the commercial aspects of sustainability.

---

or are part of collective dynamics, by examining the effect of their collective actions on other actors in the value chain or on other products.

[2] Guinea had no 'public board' type of structure before the liberalization of the 1990s.

6.2.1 *Economic and Social Dimension*

6.2.1.1 Development of the Milk Cooperative Movement in India

The development of the milk cooperative movement in India illustrates the massive social and economic inclusion resulting from the building up of the capacity of the cooperative movement in the interests of family farmers with very limited productive capital. As Kurien (2007) writes: 'The emphasis was on production by the masses and on mass production. The secret of the union (of the cooperative movement) was to combine the wisdom of farmers with the skills and knowledge of professional managers.' Poor Indian farmers have thus benefited from technical improvements (artificial insemination; health care and disease control through the cooperative; animal feed; premiums for quality) that have transformed their livelihoods by increasing incomes and improving their families' nutritional status. Total domestic production increased from 21.2 million tonnes in 1968–1969 to 132 million tonnes in 2012–2013. Over the same period, the average per capita availability of milk has increased from 112 to 290 g. In 2003, 89 % of dairy farmers cultivated less than 2 ha and 80 % less than 1 ha.

The project's overall results can be summarized as follows:

- strengthening the control of farmers and autonomy in the milk sector, at the stages of production, collection, processing, and marketing;
- an improvement in the economic situation of families and cooperatives resulting from market access;
- increased access by small farmers to intermediate and more sophisticated technologies.

The economic environment of the milk chain has been managed under a market economy regulated by the State through careful management of imports (and donations) of milk powder and of prices paid to producers and those paid by consumers (Candler and Kumar 1998).

6.2.1.2 Alianza in the Palm Oil Chain in Colombia

In Colombia, the *alianzas productivas estratégicas* constitute a model of participatory inclusion promoted since the 1990s by the national federation of palm oil producers (Fedepalma). Ever since, the number of *alianzas* and the average area under their ambit continues to grow, reflecting the popularity of this new development model amongst the rural populations. The model is based on the participation of all producers in the *alianza*'s capital and decision making. The result is an increase in productivity thanks to technical advice. In 2014, the *alianzas* represented 21 % of the Colombian oil palm orchard (nearly 70,000 ha) and brought together 5600 smallholders (Mesa-Dishington 2015).

6.2.1.3 The Plantain Chain in Cameroon: A Multi-actor Platform

In a national food chain – of plantain – in Cameroon, the implementation of decentralized multi-actor platforms has allowed the spread of a technology requiring limited financial investments: yields have increased by 20 % and labour productivity by 15 % (higher density and longer life of the plant), with visible effects on national statistics, which are evidence of the extent of this innovation's dissemination (Temple et al. 2015).

6.2.1.4 Market Information Systems

Access to information is one of the services where collective action often seems more effective than centralized schemes imposed from above. So-called 'second generation' market information systems (MIS2G) aim to improve the conditions for market insertion of producers and market performance (price stability, equitable distribution of margins). The growth of mobile telephony today permits diverse and updated information to be provided at a low price. These information systems, which are increasingly backed or promoted by producer or trader organizations, are oriented more towards supporting decision making by the actors of the value chains rather than by policymakers. The use of these information systems is still limited, but their reach is enhanced significantly when they are combined with tools for inserting producers into the market (Galtier et al. 2014): inventory credit, pooling of supplies, contracting, improvement in market infrastructure, etc.

Beyond helping the coordination between individual market participants, how does information strengthen coordination and regulation between the value chain's collective actors and the State? The platform for dialogue and management of the rice sector provides an illustration (David-Benz et al. 2014). Established in 2005 in Madagascar and consisting of representatives of different types of actors and the State, this platform's mission is to foster dialogue and the exchange of information, and to propose measures for the rice chain's development. The platform is based on information produced by the Rice Observatory established in the same year. In a context of paucity of agricultural statistics, the provision of market data, updated and analyzed regularly, has undeniably improved the level of information available to the value chain's actors and decision makers. This information has facilitated the construction of a shared vision of the rice chain and has informed debate on the measures or interventions necessary. During the effective operating period of these two mechanisms (from 2005 to end 2008), rice prices remained relatively stable, despite a volatile international and national context. The sharing, even if only partially, of the strategies of the various key actors of the rice chain was a factor in ensuring regular market supplies that eventually benefited the consumers. However, representatives of producers were quickly marginalized within the platform, with discussions on managing imports taking place mainly between the major operators and the State. In contexts where the experience and reach of producer organizations are limited, consultation mechanisms in the form of interprofessional bodies or platforms place them in situations of sharp asymmetry in terms of negotiating capacity. For such mechanisms to truly help improve the income of producers, and consequently the sustainability of national production, it is vital to train the representatives of farmer organizations. It is not enough to merely supply them with information.

6.2.1.5 'Fair Trade' Certification

Finally, even though an equitable distribution of income is one of the fundamental principles of the cooperative movement – to be achieved through democratic principles and solidarity between members –, it is through the concept of fair trade that it finds its fullest expression. Indeed, only democratically organized producers are eligible for 'fair trade' certification. This certification involves the setting of a guaranteed minimum procurement price paid to producers (fixed by the specifications) and the distribution of a collective premium for investments that benefit all the members and the community. These investments concern education (scholarships, improving school infrastructure, school equipment and supplies), health (establishment of health centres, improving health services), and community projects (construction of ovens, toilets, roads, sanitation projects, etc.).

Nevertheless, the use of the social premium is highly dependent on the skills of the organization and the interests represented in the selection of investments. This underlines the importance of participation, the producers' level of information, and decision-making processes within their organizations. Some authors even consider that the benefits of the fair trade label – which remain controversial in some cases (Vagneron and Roquigny 2010) – depend essentially on the capabilities of these organizations (Hopkins 2000). Organizations therefore offer tools and services to strengthen economic sustainability and social inclusion. Even though these tools are meant for the producer members, they also serve to strengthen the capacity of the organizations themselves.

6.2.2 Environmental and Health Dimensions

In agri-chains in the South, the producers' collective action can help draw commercial benefits from attributes of environmental sustainability, such as the limited use of chemicals by farmers, valued by consumers mainly due to health concerns. When the reputation of a product so created is considered a common good for the community of producers or the value chain's actors as a whole, collective action can be undertaken to protect it. Collective action then expresses itself in the respect for rules. These rules may take the form of specifications monitored for compliance, which go together with a common label that conveys information on the origin of the products and on the specific production practices. Such collective action can be observed in a variety of forms, for example in Vietnam, for vegetables on the outskirts of Hanoi, litchis from Thanh Ha, or fragrant rice from Hai Hau (Moustier et al. 2010; Moustier 2009). Technical advice provided within this framework of producer organizations can also have a positive impact on reducing the use of chemical inputs (Naziri et al. 2014) and, consequently, on the health of producers. Geographical indications – known for their potential for the inclusion of small family farms and for improvements in farm incomes through higher prices of raw products due to collective quality management rules – can also have an effect at the environmental level. These collective tools are based on approaches undertaken at several levels, from the local to the national, that allow the design and implementation of inclusive systems, keeping the risks of exclusion in mind (Biénabe and Vivien 2015).

RSPO certification of oil palm smallholdings within a framework of collective action can also be a vector for guaranteeing sales and improving income for farmers, improving performance in terms of oil extraction rate for the mill (perfect ripeness and presence of loose fruit), and environmentally friendly agricultural practices. This is especially true for agro-industry in Thailand which suffers, because of competition between oil palm mills, from the dual problems of product quality and the securing of supply through intermediaries. It proposes to work directly with an organization of smallholders in a logic of participatory inclusion.

In Vietnam, in the integrated poultry sector, agrifood companies provide technical services to producers. Health management (technical advice, veterinary drugs, care for sick animals and dealing with dead ones) and optimization of productivity (poultry feed) not only benefit small producers under contract with these companies but also other local producers who are geographically close (neighbours) and/or socially close (family members). This technical support ensures local management of animal diseases which, in most cases, replaces the national public system of monitoring and control (Delabouglise et al. 2015a, b; Box 6.1).

Box 6.1 Private Health Surveillance Systems for Local Risk Management

Marisa Peyre

In Vietnam and Thailand, health information circulates among livestock farmers, creating networks based on geographic proximity, on belonging to a community, or on being part of an organized agricultural value chain. These networks, based largely on private structuring of agricultural value chains, are groups of collective action that share values and objectives: identical analysis of risk and common definition of specific management tools. They correspond to a need for an improvement of public monitoring and control systems. Indeed, the latter are defined at the national level and are poorly

(continued)

Box 6.1 (continued)

adapted to local conditions or too restrictive, given the various interests of the actors and their scales of analysis. However, in these private networks, interventions often aim at minimizing the possible effects of diseases only within the network (through the sale of sick animals or those exposed to the disease, for example), while in the public system, they aim to control or even anticipate the causes of disease and limit its spread on a wider scale.

Actors other than farmers may occupy key positions in these networks because of their professional activities (collectors of animals, sellers of veterinary products) or their social standing (village head, local authority). The functioning of these networks can thus reveal power relations and issues other than those concerning animal health (Delabouglise et al. 2015a, b). These collective actions allow local and limited management of health risks. However, the lack of communication and coordination between these local collective actions and national public systems do not allow sustainable management that would reduce the local health risk. These actions transform national disease monitoring and control systems into crisis management systems, activated only when local collective actions find themselves overwhelmed.

(*According to Delabouglise* et al. 2015a, b).

6.3 Conditions Necessary for Collective Action to Improve the Sustainability of Agricultural Value Chains

6.3.1 Differentiated North/South Trajectories

Currently, producer organizations as well as those formed around marketing and processing are present in all countries, in the North as well as in the South. It is however observed that these organizations are often more effective in OECD countries than in developing ones. The OECD countries are characterized by a strong structuring of activities into agri-chains that goes hand in hand with the development of highly productive intensive and specialized agriculture. In such contexts, collective action is a particularly effective tool for economies of scale, reduced transaction costs, and enhanced bargaining power in the economic and political spheres. These dynamics are made possible by the ability of organizations to structure their collective actions in complementary dimensions necessary for success: the farm dimension, with issues concerning the innovation process, including its financing; the territorial dimension, with the territories' specificities taken into account; and, finally, the political dimension, in the context of the rule of law.

In some agri-chains, global giants have emerged from cooperatives that progressively engaged in global competition, including for upstream functions such as financing. Most of these agricultural giants are located in the OECD countries

(WCM 2014). In some of these countries, the rise of value chain organizations has resulted in the creation of interprofessional bodies serving as an intermediary with the State. But this has taken place in tandem with the establishment and stabilization of laws and the setting up of institutions to guarantee competition regimes regulated under national legal frameworks. This is not the case in many developing countries, where these institutional and regulatory frameworks are only hesitantly emerging in an environment of multiple political and institutional changes.

6.3.2 Importance of Institutional Conditions

The differences observed in the evolution of collective actions, with some organizations' collective actions not being as effective in improving sustainability as of others, reflect the weak institutional and regulatory mechanisms that limit the growth of organizations (Maître d'Hôtel et al. 2008).

6.3.2.1 Dynamics and Capacities of Actor Organizations

The prerequisite for collective action lies in the actors' willingness to work together towards one or more goals, either on the basis of existing social relationships or by creating new bonds of solidarity. The negative experiences of the past may prove to be a block but the adoption of democratic rules of functioning can often overcome it and increase social and economic inclusion.

Often structured around an agri-chain, producer organizations can diversify their actions on several products based on their members' needs. These changes should be supported by the State and aid agencies, with the only criterion being economic, social, and environmental efficiency. They can also promote joint actions undertaken with other actors of the value chains and initiate interprofessional processes.

6.3.2.2 Dynamics Downstream of the Value Chain

Consumer demands can help bring about changes in production, encouraging the adoption of practices that are more environmentally friendly and respectful of social conditions of production, either within a fair trade framework or outside of it. They can reinforce the initiatives of first processors or downstream industries, with a stronger impact on social and economic sustainability.

6.3.2.3 Roles and Functions of the State

State interventions can play an especially strategic role through the provision of public goods conducive to the agri-chains' economic development. Nevertheless, each time that the State has sought to impose organizations, the very principle of collective action has been discredited, damage that usually takes a long time to undo. The State's role remains crucial, but only as a guarantor of the enforcement of economic and social rules and for empowering organizations through capacity building.

The ineffectiveness of some collective action is often due to a lack of institutional policy and regulatory frameworks that does not allow the autonomous expression of initiatives, or their regulation and durability over time. A large number of efforts fail not because of the nature itself of collective action but because of an inability to uphold or enforce the rule of law and a lack of recourse in the face of internal and external practices that undermine collective action.

References

Barral S (2015) Capitalismes agraires. Économie politique de la grande plantation en Indonésie et en Malaisie. Presses de Sciences Po, Paris 240 p

Biénabe E, Vivien M-D (2015) Institutionalizing geographical indications in southern countries: lessons learned from Basmati and Rooibos. World Development, forthcoming, available online at http://www.sciencedirect.com/science/article/pii/S0305750X15000881. Retrieved 23 April 2016

Candler W, Kumar N (1998) India: The dairy revolution: The impact of dairy development in India and the World Bank's contribution. A World Bank operations evaluation study. World Bank, Washington, DC, 92 p

Cheyns E, Akindes F, Adie FA (2000) La filière palmier à huile en Côte d'Ivoire 3 ans après la privatisation: état des lieux d'un procès de recomposition institutionnelle. OCL 7(2):166–171

David-Benz H, Rasolofo P, Andriamparany S (2014) La régulation des prix du riz à Madagascar: l'action conjointe de l'information et de la concertation. Cah Agric 23(4–5):295–303. doi:10.1684/agr.2014.0707 Retrieved 23 April 2016

Delabouglise A, Antoine-Moussiaux N, Phan TD, Dao DC, Nguyen TT, Truong DB, Nguyen NTX, Vu DT, Nguyen VK, Le TH, Salem G, Peyre M (2015a) The perceived value of passive animal health surveillance: the case of Highly Pathogenic Avian Influenza in Vietnam. Zoonosis Pub Health, forthcoming

Delabouglise A, Dao TH, Truong DB, Nguyen TT, Nguyen NTX, Duboz R, Fournié G, Antoine-Moussiaux N, Grosbois V, Vu DT, Nguyen VK, Le TH, Salem G, Peyre M (2015b) When private actors matter: information-sharing and surveillance of Highly Pathogenic Avian Influenza in Vietnam. Acta Tropica 2015, July, 147, 38–44. doi:10.1016/j.actatropica.2015.03.025. Retrieved 23 April 2016. Epub 2015 April 3

Galtier F, David-Benz H, Subervie J, Egg J (2014) Les systèmes d'information sur les marchés agricoles dans les pays en développement: nouveaux modèles, nouveaux impacts. Cah Agric 23(4–5):245–258. doi:10.1684/agr.2014.0715 Retrieved 23 April 2016

Hopkins R (2000) Impact assessment study of Oxfam fair trade: final report. Oxfam Fair Trade Programme, Oxford 68 p

Kurien V (2007) India's milk revolution. Investing in rural producer organizations. In: Narayan D, Glinskaya E (eds) Ending poverty in South Asia. Ideas that work. World Bank, Washington, DC, pp 37–67

Maître d'Hôtel E, Bosc PM, Egg J (2008) L'action collective et la durabilité des filières agricoles au Costa Rica. Économie rurale 1(303–305):123–135

Maître d'Hôtel E, Lemeilleur S, Bienabé E (2011) Connexion des petits exploitants à des marchés efficaces. 3e forum européen du développement rural, Palencia (Espagne), 29 March–1 April 2011, 33 p

Mesa-Dishington J (2015) Innovaciones institucionales del gremio palmicultor. Palmas 36 (3):59–71

Moustier P (2009) Gouvernance et performance des filières alimentaires au Vietnam. Économies et sociétés 43(11):1835–1855

Moustier P, Phan TGT, Dao TA, Vu TB, Nguyen TTL (2010) The role of farmer organisations supplying supermarkets with quality food in Vietnam. Food Policy 35(1):69–78

Naziri D, Aubert M, Codron J-M, Nguyen TTL, Moustier P (2014) Estimating the impact of small-scale farmer collective action on food safety: the case of vegetables in Vietnam. J Dev Stud 50 (5):715–730

Rafflegeau S, Descroix F, Kourouma H, Montagnon C (1999) Caféiculture en Guinée: bilan des résultats de la recherche. Plantations, Recherche, Développement 6(6):443–453

Ramirez M, Bernal P, Clarke I, Hernandez I, Rotolo D (2014) Distinguishing patterns of learning and inclusion through the patterns of network formation in developing agricultural clusters, *SPRU: Working paper series* 2014–20. University of Sussex, Brighton

Temple L, Touzard JM, Kwa M, Boyer J, Requier-Desjardins D (2015) Comparaison des trajectoires d'innovation pour la sécurisation alimentaire des pays du Sud. Biotechnolologie, agronomie, société et Environnement 19(1):53–61

Vagneron I, Roquigny S (2010) Cartographie et analyse des études d'impact du commerce équitable: synthèse. Une étude à l'initiative de la Plate-forme pour le commerce équitable. Cirad, Paris 27 p

WCM (World Co-operative Monitor) (2014) Exploring the co-operative economy. Report, WCM, Brussels, Belgium, 116 p

Chapter 7
Agri-Chains (or *filières*) and Food and Nutrition Security: Two Independent Concepts

Sandrine Dury

7.1 Agricultural Production and Food and Nutrition Security in International Discourse

Since the surge in international commodity prices in 2008, food security has re-emerged as a major issue for international aid, both public and private (major NGOs and foundations concerned with development). Most national agricultural policies are defined with food security as their ultimate goal. After noting the significant progress made in this regard over recent decades, the United Nations (2015) maintained priority in its new Sustainable Development Goals – which replaced the earlier Millennium Development Goals – to 'end hunger, achieve food security and improved nutrition and promote sustainable agriculture' by 2030. The first three targets of goal no. 2 are to:

- 'End hunger and ensure access by all people, in particular the poor and people in vulnerable situations, including infants, to safe, nutritious and sufficient food all year round';
- 'End all forms of malnutrition, including achieving, by 2025, the internationally agreed targets on stunting and wasting in children under 5 years of age, and address the nutritional needs of adolescent girls, pregnant and lactating women and older persons';
- 'Double the agricultural productivity and incomes of small-scale food producers, in particular women, indigenous peoples, family farmers, pastoralists and fishers, including through secure and equal access to land, other productive resources and inputs, knowledge, financial services, markets and opportunities for value addition and non-farm employment'.

S. Dury (✉)
CIRAD – UMR Moisa, Montpellier, France
e-mail: sandrine.dury@cirad.fr

E. Biénabe et al. (eds.), *Sustainable Development and Tropical Agri-chains*,
DOI 10.1007/978-94-024-1016-7_7

Furthermore, the European Commission has committed to reducing by seven million the number of malnourished children by 2025, and most of its actions to this end are those promoting agricultural development.

Irrespective of the scales of intervention and the protagonists involved, agricultural production and food security are now systematically linked in the narratives. And yet, in the literature, the nature of the links between agricultural production and food and nutrition security remains unclear. Some of these links, which can be of different types (technical, economic, environmental), have however been described and analyzed in the literature at different scales (ranging from the planet to the agricultural household, for example) (Fan and Pandya-Lorch 2012).

If agri-chains or *filières* are representations of exchanges between agricultural production and (in some cases) food consumption, they can represent one way of conceptualizing the pathways between production and consumption. From an intervention-oriented perspective, we can design agri-chains having beneficial effects on food and nutrition security. Food and nutrition security, as we define it, is achieved when every individual has sufficient food to eat and when this food is safe (food safety), balanced (nutritional quality), and conforms to his or her social and cultural preferences.[1] A study undertaken at the request of Action Against Hunger (international humanitarian NGO whose goal is to end world hunger) has stirred profound reflections on the pathways between agricultural interventions and their effects on nutrition (Dury et al. 2015). We have identified several impact pathways and a series of questions to ask if we want to analyze the effects on nutrition of an intervention in agriculture. We rely on this framework to propose some approaches to specialists and proponents of food and non-food agri-chains in all their diversity to question and assess the effects of their expansion on food and nutrition security.

7.2 The Agri-Chain Centred on an Agricultural Product

We consider here that agri-chains or *filières* are constituted around the production, exchange, and distribution of agricultural, forest, or animal products. People working in and/or on agri-chains are concerned with the technical and organizational modalities of their operation, their regulation, and their impact on sustainable development. In a perspective of action or reflection, to 'develop' or 'conceptualize' one or more agri-chains means, first and foremost, thinking of the products as defined by their materiality. Characteristics, weight, volume, shelf life, and perishability are examples of criteria for defining products. Depending on the contexts and the interests of those involved and concerned, new layers of useful analytical concepts or attributes are added, such as intangible attributes (indications of origin;

[1] My interpretation of the definition of the Committee on World Food Security 2012, p. 7.

various standards or labels reflecting the technical, health, environmental, economic, or social conditions of production or of distribution).

And yet, it is important to remember that, beyond the different conceptual definitions of the agri-chain or *filière*, at its core is a tangible and material agricultural product with all its physical characteristics.

7.3 Linkages Between Agricultural Production and Nutrition at the Country Level

Different representations, especially the impact pathways, linking agricultural production and nutrition have been proposed by Hoddinott (2012) and Ruel et al. (2013). The relationships involved are always complicated and interconnected. The empirical evidence of the effects of agricultural variables on reducing malnutrition, irrespective of whether at the global or micro scale, is scarce and still subject to discussion (Masset et al. 2012). At the macro-economic level, it seems that an increase in agricultural output (as measured by its value and its contribution to gross domestic product) does have an effect on the reduction of malnutrition in countries where production is very low (Headey 2013) and that this effect varies over different periods. Using data from 2010, Gómez et al. (2013) ranked 170 countries in four categories according to their level of agricultural productivity as measured by the value produced per agricultural worker. They compared the nutritional problems of these four categories of countries. The 38 countries with the lowest agricultural productivity (below 1000 US$ per agricultural worker in 2010) displayed high prevalences of both undernutrition (measured by child stunting under age 5) and of micronutrient deficiency (measured by deficiencies in vitamin A and iron). These countries did not present any major problem of overweight or obesity. In contrast, all the other countries had two or three problems of malnutrition due to deficiency (undernutrition and micronutrient deficiency) or due to excess (overweight and obesity). Finally, only a fourth of the best performing countries (10 out of 44 countries whose agricultural productivity is more than 12,000 US$ per agricultural worker) had no problem of malnutrition. These authors point out that an increase in agricultural productivity is certainly necessary in the least productive countries, but that it would not in any way solve the problems of malnutrition due to micronutrient deficiency and of obesity. Both of these studies refer to agricultural production and productivity in terms of economic value.

Other authors, including Smith and Haddad (2001), have focused on food availability, which reflects total agricultural production expressed by volume (and in energy units), and on imports and exports. They explain changes in the prevalence of global malnutrition (as measured by underweight with respect to age) through several variables, including changes in food availability (expressed in calories per day per person) for 63 countries for every year from 1970 to 1995. They show that, over this period, increasing food availability contributed, on average, to reducing malnutrition by a quarter, while improvements in women's education, their status, and their health environment accounted for the remaining three-fourths of this reduction. Furthermore, the effect of increased food availability was much greater in the 1970s than in the 1990s. These different results show that growth in agricultural production has a positive effect on reducing malnutrition (undernutrition) only in countries with very low agricultural productivity, but that outside these countries, the effects remain either inconclusive or even negative.

These results are highly aggregated and provide no information about the pathways followed. There exist a multitude of paths (Fan and Pandya-Lorch 2012) and a number of questions can be asked, such as:

- Does increased production lead to an increase in farmer income and thus to an increase in food purchases? Does it lead to higher prices?
- Do these changes lead to improved food security and better nutrition of urban and rural populations?
- Does the progress in agriculture improve the health environment and help cope with diseases? etc.

Some issues pertaining to agri-chains are addressed below and classified into seven categories, which reflect seven major impact pathways.

7.3.1 Incomes

Do the agri-chains distribute income and to whom? Do the most vulnerable (women and children) benefit from the income created within the agri-chains? If yes, do these people use this income to diversify their diets and/or to take care of their health? If they do not, is it due to a lack of knowledge or because a varied and healthy diet is not available in their living environment or not accessible to them? For example, numerous studies (cited by Hoddinott et al. 2015) show that livestock farmers who have started marketing their production have seen their incomes rise but their children's nutritional status has deteriorated. Nutrient loss from the sale of milk (development of the dairy chain) is not compensated for by purchases on the market.

7.3.2 Availability of a Varied Diet

Do the agri-chains compete with the production of a variety of foods available to all? Does not the promotion of a single large agri-chain jeopardize the availability of and access to a varied diet (at the scale of a national population)? The Indian example, described by Dorin and Landy (2009) and Deaton and Drèze (2010), highlights the risk of concentrating on just some products (cereals) at the expense of other food crops (legumes). Even though India escaped famine thanks to technology and a proactive policy between 1970 and 1990, the nutritional situation over the same period did not improve or even deteriorated. The nutrient content of foods, on average, declined and the prevalence of anaemia among women increased (Welch and Graham 1999).

7.3.3 Prices

Do the agri-chains change price ratios of food products? And for whom? Do they allow urban and other populations (including non-agricultural rural ones) who depend on markets for food to obtain a varied and high quality diet at all times? Are the prices regulated?

7.3.4 Empowerment of Women

What are the effects of different agri-chains on equality between men and women? On the capacity of women to manage their incomes, allocate their time, and decide on their activities? Most studies on food security highlight the essential role of women (Kimura 2013, for example). Some agri-chains improve, while others limit, women's access to incomes, status, and resources that would make them more self-reliant. Bain (2010) showed, for example, how the growth of fruit and vegetable export chains in Chile and the establishment of corporate responsibility standards have undermined much of the female workforce, which has now become invisible and informal, while consolidating the status of a few privileged employees.

7.3.5 Women's Labour

A critical variable in food and nutrition security of individuals has to do with 'care'. The time that individuals have to care for themselves, their children, the elderly, etc., depends on the time available to them and this in turn is often constrained by their productive activities. In the cotton growing areas of West Africa, for example, the growth of the cotton chain and the mechanization of ploughing have led to the expansion of the area under cultivation. The non-mechanized tasks (weeding, harvesting, etc.) are still mainly undertaken by women and thus women's labour time has increased (Girard and Dugué 2009). This is one possible explanation for the 'paradox of Sikasso', which refers to the Malian region where both high cereal production and a very high prevalence of malnutrition are observed (Dury and Bocoum 2012).

7.3.6 Health Risks

The health environment is a key factor in food security. Livestock chains and peri-urban vegetable cultivation continually face it as a challenge. What are the methods used to prevent contamination of food, water, and the environment?

7.3.7 Exclusion

Does the growth of an agri-chain deprive certain populations (the most vulnerable) of access to resources (land, forest, trees, etc.)?

Answering these several questions, on a case-by-case basis, helps determine whether the development of a particular agri-chain – large or small, short or long – can favour or, on the contrary, be detrimental to food and nutrition security of those concerned with the agri-chain, either directly or indirectly (through the territory or prices).

References

Bain C (2010) Structuring the flexible and feminized labor market: global GAP standards for agricultural labor in Chile. Signs: J Women C Soc 35(2):343–367

Committee on World Food Security (2012) Coming to terms with terminology. Food security. Nutrition security. Food security and nutrition. Food and nutrition security. Thirty-ninth session. Point V.a., CFS 2012/39/4. Rome, CFS, 14 p

Deaton A, Drèze J (2010) From calorie fundamentalism to cereal accounting. Econ Pol Wkly 45 (47):87–92

Dorin B, Landy F (2009) Agriculture and food in India. A half-century review from independence to globalization. Manohar Publishers, New Delhi 280 p

Dury S, Bocoum I (2012) Le paradoxe de Sikasso (Mali): pourquoi « produire plus » ne suffit-il pas pour bien nourrir les enfants des familles d'agriculteurs? Cah Agric 21(5):324–336

Dury S, Alpha A, Bichard A (2015) The negative side of the agricultural-nutrition impact pathways: a literature review. World Food Pol 2(1):78–100

Fan S, Pandya-Lorch R (eds) (2012) Reshaping agriculture for nutrition and health. Washington, DC, IFPRI 230 p

Girard P, Dugué P (2009) La motorisation est-elle utile aux exploitations familiales du bassin cotonnier malien? Grain de sel 48:17–18

Gómez MI, Barrett CB, Raney T, Pinstrup-Andersen P, Meerman J, Croppenstedt A, Carisma B, Thompson B (2013) Post-green revolution food systems and the triple burden of malnutrition. Food Policy 42:129–138

Headey DD (2013) Developmental drivers of nutritional change: a cross-country analysis. World Dev 42:76–88

Hoddinott J (2012) Agriculture, health, and nutrition: toward conceptualizing the linkages. In: Pandya-Lorch R, Fan S (eds) Reshaping agriculture for nutrition and health. International Food Policy Research Institute (IFPRI), Washington, DC, pp 13–20

Hoddinott JF, Headey DD, Dereje M (2015) Cows, missing milk markets, and nutrition in rural Ethiopia. J Dev Stud 51(8):958–975

Kimura AH (2013) Hidden hunger. Gender and the politics of smarter foods. Cornell University Press, Ithaca (NY)/Londres 226 p

Masset E, Haddad L, Cornelius A, Isaza-Castro J (2012) Effectiveness of agricultural interventions that aim to improve nutritional status of children: systematic review. Br Med J 344:d8222

Ruel MT, Alderman H, Maternal and Child Nutrition Study Group (2013) Nutrition-sensitive interventions and programmes: how can they help to accelerate progress in improving maternal and child nutrition? Lancet 382(9891):536–551

Smith LC, Haddad L (2001) How important is improving food availability for reducing child malnutrition in developing countries? Agric Econ 26(3):191–204

United Nations (2015) Transforming our world: The 2030 Agenda for sustainable development. United Nations A/70/1, 41 p

Welch RM, Graham RD (1999) A new paradigm for world agriculture: meeting human needs: Productive, sustainable, nutritious. Field Crop Res 60(1–2):1–10

Part II
Agri-Chains as a Framework for Innovation in the Face of Challenges of Sustainable Development

Chapter 8
Contribution of Research to Innovation Within Agri-Chains

Aurélie Toillier and Luc de Lapeyre de Bellaire

Historically, the linear model of knowledge creation and technology transfer has dominated the conceptions of agricultural development (Hermans et al. 2013). Agricultural knowledge originated within universities and research centres and was disseminated through advisory and support mechanisms, funded primarily by States or by development projects driven by international cooperation. The assumption was that the technologies proposed by the researchers were based on an optimal comprehension of the dynamics of production systems and agri-chains or filières, and that factors of adoption were relatively well-understood (Leeuwis and Van den Ban 2004). Even though this linear model of technology transfer has allowed an unprecedented increase in production and productivity, it has, however, been called into question. Its responses to the issues of sustainability and development pertaining to the multifunctionality of agriculture in rural areas have been found wanting (IAASTD 2009). Following the observation that agricultural innovation is not necessarily linked to the direct application of research results, new approaches focused on supporting innovation have been proposed. These approaches allow a more in-depth examination of the processes underpinning technical, social, and institutional changes in a perspective of increased sustainability of and productivity in agricultural systems (Touzard and Temple 2012). It then becomes a matter for the research community to see how its activities and its products can facilitate or initiate these changes (Röling 2009).

Much of the research conducted at CIRAD has been focused for some time now on producing knowledge on agri-chains in countries of the South, with a strong

A. Toillier (✉)
CIRAD – UMR Innovation, CEDRES, Université de Ouagadougou, Burkina-Faso
e-mail: aurelie.toillier@cirad.fr

L. de Lapeyre de Bellaire
CIRAD – UPR Geco, Montpellier, France
e-mail: luc.de_lapeyre_de_bellaire@cirad.fr

E. Biénabe et al. (eds.), *Sustainable Development and Tropical Agri-chains*,
DOI 10.1007/978-94-024-1016-7_8

commitment to capacity building of local actors. These research efforts have led to an evolution of practices, most notably with the creation of socio-technical devices and modalities of multi-actor partnerships for innovation, now essential components of most research and development projects. Their effects on innovation and development processes are, however, not always properly understood and internalized (see examples in Chap. 9). To improve understanding of their benefits and limitations in supporting innovation processes within agri-chains, we will describe the theoretical foundations of the links between innovation processes, organizations of agri-chains, and capacity building.

We present first a systemic approach to innovation which allows the reconsideration of the role of knowledge production and dissemination mechanisms in multi-actor interactive processes. We then present the theory of transition that allows a rethinking of the support for dynamics of development in the agri-chains, most notably the scaling up of local innovations. Once these concepts are presented, we specify the characteristics of innovation processes in agri-chains in the South and discuss the contributions of research.

8.1 Innovation Systems and Socio-technical Transitions: Key Elements in Rethinking Support for the Development of Agri-Chains

8.1.1 Innovation System, Capacity Building, and Incentives to Innovate

Agricultural innovation can be seen as a process of co-evolution between technical, social, economic, and institutional changes (Klerkx et al. 2012). It is now widely accepted that agricultural innovation is a result of an interactive process that involves many public actors (State services, public research institutes, universities, etc.) and private ones (businesses, banking services, etc.), even from domains of activity outside of agriculture (Alter 2005). In a comprehensive perspective, the agricultural innovation system is then defined as all individuals, institutions, organizations, and networks, formal or informal, interacting around the implementation of new technologies, products, processes, services, or forms of organization (Hall et al. 2006). In this perspective, learning plays a key role in generating innovations by helping build up individual skills or promoting organizational changes in pursuit of greater efficiency of the institutions involved (Lundvall 1992; Malerba 2002). The types of actors involved and their modes of interaction determine the efficacy of the learning processes – and whether learning is even possible – and thus of innovation (Hermans et al. 2013). The existence of trading, learning, and cooperation mechanisms (platforms, partnerships, networks) seems to be a factor for facilitating and incentivizing innovation (Spielman and von Grebmer 2006). While public research and advisory and support services are no longer regarded as the driving forces behind these mechanisms, they remain actors able to fulfil essential functions pertaining to the creation and dissemination of new knowledge, identification of the need for new technologies, orienting and guiding the design of the innovation, and building up of the capacity to innovate of the actors involved (Spielman et al. 2008).

8.1.2 Innovation Niches, Socio-technical Transitions, and Lock-In Effects

The theory of transition – a recent concept – helps explain how innovations can spread beyond the framework in which they were designed and create significant impacts on development at much larger scales (Rip and Kemp 1998; Schot and Geels 2008). Three analytical levels can be distinguished, ranging from the local to the global: innovation niches, socio-technical regimes, and the socio-technical landscape.

Innovation niches are places where new solutions are tried out in a given local context. They are considered the potential starting point for significant changes in the dominant socio-technical system, leading also to social changes. They thus correspond to incubator communities isolated from normal operation of the dominant regime and protected from potential market selection effects (Kemp et al. 1998). Disruptive innovations that lead to radical changes first appear at these microlocal scales.

The concept of the socio-technical regime allows the interweaving of social and technological effects to be translated into the definition of a set of rules, procedures, arrangements, and routines that govern the choices and activities of the different actors who are together involved in designing an innovation. For its part, the socio-technical landscape corresponds to the environment outside the influence of local and national actors (the effects of globalization, cultural or historical factors).

Transition refers to the evolution from one socio-technical regime to another. It depends on the interactions between the processes taking place at these three levels of organization, i.e., niche, regime, and landscape. A technological paradigm (such as intensive input use) can be a real lock preventing a transition, even if a more useful or more efficient technology exists (such as agroecological techniques). Studies have shown how such a lock-in situation may discourage actors from adopting organizational and production methods that do not conform to existing standards. There are even so-called self-reinforcing mechanisms that strengthen the initial technological choice, such as increasing returns of adoption, technological compatibility, and the actors' organizational structure (Meynard et al. 2013).

In working with institutionally different standards and rules, the innovation niches allow learning and the construction of economic networks capable of supporting innovation in agri-chains at the level of their different links. By structuring themselves, these niches can facilitate the unlocking and overcoming of the self-reinforcing effects of the standard socio-technical regime.

8.2 The Features of The Innovation Process in Agri-Chains in the South

In countries of the South, support for innovation processes in agri-chains must take into account four key structural elements that can represent barriers and opportunities.

8.2.1 The Value Chain's Organizational Structure

The process of emergence of an innovation within a niche requires coordination between actors. The value chain, the entirety of financial and contractual links that determine the rules for sharing value between the upstream and the downstream, is a privileged space for coordination (Fares et al. 2012). This coordination between actors of the value chain's different links can be apprehended by the degree of vertical integration, i.e., the degree of contractual, financial, or property relations between entities (Porter 1998). The integration, or quasi-integration, helps reduce information asymmetries between actors (Hennessy 1996) and can thus help overcome problems caused by the lock-in effect. Fares et al. (2012) however show that a value chain's organizational structure may also constitute in itself a lock-in mechanism when it restricts the upstream distribution of the value realized from the marketing of processed agricultural products and, in this way, may reduce the innovation capacity of some actors.

8.2.2 The Role of Businesses and Entrepreneurial Dynamics

While it is widely recognized that the entrepreneur can play a key role in innovation by creating commercial activities (Schumpeter 1934), this is always not so easy in the agrifood sector in the context of the South due to:

- difficulties in fostering entrepreneurship. The creation of a commercial activity begins with an initiative, a readiness to take on a challenge requiring a low aversion to risk. But in the South, there are few measures taken at the national level to reduce such risks and thus encourage entrepreneurship. Chiffoleau and Prevost (2012) have shown, however, that short food supply chains offer an excellent opportunity for entrepreneurship in accompaniment with social innovation in agriculture;

- difficulties entrepreneurs have in identifying and grasping existing opportunities. In some cases, they are unable to respond to changing consumer needs or to identify opportunities for modernizing farms. For example, in many African countries, investment in the adaptation of agricultural or agrifood equipment to the constraints of small family farms is sorely lacking even though the need certainly exists (Havard and Side 2013);
- difficulties in bringing products to market that meet the challenges of ecological intensification and food security at competitive prices.

8.2.3 The Importance of Territorial Anchoring

In agri-chains, territorial aspects are of foremost importance. As studies on localized agrifood systems have shown (Courlet 2002; Muchnik et al. 2007; Perrier-Cornet 2009), the leveraging of local resources and the mobilization of local actors around these resources constitute drivers of innovation. Conversely, the territory can also impose limitations on innovation. This is the case, for example, of the impacts of pesticide pollution that influence changes in agricultural practices (Jannoyer-Lesueur et al. 2012).

The relationship with all that is local and the differentiation of networks of actors intervening across territories help in the consolidation or emergence of different models of agricultural development. These models are each backed by specific coalitions of actors, sometimes also involving researchers, and serve as technical and policy references for the agri-chain's actors. These models can coexist, compete, or hybridize between themselves at different territorial levels.

8.2.4 Asymmetries Between Actors of the Agri-Chains and Public Policies

In contexts in which agri-chains are still inadequately structured or are sometimes 'captured' by some actors, public policies have an important role to play by, for example, pairing the support for farmer innovations, through capacity-building actions and measures, to incentive policies (through subsidies, taxes, regulations, etc.). Thus, in certain agri-chains where the downstream is highly industrialized, the technological constraints of these industries can exert lock-in effects on technical innovations in the upstream part, as may be the case for varietal changes or modifications of cropping systems (de Lapeyre de Bellaire et al. 2010).

8.3 Contributions of Research

The research community continues to be a major actor in the innovation process, both through its involvement in the provision of resources necessary for the innovation process (new knowledge, new products, or skills) and in the design of new socio-technical devices for supporting innovation processes.

8.3.1 Renewing Knowledge on the Functioning of Agri-Chains with the Actors

So far, there existed four ways through which research could contribute to knowledge concerning the innovation process (Morand and Manceau 2009):

- Fundamental research. It is not directly related to innovation but can enrich a knowledge base that can be mobilized for innovation (knowledge of genetic and/or physiological mechanisms of plants and animals, studies of technological processes, macro-economic or sociological trends, etc.);
- Applied research. It aims to acquire knowledge for a practical application (studies of the effect of environmental conditions, of the impact of pests and cultivation practices on the functioning of plants; effects of a process on a product's characteristics, etc.);
- Experimentation. It leads to the development of a technique or process to achieve an expected effect (development of a technique to fight a particular disease, development of a technological process for manufacturing a food product, development of an industrial prototype, etc.);
- Technological adaptation. It helps in adapting a technique or method to specific contexts (adaptation of a pest or disease control method to a particular context, adaptation of an existing technological process to a new product, adaptation of an agricultural technology to a new group of actors).

With the linear model of knowledge and technology transfer being called into question, mechanisms for co-construction with the actors of knowledge for managing new agricultural and food systems have come into focus (Barbier and Goulet 2013). The human and social sciences have thus found a more prominent place in research on agri-chains for devising and experimenting with new approaches for co-designing innovation. Participatory approaches and modelling for the design of new production systems (Andrieu et al. 2012), mechanisms for farm support that take household strategies into account (Terrier et al. 2013), and social experiments to promote inclusion (Chiffoleau and Prévost 2012) are all methodological and organizational innovations to facilitate changes in pursuit of increased sustainability. The mobilization of expert and common knowledge in these mechanisms calls for a re-examination of knowledge production processes, the tools used, and the relationships between researchers and actors of development (Fofiri et al. 2015; Goulet et al. 2015).

8.3.2 Mobilizing and Disseminating Knowledge: The Importance of Expert Appraisals at the Request of Actors of Agri-Chains

The expert appraisals undertaken by CIRAD are part of a privileged form of knowledge mobilization and dissemination and lead to a mutual enrichment between development and research. They are conducted in response to diverse requests from individuals, private companies, producer organizations, development agencies, institutional funding entities, or governments. The different actors of agri-chains find CIRAD's knowledge base structurally attractive because this knowledge has, in general, been obtained over long periods and across several agri-chains and territories. Furthermore, CIRAD is also a repository of knowledge shared and enriched through dialogue between groups of specialists (agronomists, specialists of diseases and pests, geneticists, technologists, economists, etc.), and this multidisciplinary expertise can be pooled in response to a request. CIRAD's expertise can be individual, collegial, or institutional.

The application of the knowledge base in new contexts (partners, terrains, adaptations of existing technology, etc.) typically adds back to the available experience and knowledge, and thus makes them more robust. In addition, the diagnosis undertaken usually allows scientific inquiry to be brought to bear on a reality that is undergoing continuous renewal.

Thus, through requests of the various actors of the agri-chain, the expertise available allows the research community to respond to the needs of the moment and produce immediately usable results, while using the knowledge base to renew research issues from the necessary perspective (Fig. 8.1). This feedback loop allows, at the same time, to stimulate innovation, to tailor innovative products to different contexts, and to renew the demand for innovation addressed to the researchers by the agri-chain.

8.3.3 Contributing to Capacity Building

When individuals, communities, or organizations acquire or implement new capacities and skills, changes can take place: changes in policy, practices, or products, which ultimately contribute to innovation, or even to development. Capacity building involves the acquisition of new knowledge or new skills, usually over periods long enough to allow time for learning cycles to unfold.

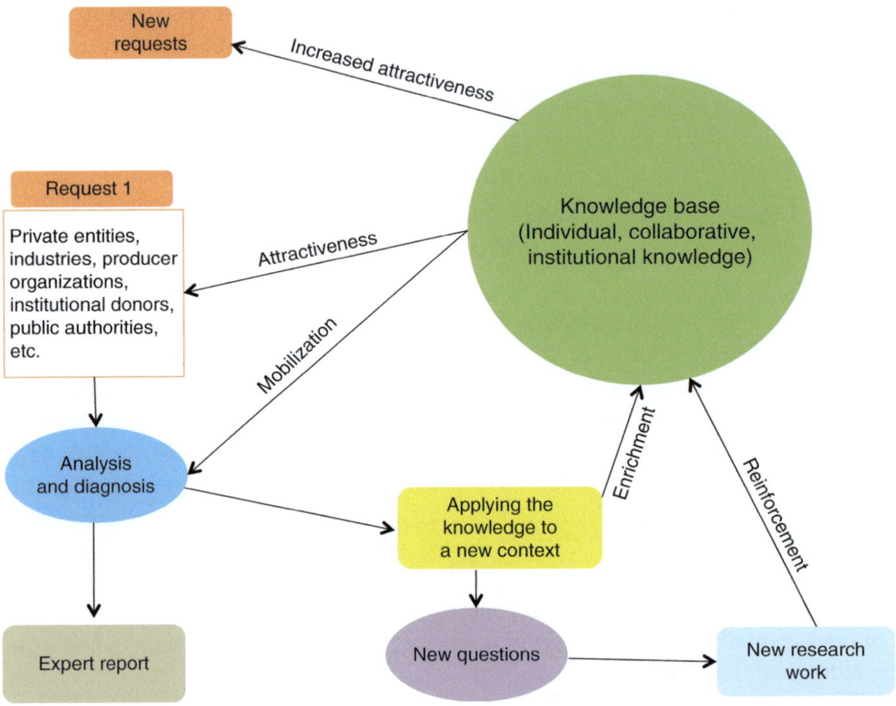

Fig. 8.1 Reciprocal contributions of expertise to research and development through production and the provision of knowledge

Research activities generate different resources that help this capacity-building process: socio-technical devices and learning materials and aids.

8.3.3.1 Socio-technical Devices

Socio-technical devices are arenas for collaboration between researchers and actors of development, supported by financial and human resources, which allow learning cycles to take place through knowledge sharing, discussion, experimentation, and evaluation, with the aim of providing answers to problems, to introduce changes, or to innovate. Examples include mechanisms for action research in partnership (Faure et al. 2010) or multi-actor platforms (Kilelu et al. 2013). However, their effectiveness is not systematic. Jatoe et al. (2013) show that intense interactions around a shared goal in a formalized framework are not necessary to the development of certain agri-chains.

8.3.3.2 Learning Materials and Aids

Learning materials and aids include scientific courses designed for students and researchers, as also professional training for agricultural advisers, technicians, and staff of development organizations and private firms (Box 8.1). In the case of CIRAD, these courses are offered in a variety of areas (plant science, crop protection, product technology, etc.) and can include technical guides (books, data sheets, CD-ROMs, films). Many of these documents have become key information sources for practitioners and are the only ones of their type available to them. Thus, for the cotton chain, a comprehensive series of technical manuals has been published and is updated regularly: identification guides to cotton pests and diseases in sub-Saharan Africa (Cauquil and Vaissayre 2004) and in Brazil; an economic approach to cotton chains (Nubukpo 2011); a manual on the quality of cotton chains in the eight West African countries belonging to the UEMOA monetary union, which covers aspects ranging from production to marketing (Crétenet et al. 2006; Chanselme et al. 2006; Amadou and Bachelier 2006; Gourlot et al. 2006; Diop and Bachelier 2006).

Box 8.1 Training of Sofitex Agricultural Advisers in Burkina Faso

Pierre Rebuffel

As part of its cotton revival programme (1996), the Burkina Faso Textile Fibre Company (Sofitex) scaled up its field presence, increasing its staff strength from 31 to 400 between 1993 and 1998. At the core of its field strategy were 130 cotton correspondents recruited with a high level of initial training (engineer level).

Between 1999 and 2002, Sofitex conducted training programmes for these cotton correspondents with the support of CIRAD in order to better respond to changing demands for support of cotton farms. This programme consisted of two components:

- training in overall farm analysis, so that agents could better apprehend the variety of producer expectations;
- an action-training component to help agents acquire methods and tools so that they could support producers in taking major farm-management decisions.

Even though the Burkinabe cotton sector was subsequently privatized (2004), the new cotton companies (Sofitex in the west of the country, Faso Cotton in the centre, and Socoma to the east) continue to rely upon these skills. They still form the basis of the support mechanism for producers supplying these companies.

8.4 Conclusion

Many innovations emerge independently of any involvement of public or private research (for example, agroecology and organic farming when they began in France, direct seeding in Brazil). Other innovations originate mainly from research: new varieties, new inputs, new sanitary control technologies, and new forms of organization. The role of the various actors, and therefore of research, in the innovation process varies, both in content and in intensity.

The innovation system approach is a framework that permits the conditions that impede or facilitate the innovation process to be analyzed. It is also conducive to a rethinking of the modalities of intervention by research based on the functions that it fulfils within innovation systems. While research does sometimes contribute significantly to innovation, it is not its sole preserve; other actors can also play a fundamental role, with or without interaction with research. It is therefore not always easy to specify the innovation that stems from research activities, especially because innovation always incorporates technical, organizational, institutional, and social dimensions.

Furthermore, every innovation does not necessarily contribute to development in exactly the way various local and national actors expect it to. Microlocal innovations, at the level of 'niches', which is the research community's preferred level of intervention, are not all viable and undergo a process of selection (Nelson 1993). The transition theory approach can be used to reveal the different possible trajectories of innovations and the multiplicity of levels at which research can intervene, as intermediary, facilitator, leader, or just a partner. In Chap. 9, we will see different examples of the ways in which research at CIRAD contributes to innovation processes within agri-chains and the various forms of partnership that this involves.

Monitoring and assessing the impacts of research on development in general, especially that of agri-chains, represent a major new challenge for cooperation for development. International public research is being increasingly called upon to show how it contributes to solving major challenges of food security, climate change, and agroecological transitions (Gaunand et al. 2015). In this perspective, CIRAD has invested its resources in developing impact assessment procedures tailored to contexts of the South and in different innovation processes in which it is involved (Triomphe et al. 2015). The knowledge produced should be able to help researchers and other stakeholders cast a critical and reflective look at the effectiveness of practices of research in partnership and, at the same time, improve the contribution to the processes of innovation themselves.

References

Alter N (2005) L'innovation ordinaire, 2e édition. Paris, PUF, 480 p

Amadou SA, Bachelier B (2006) Manuel qualité pour les filières cotonnières UEMOA. 3. Standards « Afrique » de qualité du coton fibre. Unido, Vienna 39 p

Andrieu N, Dugué P, Le Gal PY, Rueff M, Schaller N, Sempore A (2012) Validating a whole farm modelling with stakeholders: evidence from a West African case. J Agric Sci 4(9):159–173
Barbier JM, Goulet F (2013) Moins de technique, plus de nature. Pour une heuristique des pratiques d'écologisation de l'agriculture. Nature Sci sociétés 21:200–210
Cauquil J, Vaissayre M (2004) Principaux ravageurs et maladies du cotonnier en Afrique au sud du Sahara. Montpellier, Cirad 60 p
Chanselme JL, Kinré H, Bachelier B (2006) Manuel qualité pour les filières cotonnières UEMOA. 2. Égrenage du coton-graine. Unido, Vienna 71 p
Chiffoleau Y, Prévost B (2012) Les circuits courts, des innovations sociales pour une alimentation durable dans les territoires. Norois 224(3):7–20
Courlet C (2002) Les systèmes productifs localisés. Un bilan de la littérature. In: Torre A (ed) Le local à l'épreuve de l'économie spatiale. Agriculture, environnement, espaces ruraux, vol 33. Inra éditions, coll. Études et recherches sur les systèmes agraires et le développement, Versailles, pp 27–40 http://prodinra.inra.fr/record/195184, retrieved 30 April 2016
Crétenet M, Dessauw D, Bachelier B (2006) Manuel qualité pour les filières cotonnières UEMOA. 1. Production de coton-graine de qualité. Unido, Vienna 75 p
de Lapeyre de Bellaire L, Fouré E, Abadie C, Carlier J (2010) Black Leaf Streak Disease is challenging the banana industry. Fruits 65(6):327–342
Diop AM, Bachelier B (2006) Manuel qualité pour les filières cotonnières UEMOA. 5. Pratiques du commerce de la fibre de coton. Unido, Vienna 75 p
Fares M, Magrini MB, Triboulet P (2012) Transition agroécologique, innovation et effets de verrouillage : le rôle de la structure organisationnelle des filières. Le cas de la filière blé dur française. Cahiers Agricultures 21(1):34–45
Faure G, Gasselin P, Triomphe B, Temple L, Hocdé H (eds) (2010) Innover avec les acteurs du monde rural : la recherche-action en partenariat. Versailles/Paris/Gembloux (Belgium), Éditions Quæ/CTA/Presse agronomique de Gembloux, collection Agricultures tropicales en poche, 224 p. (English version: 2014. Innovating with rural stakeholders in the developing world – Action research in partnerships. CTA LM Publishers, 224 p.)
Fofiri Nzossié EJ, Temple L, Ndjouenkeu R (2015) La contribution de la recherche universitaire à la formation d'un système sectoriel d'innovation agro-alimentaire au Cameroun. Innovations 47(2):55–77. doi:10.3917/inno.047.0055 Retrieved 30 Apr 2016
Gaunand A, Hocdé A, Lemarié S, Matt M, De Turckheim E (2015) How does public agricultural research impact society? A characterization of various patterns. Res Policy 44(4):849–861
Goulet F, Sauvegrain SA, Arciniegas L, Bricas N (2015) Innover de façon plus juste ? De la place des connaissances dans un dispositif de coopération sciences sociales – industrie. Innovations 47(2):33–54
Gourlot J-P, Gawrysiak G, Bachelier B (2006) Manuel qualité pour les filières cotonnières UEMOA. 4. Classement de la fibre de coton. Unido, Vienna 51 p
Hall A, Janssen W, Pehu E, Rajalahti R (2006) Enhancing agricultural innovation: how to go beyond the strengthening of research systems. World Bank, Washington DC 135 p
Havard M, Side SC (2013) Les dynamiques de mécanisation de la production et de la transformation agricoles en Afrique de l'Ouest. In: Blin J, Mouras S, Wadre A, Voron A (eds) 4e Conférence biocarburants/bioénergies. Quel bilan et quelles voies d'avenir pour les biocarburants et les bioénergies en Afrique ? 2IE/Cirad/Ministère des Mines et de l'énergie du Burkina Faso/Sud sciences et technologies, Ouagadougou, pp 109–120
Hennessy DA (1996) Information asymmetry as a reason for food industry vertical integration. Am J Agric Econ 78(4):1034–1043. doi:10.2307/1243859 retrieved 30 April 2016
Hermans F, Stuiver M, Beers PJ, Kok K (2013) The distribution of roles and functions for upscaling and outscaling innovations in agricultural innovation systems. Agric Syst 115:117–128
IAASTD (2009) In: McIntyre BD, Herren HR, Wakhungu J, Watson RT (eds) Agriculture at a crossroads. A synthesis of the global and sub-global IAASTD reports. Island Press, Washington, DC 95 p

Jannoyer-Lesueur M, Cattan P, Monti D, Saison C, Voltz M, Woignier T, Cabidoche YM (2012) Chlordécone aux Antilles : évolution des systèmes de culture et leur incidence sur la dispersion de la pollution. Agronomie, environnement et sociétés 2(1):45–58

Jatoe JBD, Lankoandé DG, Sumberg J (2013) Does rapid agricultural growth require a system of innovation? Evidence from Ghana and Burkina Faso. Future Agricultures Consortium Working Paper 058, 14 p

Kemp R, Schot J, Hoogma R (1998) Regime shifts to sustainability through processes of niche formation: the approach of strategic niche management. Tech Anal Strat Manag 10(2):175–198

Kilelu C, Klerkx L, Leeuwis C (2013) Unravelling the role of innovation platforms in supporting co-evolution of innovation: Contributions and tensions in a smallholder dairy development programme. Agric Syst 118:65–77

Klerkx L, van Mierlo B, Leeuwis C (2012) Evolution of systems approaches to agricultural innovation: concepts, analysis and interventions. In: Darnhofer I, Gibbon D, Dedieu B (eds) Farming systems research into the 21st Century: the new dynamic. Springer, Dordrecht/New York, pp 457–483

Leeuwis C, Van den Ban A (2004) Communication for rural innovation: rethinking agricultural extension, 3rd edn. Blackwell Science, Oxford 428 p

Lundvall BA (1992) National innovation systems: towards a theory of innovation and interactive learning. Pinter, London 404 p

Malerba F (2002) Sectoral systems of innovation and production. Res Policy 31(2):247–264

Meynard JM, Messéan A, Charlier A, Charrier F, Fares M, Le Bail M, Magrini MB, Savini I (2013) Freins et leviers à la diversification des cultures. Étude au niveau des exploitations agricoles et des filières. Synthèse du rapport d'étude, Inra, 52 p

Morand P, Manceau D (2009) Pour une nouvelle vision de l'innovation. Ministère de l'Économie, de l'industrie et de l'emploi. La documentation française, Paris 106 p

Muchnik J, Requier-Desjardins D, Sautier D, Touzard JM (2007) Les systèmes agroalimentaires localisés (SYAL): introduction. Économies et sociétés 29:1465–1484

Nelson RR (ed) (1993) National innovation systems: a comparative analysis. New York, Oxford University Press, 542 p

Nubukpo K (2011) L'improvisation économique en Afrique de l'Ouest : du coton au franc CFA. Karthala, Paris, 144 p

Perrier-Cornet P (2009) Les systèmes agroalimentaires localisés sont-ils ancrés localement ? Un bilan de la littérature contemporaine sur les SYAL. In: Aubert F, Piveteau V, Schmitt B (eds) Politiques agricoles et territoires. Éditions Quæ, coll. Update Sciences & Technologies, Versailles, pp 49–68

Porter ME (1998) The competitive advantage of nations. Free Press, New York, 896 p

Rip A, Kemp R (1998) Technological change. In: Rayner S, Malone L (eds) Human choice and climate change, vol. 2. Resources and technology. Battelle Press, Columbus (Ohio), pp 327–399

Röling N (2009) Pathways for impact: scientists' different perspectives on agricultural innovation. Int J Agric Sustain 7(2):83–94

Schot J, Geels FW (2008) Strategic niche management and sustainable innovation journeys: theory, findings, research agenda, and policy. Tech Anal Strat Manag 20(5):537–554

Schumpeter JA (1934) The theory of economic development. Harvard University Press, Cambridge (Mass.)

Spielman D, von Grebmer K (2006) Public-private partnerships in international agricultural research: an analysis of constraints. J Technol Transf 31(2):291–300

Spielman D, Ekboir J, Davis K, Ochieng CM (2008) An innovation systems perspective on strengthening agricultural education and training in sub-Saharan Africa. Agric Syst 98(1):1–9

Terrier M, Gasselin P, Le Blanc J (2013) Assessing the sustainability of activity systems to support households' farming projects. In: Marta-Costa AA, da Silva ES (eds) Methods and procedures for building sustainable farming systems. Springer, Dordrecht, pp 47–61

Touzard JM, Temple L (2012) Sécurisation alimentaire et innovations dans l'agriculture et l'agroalimentaire : vers un nouvel agenda de recherche ? Une revue de la littérature. Cahiers Agricultures 21(5):293–301. doi:10.1684/agr.2012.0577 retrieved 30 April 2016

Triomphe B, Barret D, Clavel D, Dabat MH, Devaux-Sparatakis A, Faure G, Hainzelin E, Mathé S, Temple L, Toillier A (2015) Towards a generic, comprehensive and participatory approach for assessing the impact of agricultural research in developing countries. Communication at the international symposium Societal Values of Agricultural Research, 3–4 November, Paris, Inra, 27 p

Chapter 9
Agri-Chains and Partnership Approaches to Research

Aurélie Toillier and Luc de Lapeyre de Bellaire

In Chap. 8, we explored the relationships between the processes of innovation and the development of agri-chains. The research community is an important actor in this innovation process, with its main contribution being the generation of new knowledge on biotech, economic, and socio-organizational processes. It is also active in the mobilization and dissemination of knowledge in order to build up the actors' capacity to innovate.

The aim of this chapter is to examine the role of research in different innovation processes associated with agri-chains in pursuit of an improved understanding of its relevance to the issues of sustainable development.

To this end, we rely on a wide variety of research experiments conducted by CIRAD to highlight the different types of partnerships concerning innovation that the research community establishes with the actors of the agri-chains. We also explore the issues these partnerships raise about the design and dissemination of innovations.

Using two in-depth case studies of contrasting situations, in which the role of research has evolved over time, we show the growing importance acquired by the partnership in supporting innovation processes and facilitating their impacts on development.

The historical presence of the banana chain on French territory in the West Indies has engendered a relationship between CIRAD and this agri-chain's actors that has endured for over 60 years. The development of the banana chain in a fragile island environment and the specificity of French legislation in relation to other

A. Toillier (✉)
CIRAD – UMR Innovation, CEDRES, Université de Ouagadougou, Burkina-Faso
e-mail: aurelie.toillier@cirad.fr

L. de Lapeyre de Bellaire
CIRAD – UPR Geco, Montpellier, France
e-mail: luc.de_lapeyre_de_bellaire@cirad.fr

E. Biénabe et al. (eds.), *Sustainable Development and Tropical Agri-chains*,
DOI 10.1007/978-94-024-1016-7_9

production areas have continually fuelled the need for innovations and thus turned this territory into a particularly fecund laboratory.

The emerging organic cotton chain in Burkina Faso, which is based on a small family farms, has required technical, organizational, and methodological innovations for which the research community, including CIRAD with its long history of support of Burkina Faso's cotton chain, has been mobilized.

9.1 The Different Types of Partnerships Formed by CIRAD to Accompany Innovation

At the systemic level, an agri-chain consists of distinct groups of actors. Thus, a first group works along the value chain: upstream, including suppliers of inputs (seeds, fertilizers, phytosanitary products, packaging materials, agricultural machinery, irrigation equipment, etc.), access to credit (banking system), the labour force; producers, whether independent or organized; the downstream processing sector (industries, ripening facilities, etc.); and the downstream marketing sector (buyers, distributors, etc.). On this linear dimension of the value chain, a second group of actors is overlaid (States, national and international institutions) who create and enforce the standards and operating rules that apply to the agri-chains.

The need for innovation may arise as a request expressed by all of these groups of actors or, if coordination within agri-chains is weak – as is the case often in the South –, by a subset of these actors. The links between the research community and the agri-chain's actors materialize through these requests and are organized in more or less integrated and formalized frameworks, ranging from a bilateral partnership between a research entity and a single actor to a broader partnership within multi-actor platforms. Research, when it pertains to technical matters, is especially involved in the stages of design of the innovation, which lead to the formalization of partnerships for allocating resources and dividing responsibilities. Formalized partnerships for the dissemination of innovations are less common. The role of the research community may also vary in nature and intensity.

Table 9.1 illustrates the diversity of partnerships that can exist between the research community and these actors and provides examples of CIRAD's involvement in them.

9.1.1 Bilateral Partnerships

CIRAD has entered into many bilateral partnerships with private groups for the design of technical innovations (Boxes 9.1 and 9.2). These partnerships were created in response to direct requests from the private actors concerned. Innovation demands here confront the economic environment of these actors, who participate

Table 9.1 Different types of partnerships for innovation between the research community and actors of the agri-chains

	Partners	Examples	Type of innovation	References
Bilateral partnerships (Research and development)	Private groups	Industrial producer of yeast (Bionext/ Lesaffre) and with Gembloux-Agrobiotech	Developing a biological control solution for a banana storage disease	Bastiaanse et al. (2010)
		Industrial producer of organic matter (Cifre thesis)	Effect of different types of organic matter on the biological control of parasitic nematodes of the banana tree	Tabarant et al. (2011)
		PT Smart (Indonesia), oil palm plantation company	Multidisciplinary research and training on the sustainability of palm oil production systems: fertilization management, soil conservation, integrated pest management, ecosystem services, waste recycling, agri-environmental indicators	Caliman (2011)
		Ecom (green coffee trader), 2003–2018	Transfer of somatic embryogenesis technology of Arabica F1 hybrids – Creation of two micropropagation laboratories (Nicaragua, Mexico)	Etienne et al. (2012)
		Senegalese Sugar Company (CSS), Richard-Toll, Senegal	Integrated sugarcane pest management strategies against Lepidoptera stem borers, advice and recommendations on weed control	Goebel (2013), Marnotte (2015)
	Producer organizations	*Confédération paysanne du Faso* (Burkina Faso peasant confederation) – Assessing the impact of warrantage on agricultural productivity and household food security	A new warrantage tool (interest for banks: offering secured loans; interest for producers and their organizations: funding of counter-season activities, higher market valuation of their products, protection of	Ghione et al. (2013)

(continued)

Table 9.1 (continued)

	Partners	Examples	Type of innovation	References
			part of their production through storage in a community warehouse until the lean period)	
		Ancupa, Ecuador, *Asociacion Nacional de cultivadores de palma aceitera* – Agronomic diagnosis of the performance of village oil palm farms in Quinindé region	Reduce yield gaps between industrial plantations and small village plantations – Redesign, in association with Ancupa, the technical advice provided to smallholders faced with the risks of the spread of heart rot disease	Rafflegeau et al. (2015)
		Réunion Island fruit cooperative –Fruits and vegetables	Design of innovative cropping systems	Dorey (2014)
	International organizations	Promecafe (IICA) – Regional Cooperative Program for the Technological Development and Modernization of Coffee Cultivation in Central America	Creation and dissemination of new F1 hybrid varieties of Arabica in Central America	Bertrand et al. (2012)
	National institutions	Research partnership with Sodecoton in Cameroon	Cotton breeding programme which has helped improve yields, quality of production, and producer incomes	
Partnership platforms	Village innovation platform	Local populations, POs, and NGOs	Design of technical itineraries and territorial management methods by adapting the principles of conservation agriculture	Tittonell et al. (2012)
	Platform for innovative cropping systems	UGPBAN, Union of banana producer groups of Guadeloupe and Martinique (Banamart, LPG) and Tropical Technical Institute	Cropping systems that incorporate cover crops	Dorel et al. (2013)
	Multi-actor international platform	International business-to-business platform: Roundtable on Sustainable Palm Oil (RSPO)	Establishment of voluntary sustainability standards for the palm oil chain	Djama and Verwilghen (2012)

not only in the definition of specifications but also in the evaluation and selection of the innovations of which they are usually the end users. A bilateral partnership can be formed with different types of actors:

- upstream actors such as suppliers of inputs, agricultural machinery, seeds, etc.
- production actors, most often with large plantation groups (Box 9.1 and Table 9.1);
- downstream actors (processing and storage industries, distributors, etc.) (Box 9.2).

Such a partnership is formed within a contractual framework of research collaboration agreements, one-off contracts, or collaborations on theses and/or vocational training. The issues concern the generation of knowledge through basic research, applied research, experimental development, or technological adaptation (Boxes 9.1 and 9.2).

Box 9.1 A Structuring and Evolving Partnership with a Private Company to Accompany Changes in the Palm Oil Chain in Indonesia

Alain Rival

For almost two decades now, CIRAD has been developing, with the Indonesian plantation company PT Smart, a type of partnership whose form has evolved in tandem with recent developments in the palm oil chain. Originally created for the provision of consultancy services by one of CIRAD's specialized research units in order to support and improve the management of fertilization, the partnership evolved rapidly in attempting to respond to new research issues and by incorporating new scientific themes for supporting the company in its efforts to transition towards sustainable production systems: waste recycling, typology of family farming, ecosystem services, agri-environmental indicators, modelling of plant functioning, etc. In this way, the partnership triggered truly multidisciplinary dynamics while facilitating the immediate transfer of innovations conducive to good agronomic practices within the company, its satellites, and associated smallholders (over 50,000 of them). Over 15 % of fertilization at PT Smart is now provided by the composting of oil mill effluents.

Furthermore, by providing privileged access to land, diversified agricultural systems, and very long-term data series, the CIRAD/PT Smart partnership has played a significant role in the training of students, plantation executives, and young researchers. The Libo research station, located near Pekanbaru in Riau Province (east-central Sumatra) has become an international research centre boasting high-level infrastructure, as evidenced by numerous theses and publications. This facility attracts not only CIRAD

(continued)

Box 9.1 (continued)

researchers and PhD students but also colleagues from prestigious research centres, most notably via the ongoing Befta project (Biodiversity and Ecosystem Function in Tropical Agriculture) being conducted with the universities of Cambridge, Oxford, and Southampton (http://oilpalmbiodiversity.com/).

Aware of the growing importance to industry of environmental issues and the lacunae of knowledge in many scientific fields, PT Smart has been organizing, in partnership with WWF and CIRAD, an 'International Conference on Oil Palm and Environment' every 2 years since 2006 (www.icope-series.com).

Box 9.2 A Partnership with a Private Group for Plant Breeding in Brazil

Jérôme Sainte-Beuve

In 1984, CIRAD launched a research programme in Guyana on combating South American leaf blight (SALB) of rubber trees through genetic methods, based on a multigenic tolerance against *Microcyclus ulei* associated with cost-effective rubber production. The rubber varieties grown in Asia and Africa are all highly vulnerable to SALB and any accidental introduction of this fungus in these regions would be catastrophic. To develop rubber cultivation in Brazil and aware of the global challenge posed by SALB, the Michelin company decided in 1992 to launch an ambitious research programme in partnership with CIRAD to counter this threat.

Two research axes were selected: virulence of the fungus and population genetics; and disease epidemiology and study of the genetic basis of resistance to SALB of certain rubber varieties. This multidisciplinary effort is being undertaken by teams located in Brazil (plant breeders), Guyana (phytopathologists), and in Montpellier (molecular geneticists). To date, more than 30,000 genotypes have been created by crossing and are currently being evaluated in the field in different Brazilian regions.

In addition, joint studies at the programme's three experimental sites have led to an improved understanding of the transmission of resistance to SALB from one generation to the next. This has helped in making new tools available to plant breeders to undertake selection based on more reliable and faster markers.

(continued)

Box 9.2 (continued)

In an international context marked by a lack of research on SALB, this collaboration between CIRAD and Michelin has led to the selection of 12 clones exhibiting good tolerance to the disease in the environmental conditions prevailing in the Amazon basin and Bahia State. These clones were made available to other producing countries for evaluation, as part of an international exchange agreement for clones under the aegis of the IRRDB (International Rubber Research and Development Board).

Apart from such partnerships with private groups, CIRAD has also entered into important collaborations with producer groups (cooperatives, producer organizations, federations, etc.). This form of interaction has proven very fruitful in the French overseas departments, where some well-structured tropical agri-chains are active (Table 9.1): sugarcane in Réunion, Martinique, and Guadeloupe; banana in Martinique and Guadeloupe; and pineapple in Martinique, Guadeloupe, and Réunion. A similar situation prevails in West Africa, most notably with cotton producers in Mali, Burkina Faso, Benin, and Cameroon. Very often, these are long-term partnerships, part of projects combining both private funding (from parafiscal levies) and public funding.

Furthermore, in cases where farmers are poorly organized, CIRAD enters into partnership with national entities (Ministries of Agriculture and Research, research institutes) or international ones (European development funds, FAO, international organizations, NGOs) (Box 9.3).

Box 9.3 A Partnership with National Institutions to Support Cotton Farmers in Cameroon

Bruno Bachelier

The Cameroon Institute of Agricultural Research for Development (IRAD) has been undertaking research on cotton continually since the 1950s, originally in partnership with the French Institute for Research on Cotton and Exotic Textiles (IRCT), which later became part of CIRAD.

Beginning in the 1990s, the public funding of Cameroonian cotton research became difficult and the Cotton Development Corporation of Cameroon (Sodecoton), a major beneficiary of research results, was asked to participate in its funding.

Since 2009, research topics are jointly chosen and resources (human, material, financial) are provided by Sodecoton, IRAD, and CIRAD as part

(continued)

Box 9.3 (continued)

of a tripartite framework agreement on scientific and technical cooperation. This framework also provides a basis for the evaluation of research results.

A review of the 2009–2014 period highlights in particular the effectiveness of plant breeding, which has resulted in (1) the popularization with farmers of three new local cultivars (IRMA L457, L484, and Q302) and (2) requests for seeds of these cotton varieties from other countries.

9.1.2 Partnership Platforms

The need for greater incorporation of local knowledge and practices in the design of innovations has led to more extensive partnerships with producers (whether part of an association or not) and the inclusion of local village populations as partners. Their contributions are made mainly within innovation platforms, usually at the local levels (Boxes 9.4 and 9.5 and Table 9.1).

Box 9.4 Management Advice for Family Farms: A Methodological and Organizational Innovation

Aurélie Toillier

Management advice for family farms (MAFF) is based on learning and decision-making processes, following the principles of the management sciences. It aims to build up the capacity of farmers to better manage their farm resources (land, labour, inputs, cash flow, crops, livestock herds) and their off-farm and non-agricultural activities. Participatory approaches are used to allow participants to undertake themselves analyses of their practices, at the various dimensions of their farms (production, processing, marketing, etc.), while taking into account the stages of the management cycle (analysis, planning, monitoring, adjustment, and evaluation) as well as their economic and social environment. MAFF is based on the use of decision-making tools that help farmers analyze their technical and economic results mainly on the basis of farm data that they themselves record. Therefore, farmers acquire a vision and a new and comprehensive understanding of their entire farm and its activities. They become capable of improving their situation on their own, through the initiation and development of new projects or improvements in their farming, managerial, and/or social practices. MAFF can thus be adapted to various agroecological, institutional, and organizational contexts. It complements other advisory approaches, such as specialist advice or support for collective action (Faure et al. 2015).

Furthermore, CIRAD is also present in multi-actor coordination and intermediation mechanisms (Box 9.4), in view of facilitating the networking and the creation of media for disseminating innovations at regional or national scales. These involve entering into partnerships with entities that support and accompany innovation (training centres, consultancy services, advisory and support services).

Box 9.5 The Role of a Partnership with Local People in Incorporating Local Knowledge and Practices in the Development of a New Method of Monitoring

Flavie Goutard

Since 2010, CIRAD is working in Southeast Asia on participatory approaches involving pastoral communities in pursuit of an integrated management of health risks that uses innovative methods of monitoring and control. These approaches allow the leveraging of local knowledge (Binot et al. 2015) and the taking into account of socio-economic constraints (Calba et al. 2015; Binot and Morand 2015) and of various local perspectives (Delabouglise et al. 2015; Calba et al. 2014) in order to propose solutions that are not only effective but also acceptable to the actors.

9.2 A Few Key Determinants of the Contribution of Research to Innovation Processes

In this section, we present the long trajectories of innovations within two agri-chains to which CIRAD has contributed at different stages and the resulting climate of trust and shared understanding that has been built up.

9.2.1 The Export Banana Chain in the French West Indies: Towards an Agroecological Conversion of Intensive Monocultures

Banana cultivation began in the late 1930s in the West Indies and within a decade, starting in the late 1940s, research began to be undertaken to accompany this nascent agri-chain. Research stations were set up at that time by French agricultural institutes which would later be integrated into CIRAD (Lassoudière 2012). Their existence on French soil, in a tropical location and a stable environment, has provided a basis for uninterrupted agricultural research to be conducted to support this agri-chain for a period of over 70 years and has even led to an increased allocation of resources to research. These resources have been mobilized over long time spans, in terms of personnel and structures, in the form of laboratories for agriculture, phytopathology, nematology, entomology, genetics and molecular biology, soil analysis, and post-harvest physiology. In addition, two research stations, several experimental plots on smallholder farms, and observation networks were set up. A collection of genetic resources was created and production basins were instrumented to assess the impacts of pollution. Such an investment was made possible by the high degree of organization of the banana chain in Martinique and Guadeloupe, which facilitated demands for innovation and helped mobilize the necessary funding (Maillard 1991; Lassoudière 2010). The relationship underpinning this partnership has gone through three major phases of innovation over the years and the partnership itself has steadily grown around innovation platforms with the aim of putting this agri-chain firmly on a trajectory of sustainable agriculture. Several specific aspects have been critical to this development: increased productivity and international economic competition, a fragile island environment, the stringency of European legislation, and the challenge of reducing pesticide use.

9.2.1.1 The Revolution of the Cavendish Variety and the Advent of Intensive Monocultures (1960–1990)

From the early 1960s, a varietal revolution had perforce to take place as banana plantations cultivated with the Gros Michel variety were decimated by a fusarium wilt (Panama disease, race 1). A major varietal innovation was the result: a complete replacement of the Gros Michel variety with the Cavendish variety, resistant to Panama disease. This varietal conversion took place in a context of growing international trade in bananas and increased market competition, especially with banana cultivated in the so-called 'dollar zone' in Latin America. This situation resulted in the establishment of intensive monoculture systems to achieve quality objectives defined in this export agri-chain's standards (packaging of fruits in cartons, storage compatible with artificial ripening, reduction in cosmetic defects and storage disease) and the high yields necessary for the agri-chain's economic survival. Ecophysiological knowledge about this new cultivar largely contributed to its growth, by optimizing plant behaviour within production systems (fertilization, stand management, storage physiology, etc.). At the same time, research came up with management methods and tools for these systems to minimize the use of fertilizers and pesticides (Lassoudière 2012). A notable example of this contribution was the design of sustainable ways to combat yellow Sigatoka disease using control strategies based on early-warning systems, which has significantly reduced the number of fungicide treatments in the French West Indies to just 5 or 6 applications on average against 20 in other production areas (Ganry and Laville 1983; Ganry et al. 2011).

9.2.1.2 The Limitations of Intensive Monocultures and the First Innovations in Cropping Systems (1990–2005)

In the late 1980s, the first signs of declining yields began to appear in the French West Indies. Diagnostic surveys conducted on the two islands confirmed the cause: soil parasitism (phytophagous nematodes and banana weevil) that had spread in these intensive monocultures in spite of chemical control (Dorel and Perrier 1990). A first innovation in cropping systems was therefore a proposal to break up the monoculture with an adequate fallow period to clean up the soil, and then replant healthy banana trees originating from *in vitro* culture to prevent new infections by infested planting material (Ternisien and Melin 1989). It took a relatively long time – some 10 years – with many experiments conducted in partnership to validate this approach, understand its limitations, and propose technical adjustments so that it could be appropriated by the producers (Marie et al. 1993; Dorel and Risède 2004). Over time, this innovation spread widely and supporting measures were put in place. During this period, a programme for the genetic improvement of the banana was also launched at the Neufchâteau research station in Guadeloupe to create varieties resistant to Sigatoka disease, which is the cause for the greatest amount of pesticide use in banana cultivation (Risède et al. 2010).

9.2.1.3 Awareness of the Environmental Impacts, Reduction in the Use of Pesticides, and Agroecological Conversion in the Context of the Sustainable Banana Plan

In the mid 2000s, several concomitant factors combined to call into question existing production systems and, in particular, the use of pesticides: the discovery of a long-term soil contamination by an organochlorine insecticide – chlordecone – and an improved understanding of the spread of pollution in the environment through the hydrographic network (Cabidoche et al. 2005, 2009; Saison et al. 2008). At the same time, European and French legislation on phytosanitary products became more stringent (de Lapeyre de Bellaire et al. 2009) in the context of the Grenelle Environment Forum (2007) and the establishment of the Ecophyto plan. This legislative and regulatory context prompted the profession (UGPBAN), CIRAD, governments, and the European Commission to come up with an innovation plan, called the Sustainable Banana Plan (2008–2013), whose objective was to propose more ecological modes of banana production, based on reducing pesticide use and adverse environmental impacts. This innovation plan gave birth to the Tropical Technical Institute and two partnership platforms: one platform for the design and evaluation of innovative cropping systems that incorporate cover crops and the other a plant breeding platform for the selection and evaluation of new hybrid banana varieties resistant to Sigatoka disease. This innovative plan has led to significant advances (considerable reduction in pesticide use, beginning of the adoption of new multi-species cropping systems incorporating cover plants,

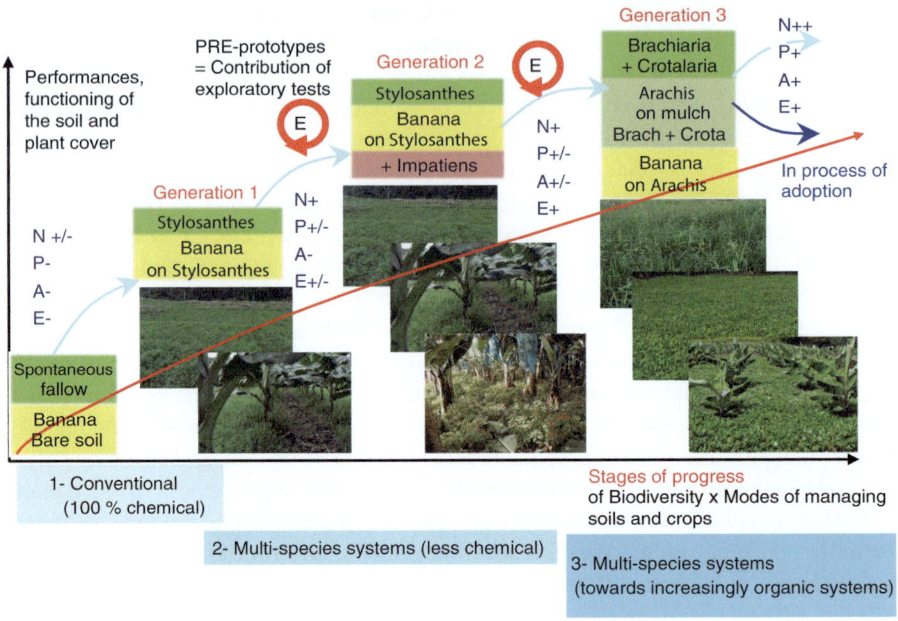

Fig. 9.1 Evolutions of cropping systems incorporating cover crops (Based on Dorel et al. 2014)

Fig. 9.1). A second plan is being prepared, to address new constraints faced by the agri-chain (black leaf streak disease, a new and more aggressive strain of Sigatoka, and the ban on aerial spraying, Guillermet et al. 2013). The Sustainable Banana Plan 2 has the objective of involving all the actors of the agri-chain and representatives of civil society in the design and evaluation of agricultural innovations in order to find the best compromise between all three dimensions of sustainability (economic, environmental, and social) of production systems and to integrate them.

From the initial system of replanting after a spontaneous fallow on bare soil (1) to the more elaborate system (3) of Bracharia + Crotalaria fallow/planting of *Arachis* on mulch/planting of banana trees on the *Arachis* cover, the services provided become increasingly important and the use of inputs is greatly reduced.

9.2.2 The Cotton Chain in Burkina Faso: Contribution of Research to a Service Innovation for Supporting the Conversion to Organic Production

9.2.2.1 Organic Cotton, Genetically Modified Cotton, and Conventional Cotton: Transformations and Innovations in the Cotton Chain

In Burkina Faso, cotton is one of the few cash crops grown exclusively for export. Cotton production in the country has followed a classic trajectory of intensification: chemical inputs, animal traction, mechanization, high-yield varieties, credit schemes, and cooperatives. It enjoyed the support of national research entities for the creation of improved varieties and adaptation of technical itineraries, all under the impetus of the Burkina Faso Textile Fibre Company (Sofitex), which ensured the dissemination and adoption of research recommendations. In this 'green revolution' logic, the use of biotechnology was advocated by experts to raise the productivity and competitiveness of the West African cotton chains (Baffes 2007). Thus, the marketing of transgenic cotton was approved in 2006, paving the way for the design of new production systems that were conceived to be more productive and less reliant on insecticides, have lower production and labour costs, have a smaller environmental impact, and lead to fewer adverse health effects for workers. Transgenic cotton started being grown in 2008 and now accounts for almost 80 % of surface areas under cotton cultivation.

At the same time, the first programme for organic and fair trade cotton production was launched, in 2003, with the support of a Swiss NGO, in partnership with the National Union of Cotton Producers of Burkina Faso (UNPCB). It started with 72 producers but now reaches 8000 farmers, often marginalized in the conventional production system – women, young people who are not farm heads, farmers whose farms were deemed insolvent by cotton producer groups – and who therefore did not benefit from receiving inputs from cotton companies. The establishment of the organic and fair trade cotton chain necessitated innovations at various levels: production techniques (use of green fertilizers, manure, and natural pesticides), input supply channels, producer organizations in the villages for the use of the fair trade premium, and advisory and support services to implement the certification process.

Even though it still represents less than 1 % of production (6100 tons in 2013), organic cotton, whose profitability is 20–40 % higher than that of genetically modified cotton, is slowly hewing an alternate path. Burkina Faso is now the tenth largest producer of organic cotton in the world.

For all three types of cotton – genetically modified, organic, or conventional –, it has been shown that if producers are inadequately informed and supported in their efforts to master the technical innovations proposed to them, the yield and quality of the cotton produced suffer and thus the expected economic, social, or environmental benefits fail to materialize (Renaudin et al. 2012). And yet these changes in the cotton chain took place in a context in which, since the 1990s, funding available for research and advisory and support services has continually declined (Vaissayre et al. 2006). Thus, the deficiency in the dissemination of knowledge to farmers largely explains why genetically modified cotton is falling short of its expectations (Renaudin et al. 2012) and why the organic cotton chain remains fragile, with low yields and inconsistent quality.

9.2.2.2 A Partnership Between Research Entities and Producer Organizations for the Development of Services Adapted to Their Members for the Production of Organic Cotton

The strengthening of the organic cotton chain is partly dependent on:

- the creation, with the producers, of technical and economic reference bases, adapted to their means and contexts of production, that meet specifications of organic agriculture and fair trade certification bodies;
- an improved knowledge of the target farmers and of the differentiated support necessary for the different types of farmers;
- a territorial approach to controlling the expansion of organic cotton production to avoid interfaces with genetically modified cotton and conventional cotton and to improve the efficiency of organic practices.

Profound changes in the design of support and advisory services are therefore necessary. Indeed, these services are still very focused on the transfer of technology packages in accompaniment with collective field or classroom training, but over such short time frames that these efforts are hardly conducive to true learning. CIRAD has been involved for almost 30 years now in the design, testing, and dissemination of new methods of providing advice to farmers (Box 9.4) based on the principles of management advice for family farms (Faure et al. 2004, 2015). With its proven expertise in projects for providing support to producer organizations and the cotton chain, CIRAD has been working in this field in Burkina Faso since the 1990s, initially in partnership with INERA and Sofitex, and later within the framework of a project to provide support to farmer organizations (PA-OPA) from 1996 to 2000 and a project to build up the cotton chain in Burkina Faso (PRFCB) between 2009 and 2011, in partnership with UNPCB. These experiences over the long term have led, on the one hand, to an appropriation of the approach by the agri-chain's actors and, on the other, to a strengthening of their capacity to fine tune this approach as needed.

Thus, in 2013, UNPCB made a request for methodological support to customize CIRAD's approach for providing support to producers to the specific requirements of organic cotton farmers, and to redesign the service so that it was not dependent on external funding. To meet the challenges of the organic cotton chain, advisory services destined for organic cotton producers had therefore to perform several functions: training in organic farming practices, providing tools to help in the financial management of the farm, helping in reorganizing farm labour and production, and helping identify strategies to deal with low yields and/or to compensate for loss of income. This required very different kind of activities, some of which were alien to traditional agricultural extensions services (Box 9.4): training; organizing forums for collective exchanges in the form of discussions of experiences between peers, and between farmers who have converted to organic farming and those intending to do so, in order to capitalize on and develop new technical references; providing advice to individuals or target groups to address technical, economic, and organizational problems; and encouraging experiments by farmers themselves, even helping them create model farms.

Another major concern of these producer organizations is the internalization of the their advisers' skills which allows them to be less dependent on the context and better respond to their members' needs (Toillier et al. 2015). Indeed, UNPCB's request could not be met by suggesting a new way of doing things. What was required instead was the co-construction with the producer organization of new ways of designing and orienting farm support and new advisory tools that could be configured to match the skills of the advisers. The establishment of mechanisms at the level of the producer organization to govern support and advisory services, appropriation of advisory tools by the advisers, and the training of these advisers appear as key factors in improving the performance of advisory services. This diagnosis has led to the design of original teaching aids and the training of 60 advisers who are currently testing these new tools. The main impacts of these methodological and organizational innovations consist of stronger management capacities – at the level both of the individual producer (de Romémont et al. 2014) and of producer organizations – which are key for undertaking changes (Toillier 2014).

9.3 Conclusion

Research undertaken in partnership that leads to innovation relies on the decision taken jointly by several organizations or actors to pursue a specific research and development programme, or simply to establish, over a relatively long period, some degree of cooperation amongst themselves in the form of co-production of knowledge, technologies, methods, or even skills. These partnership agreements can take many different forms but very often, because of their *ex ante* nature, they do not adequately address issues of appropriation and dissemination of innovations. As a consequence, other mechanisms are needed, once the innovation is designed, to ensure the scaling up required for a wider validation and appropriation of the innovation.

The long-term commitment of the research community and the availability of human and financial resources at key stages, most notably in the form of expertise and skills, appear critical to the success of the innovation process. While earlier the focus was squarely on the design of new technologies, the capacity building of actors so that they can participate all through the innovation process is now becoming an increasingly important activity. This has led to changing forms of partnerships and the role of expertise, and a relook at issues of knowledge production.

References

Baffes J (2007) The "cotton problem" in West and Central Africa: the case for domestic reforms. Economic Development Bulletin (Cato Institute), 11, 4 p. www.cato.org/pubs/edb/edb11.pdf. Retrieved 13 May 2016

Bastiaanse H, de Lapeyre de Bellaire L, Lassois L, Misson C, Jijakli MH (2010) Integrated control of crown rot of banana with *Candida oleophila* strain O, calcium chloride and modified atmosphere packaging. Biol Control 53(1):100–107

Bertrand B, Montagnon C, Georget F, Charmetant P, Etienne H (2012) Création et diffusion de variétés de caféiers Arabica: quelles innovations variétales ? Cahiers Agricultures 21 (2–3):77–88

Binot A, Morand S (2015) Implementation of the one health strategy. Lessons learnt from Community-Based Natural Resource Programs for communities' empowerment and equity within an ecohealth approach. In: Morand S, Dujardin JP, Lefait-Rollin R, Apiwathnasorn C (eds) in pressSocio-ecological dimensions of infectious diseases. Springer, Singapore

Binot A, Duboz R, Promburom P, Phimpraphai W, Cappelle J, Lajaunie C, Goutard FL, Pinyopummintr T, Figuié M, Roger FL (2015) A framework to promote collective action within the one health community of practice: using participatory modelling to enable interdisciplinary, cross-sectoral and multi-level integration. One Health 1:44–48

Cabidoche YM, Clermont-Dauphin C, Lafont A, Cattan P, Achard R, Caron A, Chabrier C (2005) Stockage dans les sols à charges variables et dissipation dans les eaux de zoocides organochlorés autrefois appliqués en bananeraies aux Antilles : relation avec les systèmes de culture. In: Pesticides : comment réduire les risques associés ? Avignon, 14–16 November 2005 (Ministry of Ecology and Sustainable Development). Colloque « Pesticides : comment réduire les risques associés ? », 14–16 November, Avignon (France), 6 p

Cabidoche YM, Achard R, Cattan P, Clermont DC, Massat F, Sansoulet J (2009) Long-term pollution by chlordecone of tropical volcanic soils in the French West Indies: a simple leaching model accounts for current residue. Environ Pollut 157(5):1697–1705

Calba C, Ponsich A, Nam S, Collineau L, Min S, Thonnat J, Goutard FL (2014) Development of a participatory tool for the evaluation of village animal health workers in Cambodia. Acta Trop 134:17–28

Calba C, Antoine-Moussiaux N, Charrier F, Hendrikx P, Saegerman C, Peyre M, Goutard FL (2015) Applying participatory approaches in the evaluation of surveillance systems: a pilot study on African swine fever surveillance in Corsica. Prev Vet Med 122(4):389–398

Caliman JP (2011) Palmier à huile : le management environnemental des plantations: le cheminement de PT. Smart. Ol, Corps Gras, Lipides 18(3):123–131

de Lapeyre de Bellaire L, Essoh NJ, Abadie C, Chabrier C, Blanco R, Lescot T, Carlier J, Côte F (2009) Is chemical control of Mycosphaerella foliar diseases of bananas sustainable? ISHS Acta Hortic 828:161–170

de Romémont A, Faure G, Macombe C (2014) Can management advice to small-scale farmers trigger strategic thinking? In: Aenis T, Knierim A, Riecher MC, Ridder R, Schobert H, Fischer H (eds) 1–4 April 201411th European IFSA Symposium, "Farming systems facing global challenges: capacities and strategies". IFSA, Berlin, pp 396–407

Delabouglise A, Dao TH, Truong DB, Nguyen TT, Nguyen NTX, Duboz R, Fournié G, Antoine-Moussiaux N, Grosbois V, Vu DT, Nguyen VK, Le TH, Salem G, Peyre M (2015) When private actors matter: information-sharing and surveillance of highly pathogenic avian influenza in Vietnam. Acta Trop 2015(147):38–44

Djama M., Verwilghen A.. 2012. Figures de l'expertise dans un dispositif de normalisation en agriculture durable. L'exemple de la certification « huile de palme durable » (RSPO). In: (Alphandéry P, Djama M, Fortier A,Fouilleux E, dir.) Normaliser au nom du développement durable. Éditions Quæ, Versailles, pp 175–192

Dorel M, Perrier X (1990) Influence du milieu et des techniques culturales sur la productivité des bananeraies de Guadeloupe. Enquête-diagnostic. Fruits 45(3):237–244

Dorel M, Risède JM (2004) Fiches techniques : « La jachère raisonnée » et « La conduite au champ des vitroplants de bananier ». In: Le manuel du planteur de Guadeloupe. Cirad, Montpellier

Dorel M, Meynard C, Lakhia K, Ramassamy M, Tobal M, Ramdaya D, Racel C, Saint-Hilaire R, Magen P, Achard R, Rosalie E, Telle N, Birba O, Marville E, Ornem G, Aliker M, Hubervic C, Rieux R, Bastol C (2013) Plan banane durable Guadeloupe-Martinique. Plateforme d'évaluation de Systèmes de culture innovants 2008–2013, Rapport d'exécution année 2012. Cirad, Montpellier, 62 p

Dorel M, Tran Quoc H, Anselm R, Satney M (2014) The platform "Innovative banana cropping system", an original tool for designing and evaluating cropping systems. Closing seminar of the project Interreg Banane Durable Caraïbe, Saint Lucia, 13–16 October.

Dorey E (2014) Modélisation de l'élaboration du rendement et de la qualité de l'ananas « Queen Victoria ». Application à la conception de systèmes de culture durables à La Réunion, Doctoral thesis. Montpellier, University of Montpellier 2, 153 p

Etienne H, Bertrand B, Montagnon C, Bodadilla LR, Dechamp E, Jourdan I, Alpizar E, Malo E, Georget F (2012) Un exemple de transfert de technologie réussi dans le domaine de la micropropagation : la multiplication de *Coffea arabica* par embryogenèse somatique. Cahiers Agricultures 21(2–3):115–124

Faure G, Dugué P, Beauval V (2004) Conseil à l'exploitation familiale, Expériences en Afrique de l'Ouest et du centre. Gret/Cirad, Paris/Montpellier, 128 p

Faure G, Pautrizel L, de Romémont A, Toillier A, Odru M, Havard M (2015) Management advice for family farms to strengthen entrepreneurial skills. Note 8: GFRAS good practice notes for extension and advisory services. GFRAS, Lindau (Switzerland), 4 p

Ganry J, Laville E (1983) Les cercosporioses du bananier et leurs traitements. 1. Traitements fongicides. 2. Avertissement. Fruits 38(1):3–20

Ganry J, Fouré E, de Lapeyre de Bellaire L, Lescot T (2011) An integrated approach to control the black leaf streak disease (BLSD) of bananas, while reducing fungicide use and environmental impact. In: Dhanasekaran D, Thajuddin N, Panneerselvam A (eds) Fungicides for plant and animal diseases. InTech, Rijeka, pp 193–226

Ghione A, Kambou F, Le Cotty T, Maître d'Hôtel E, Malnoury G (2013) Le warrantage paysan, un outil de protection des ressources. Grain de sel 59-62:17–18

Goebel R (2013) Rapport de mission au Sénégal du 18 au 22 novembre 2013. Expertise pour le compte de la Compagnie sucrière sénégalaise (CSS – Richard-Toll) sur les attaques de foreurs des tiges de canne à sucre. Cirad, Montpellier, 13 p

Guillermet C, Fouré E, Dorel M, Lescot T, de Lapeyre de Bellaire L (2013) Growing cavendish bananas without chemical control of black leaf streak disease. In: Borges AL, Lichtemberg L (eds) Proceedings of XXth International Meeting. Acorbat, Fortaleza, p 159

Lassoudière A (2010) L'histoire du bananier. Versailles, Éditions Quæ, 384 p

Lassoudière A (2012) Le bananier. Un siècle d'innovations techniques. Éditions Quæ, Versailles, 336 p

Maillard JC (1991) Le marché international de la banane, étude géographique d'un « système commercial ». Presses universitaires de Bordeaux, Bordeaux, 454 p

Marie P, Dave B, Côte F (1993) Utilisation des vitroplants de bananiers aux Antilles françaises : atouts et contraintes. Fruits 48(2):89–94

Marnotte P (2015) Rapport de mission de malherbologie à la Compagnie sucrière sénégalaise à Richard-Toll (Sénégal) du 23 au 31 mai 2015. Cirad, Montpellier, 14 p

Rafflegeau S, Rioualec AL, Winterhalter J, Burgos R, Dubos B, Noboa I, Pech M (2015) Los desafío del desarrollo para los palmicultores de Quinindé y la Concordia. Revista Institucional Palma 26:30–34

Renaudin C, Pelc H, Opois J (2012) Cotonnier génétiquement modifié : l'expérience d'une campagne agricole au Burkina Faso. Cahiers Agricultures 21(6):387–394

Risède JM, Lescot T, Cabrera CJ, Guillon M, Tomekpé K, Kema GHJ, Côte F (2010) From science to field. Banana case study: guide Number 1. Challenging short and mid-term strategies to reduce pesticides in bananas. Endure, Paris, 8 p

Saison C, Cattan P, Louchart X, Voltz M (2008) Effect of spatial heterogeneities of water fluxes and application pattern on cadusafos fate on banana-cultived andosols. J Agric Food Chem 56 (24):11947–11955

Tabarant P, Villenave C, Risède JM, Roger-Estrade J, Thuriès L, Dorel M (2011) Effects of four organic amendments on banana parasitic nematodes and soil nematode communities. Appl Soil Ecol 49(1):59–67

Ternisien E, Melin P (1989) Études des rotations culturales en bananeraies. Fruits 44 (7–8):373–383

Tittonell P, Scopel E, Andrieu N, Posthumus H, Mapfumo P, Corbeels M, van Halsema G, Lahmar R, Lugandu S, Rakotoarisoa J, Mtambanengwe F, Pound B, Chikowo R, Naudin K, Triomphe B, Mkomwa S (2012) Agroecology-based aggradation-conservation agriculture (ABACO): Targeting innovations to combat soil degradation and food insecurity in semi-arid Africa. Field Crop Res 132:168–174

Toillier A (2014) Invisible technologies: how decisive are the contributions of capacity development interventions to ecological intensification? Keynote at the International conference Agroecology for Africa, 3–7 November, Antananarivo

Toillier A, Chander M, Faure G, Somé P, Havard M (2015) The Role of Producer Organisations in Rural Advisory Services. Note 12. GFRAS Good Practice Notes for Extension and Advisory Services. GFRAS, Lindau

Vaissayre M, Ochou OG, Herna SAO, Togola M (2006) Quelles stratégies pour une gestion durable des ravageurs du cotonnier en Afrique subsaharienne ? Cahiers Agricultures 15 (1):80–84

Chapter 10
Design of Cropping Systems and Ecological Intensification

Luc de Lapeyre de Bellaire, Bruno Bachelier, Marc Dorel, François-Régis Goebel, Hervé Guibert, Olivier Husson, Patrick Jagoret, Pascal Marnotte, Jean-Michel Risède, and Éric Scopel

Conventional agriculture systems are still largely based on the phase of agricultural modernization initiated in the 1960s (the Green Revolution). Their objective is to provide agricultural supply chains with standardized products and rationalize the different production stages as best as possible by optimizing costs, especially that of labour. In these systems, recourse to external inputs and their proper use are key to the adaptation of cropping systems. The design process is often simplified and linear: technologies are adjusted (optimal dosage rates, varietal choices, scheduling of cycles, etc.) depending on the primary environmental constraints, and these adjustments are then recommended to producers in a very top-down pattern. In this chapter, we present the limits of sustainability of these conventional systems, and how the work undertaken by CIRAD and its partners from the South in pursuit of an agroecological transition addresses the biological, ecological, and biophysical mechanisms that underpin agricultural systems. We also discuss the processes of

L. de Lapeyre de Bellaire (✉) • J.-M. Risède
CIRAD – UPR Geco, Montpellier, France
e-mail: luc.de_lapeyre_de_bellaire@cirad.fr; jean-michel.risede@cirad.fr

B. Bachelier • F.-R. Goebel • H. Guibert • P. Marnotte • É. Scopel
CIRAD – UPR Aida, Montpellier, France
e-mail: bruno.bachelier@cirad.fr; francois-regis.goebel@cirad.fr; herve.guibert@cirad.fr; pascal.marnotte@cirad.fr; eric.scopel@cirad.fr

M. Dorel
CIRAD – UPR Geco, Neufchâteau, Guadeloupe
e-mail: marc.dorel@cirad.fr

O. Husson
CIRAD – UPR Aida – Africa Rice, Cotonou, Benin
e-mail: olivier.husson@cirad.fr

P. Jagoret
CIRAD – UMR System, Montpellier, France
e-mail: patrick.jagoret@cirad.fr

E. Biénabe et al. (eds.), *Sustainable Development and Tropical Agri-chains*,
DOI 10.1007/978-94-024-1016-7_10

innovation and institutional regulation that have to be implemented with the actors for the design, adoption, and dissemination of these innovative systems. While Chaps. 8 and 9 cover the management of the innovation process and the experience of CIRAD in partnering with agri-chains' actors, this chapter presents the issues that the research community has to address concerning the agroecological transition of production systems and agri-chains. As in the previous chapter, we rely here too on the work done on the export banana chain in the French West Indies in partnership with West Indian producer groups and the Tropical technical institute, as well as on the cotton chain in northern Cameroon in partnership with IRAD (Institute of Agricultural Research for Development) and Sodecoton (Cotton Development Company of Cameroon). We also draw lessons from the rice chain in Madagascar in partnership with FOFIFA (National Centre for Applied Research on Rural Development), GSDM (Direct Seeding Association of Madagascar), the NGO Tafa (Land and Development), and the University of Antananarivo; and on the sugarcane chain in Réunion in partnership with eRcane within the framework of RITA Canne (Agricultural Innovation and Dissemination Network in French overseas departments).

10.1 Limitations of Conventional Practices When Confronted by Current Challenges

In this first section, we review the limitations of conventional practices in terms of sustainability through their effects on the biophysical environment (plot and territory), on all the communities of living organisms (pests as well as useful communities), and on the health aspects of crop production.

10.1.1 The Limitations of Tillage

The first objective of tillage in conventional systems is to use mechanized means to ensure a good soil structure prior to planting a crop: preparation of seedbeds or planting furrows. The soil is mechanically prepared before planting the crop mainly by using tools with discs or teeth. The loosening of the soil resulting from the use of these tools over the fields is, however, short-lived since these loosened layers of soils quickly recompact (Dorel 1994). In addition, these tools frequently result in the formation of structural discontinuities and shallow compacted zones (plough pans) that limit root growth and reduce the plant's access to soil mineral and water resources (Dorel et al. 2000). The inputs of water and fertilizers must then be increased to sustain a high level of productivity.

Tillage also disturbs the distribution of organic matter in the cropping profile and accelerates its rate of mineralization. It also increases the risk of nutrient losses through leaching and erosion, as well as the risk of loss of organic matter if additions are smaller than the losses due to mineralization. Tillage has a negative impact on the structure and diversity of soil organism communities (microflora, and, in particular, micro and macro fauna) as it destroys habitats and decreases food resources. This is especially true for communities of free-living nematodes (Dorel et al. 2010), with disruption in the functions they are associated with (stabilization of soil structure, regulation of soil parasitism, nutrient recycling). Finally, the bare soil is exposed to severe erosion (wind, tillage, or through runoff) in the tropics and its surface can harden and even form a slaking crust.

The second objective of tillage, weed control prior to planting, is equally short term. Tillage can actually increase weed pressure in the medium term by placing weed seeds in conditions that are favourable for germination. Furthermore, the use of disk harrows promotes the splitting of rhizomes and stolons and increases the vegetative propagation of weed plants.

Finally, tillage entails a large energy cost and the consequential emission of greenhouse gases when mechanized equipment that runs on fossil fuels is used.

10.1.2 Limitation of the Use of Fertilizers for Plant Nutrition

In most conventional systems, the quantity of crop residues is insufficient to maintain the soil carbon stock at a level to ensure fertility, especially in tropical areas where mineralization occurs rapidly. Moreover, this biomass is often exported or burned, which results in a low level of carbon returned to the soil, leading to a steady decline in the capacity of the soil to store nutrients. A considerable amount of nutrients required by crops therefore have to be provided by the farmer. Losses due to surface runoff and leaching are also high due to the high solubility of certain fertilizers, the rapid mineralization of organic matter, and slow rooting. Nutrient loss also takes place due to soil erosion and volatilization of soils exposed to high temperatures (Husson et al. 2013). Consequently, the level of fertilization often has to greatly exceed the exports through harvests (Dorel et al. 2008).

Furthermore, the manufacturing process for mineral fertilizers emits large quantities of greenhouse gases. Indeed, nitrogen fertilization accounts for more than a quarter of the carbon footprint for the banana crop (Lescot 2012).

10.1.3 Limitations of Pesticide Use in Pest Management

Pests affect crop yields in several ways: competition for resources, consumption of plant tissue, consumption of plant assimilates, and reduction of photosynthesis efficiency (Bancal et al. 2006). They can also affect the crop quality. Pesticides remain the most widely used method to control pests in conventional systems. Indeed, the use of pesticides is often a guarantee of favourable results for farmers who seek to limit risk factors. These practices are, however, far from sustainable and cause numerous undesirable environmental, technical, and socio-economic impacts.

10.1.3.1 Environmental Impacts

The use of pesticides in agricultural systems has progressively contaminated the different environmental components: atmosphere, soil, surface water, and groundwater (Charbonnier et al. 2015). Two realities can make these problems particularly acute in tropical agricultural systems: legislation that is less stringent than in the countries of the North (de Lapeyre de Bellaire et al. 2009) and the tropical conditions that favour the development of pests without breaking their life cycles. Thus, in the case of black Sigatoka disease in banana, pest control is a systematic process carried out weekly in many countries as the inoculum is present throughout the year, and new leaves vulnerable to the infection sprout every week on each banana plant. With over 60 kg/ha/year of active material used, this is the main use of pesticide on this crop. Finally, the insecticides have significant impact on natural enemies of pests (parasitoids and predators), which, in Africa, are already weakened by the harsh environment, especially in the Sudano-Sahelian zone. Pesticides reduce biodiversity and obstruct ecosystem services such as pollination by bees.

10.1.3.2 A Major Technical Limitation: The Emergence of Resistant Individuals

The emergence of resistant individuals who can no longer be controlled constitutes a major technical limitation in the use of certain pesticides. The different stages in the emergence of resistant individuals are well known (zur Wiesch et al. 2011). The speed of this process depends on factors that are based on a combination of agricultural practices, genetic medium of the resistance, and the characteristics of the pest, especially its dispersal ability (Milgroom et al. 1989).

Several publications have explored the phenomena of resistance developed by bollworm caterpillars (for example, *Helicoverpa armigera* or *Spodoptera littoralis* to mention only two such insects) to insecticides in cotton cultivation. In Africa, where insecticides have long been used in cotton cultivation, bollworm caterpillars have become resistant to pyrethinoids, while phloem-feeding insects (aphids, whiteflies) have become resistant to organophosphates.

In the case of the black Sigatoka disease of the banana, CIRAD developed reasoned control strategies based on warning systems, which helped greatly reduce the number of treatments needed. These strategies rely on the use of systemic fungicides with high curative effects (de Lapeyre de Bellaire et al. 2009). The emergence of resistant strains in Cameroon and parts of Côte d'Ivoire resulted in the discontinuation of the reasoned control process (12–14 treatments/year) in favour of a systematic strategy based on a weekly application of contact fungicides (40–50 treatments/year).

Some cases of resistance to herbicides used in cotton cultivation were also observed. Conventional agriculture is adapting to such resistance by adopting new and ever more effective active materials which are, however, costlier and often more polluting.

10.1.3.3 Economic Limitations

While the use of pesticides was originally designed to achieve an economical optimum that limited the risk of crop loss or decline in quality, the implementation of chemical pest control can turn out to be very expensive in some cases, especially when resistance to pesticides requires making new technical choices. Thus, the weekly systematic control of black Sigatoka disease requires application logistics (aircrafts, helicopters) and additional labour (weekly leaf-stripping) that are not only very expensive but offer no guarantee against crop loss (de Lapeyre de Bellaire et al. 2009).

Weed control is also a major factor and expense in technical itineraries practiced in the tropics. For example, 50 % of labour in rice cultivation is devoted to weeding. This work is difficult to carry out under proper conditions, which often makes weeding one of the main limiting factors. Parasitic species like striga only exacerbate the problem, especially for cereal crops in the Sudano-Sahelian zones. In case resistance to herbicides is observed, manual weeding of the plots is often the only affordable solution, despite this option being very labour intensive.

10.1.3.4 Increasingly Stringent Legislation

Pesticide legislation has become increasingly stringent in the countries of the North in recent years, and the use of many pesticides has been discontinued or banned. Although corresponding legislation in most tropical countries is currently less stringent than in Europe or the United States, there are other factors that limit the use of pesticides, especially for use on export crops (de Lapeyre de Bellaire et al. 2009). First, maximum residue limits (European Union standards, for example) in plant products are applicable at the point of entry into the importing countries, often in the countries of the North. Second, the specific standards imposed by agrifood trading intermediaries reflect the concerns of consumers regarding the potential impacts of pesticides on health in the countries of production and on the environment in the areas of production.

10.1.4 A Very Poor Utilization of Water Resources

Conventional practices that include tillage and bare soils cause heavy runoff, reduced water infiltration, and high evaporation due to the elevated temperature of the surface soil. These practices are not conducive to soil water storage. Furthermore, the frequent construction of plough pans hinders deep rooting in plants. The volume of soil that comprises the root zone is small and consequently the useful water reserve remains low. In such conditions, water supply to plants is strongly dependent on the regularity of inflows. Plant growth becomes very quickly limited in rainfed agriculture without access to irrigation, especially during sensitive crop stages such as flowering. The water supply in conventional systems is thus heavily dependent on the irrigation capacity, which results in high water consumption, even as this resource is getting increasingly scarce.

10.2 New Ecological Intensification Practices

The inadequacies of conventional practices when confronted by current challenges have led CIRAD, along with its partners in the South, to devise and implement new ecological intensification practices for different agricultural supply chains to initiate their conversion to agroecological practices. The introduction of a functional vegetal biodiversity (cover crops) in cropping systems is one of the cornerstones of this conversion. In fact, these plants provide services that were earlier obtained from various external inputs.

10.2.1 Organic Tillage and Restoration of Soil Biology

The elimination of tillage, a practice that is very harmful to soil fauna and microflora and their habitat, is a major step towards rebuilding biological activity in the soil. In conservation agriculture such as the innovative banana cropping systems of the Sustainable Banana Plan, the use of cover crops and inputs of exogenous organic matter ensure the creation and maintenance of good soil structure:

- soil aeration and restructuring are achieved by macrofauna (environmental engineers such as earthworms, ants, collembolae, macroarthropods, etc.) and by the strong root systems of cover crops used in associations and rotations;
- root systems that are very dense at the surface also act like a flexible and resilient framework that limits the impact of compaction resulting from the use of heavy farm machinery;
- the incorporation of organic material of varying quality (from sugars and easily decomposable simple proteins to lignin composed of large molecules that

decompose slowly) in the soil fuels the various reservoirs of soil organic matter that are more or less stable. This organic matter promotes the formation and stabilization of aggregates through intense biological activity (Husson et al. 2013).

10.2.2 Nutrient Cycling and Organo-mineral Management of Crops

The use of cover crops rapidly increases soil fertility and, in particular, the quantity of nutrients available to crops. These systems are designed to ensure a regular and balanced supply of nutrients to the crops by a steady decomposition and mineralization of biomass. Even though biomass is replenished regularly, the quality of the inputs is very varied, which results in different rates of mineralization. A leguminous ground cover mineralizes rapidly and restores nutrients to the soil (especially nitrogen) that can be directly used by the following crops from the beginning of their cycle. Conversely, grasses with a high C:N (carbon-nitrogen) ratio and with high lignin and polyphenol content decompose slowly and release their nutrients later, thus supplying nutrients to subsequent crops several months after their contribution to the plant litter. The absence of tillage also helps protect the upper layers of the litter. Finally, the microflora present in these undisturbed systems (bacteria, mycorrhizae, and trichodermas) increases the amount of nutrients available to plants (very large increase in the area of exchange, as in symbiotic associations with mycorrhizae, solubilisation by bacteria, etc.) (Husson et al. 2013).

10.2.3 Improvement in Water Efficiency

A permanent cover substantially reduces runoff and losses due to evaporation. Moreover, infiltration is rapid due to a good macroporosity, created and maintained through agroecological practices. This rapid infiltration prevents clogging during heavy rainfall ('flush' effect). Microporosity, which is also created and maintained through agroecological practices, leads to a large storage capacity and, along with good infiltration, helps the formation of a significant water reserve. This water reserve is easily accessible to plants with deep root systems in soil that is well structured. The use of deep water (during dry periods) and limited runoff losses (periods of heavy rain) present a buffer against climatic hazards (Husson et al. 2013).

10.2.4 Erosion Control

Measures to control erosion include the planting of rows of herbaceous or shrubby species that break the flow of water, thus decreasing the possibility of water flowing fast enough to cause severe erosion. However, these plant barriers have drawbacks: loss of productivity, possible competition with crops (shade), and the maintenance they require. Permanent cover offers the advantage of protecting the soil, both against the impact of falling water drops (high kinetic energy) and against wind and/or water erosion.

10.2.5 Biological Pest Regulation

Different methods of pest control, other than the use of pesticides, are being researched and can be combined into an integrated management strategy, since none of these solutions, when implemented independently, are as efficient as pesticides.

10.2.5.1 Biological Regulation of Pest Populations to Limit the Damage They Cause

Biocontrol methods (use of living organisms and natural or mineral substances) are preferred mechanisms of biological regulation, along with appropriate farming practices to regulate populations of pests or limit the damage they cause.

10.2.5.1.1 The Introduction of a Functional Plant Biodiversity

Species that do not host pests can be deployed spatially or temporally.

Use of resistant varieties Resistance to pests can either be total or partial, limiting not only the damage, but also the spread of pests. However, such varieties are not always available, as in the case of banana systems.

Rotations of crops or cover crops allow for a fallow period, a technique that is relatively effective in controlling soil pests. The practice is now commonly used to combat plant-parasitic nematodes on banana prior to replanting plots using healthy planting material from in vitro cultures.

Crop associations or the use of cover crops The use of non-host genotypes in the cropping system has a direct effect on the life cycle of pests via two essential mechanisms: barriers for dispersal resulting from an increase in the distance between sensitive plants, and a decrease in the abundance of pests in the plot (Mundt 2002). The introduction of plant biodiversity also modifies the spatial structure of the canopy, which can hinder the spread of pests (interception, modification of physical parameters such as rain or wind) (Boudreau 2013). These changes in canopy structure may also promote micro-climatic effects (shading, wetness duration, etc.) and the partitioning of the agricultural system into several distinct habitats. Thus, trap plants that attract insects (*Arachis pintoï* attracts bugs, millet attracts crickets, and *Pennisetum* attracts borers) discourage crop predators (push-pull approach). Moreover, the introduction of cover crops can alter food webs and improve the natural control of pests, as in the case of the banana weevil or plant-parasitic nematodes. The presence of a permanent cover can also reduce the incidence of fungal diseases of cereals, such as fusarium wilt, root rot (*Rhizoctonia* sp.) and damping off (*Pythium* sp.), as a result of the microbial activity induced in these soils by the organic material. Finally, some plants produce substances that are toxic or repulsive for pests. Rice cultivators in Madagascar make use of several species: hairy vetch and fodder radish are grown to control white grub populations, and members of the *Desmodium* genus are known to repel borers.

The case of agroforestry Agroforestry is defined as the simultaneous or sequential association of trees and annual crops or livestock in a single plot, with the aim of obtaining goods and services useful to people (Torquebiau 2000). These systems put ecological principles based on biodiversity and the interactions between species – whether positive, negative, or neutral – into practice. Agroforestry systems exhibit a higher overall productivity per unit area, better control of pest pressure, increased environmental services (biodiversity conservation, maintenance of soil fertility, carbon sequestration, etc.), higher profitability, and increased sustainability. Such is the case of cocoa agroforestry systems in Cameroon's Centre Region. The lifespan of the cocoa trees in these systems often exceeds 50 years (Jagoret et al. 2011), which is much higher than the 20–30 years for cocoa grown as a monoculture. In addition, the yield of the former is similar to, or even greater than, the latter. These systems also allow cocoa to be grown in savannah zones, usually regarded as unfavourable for cocoa cultivation (Jagoret et al. 2012). The intercropping of cocoa with several fruit and forest species and their organization and management in space and time helps maintain, or even restore, soil fertility by stabilizing the biomass recycling and nutrient cycling processes. Even without the use of chemical fertilizers, the level of organic matter content in cocoa farms set up in forests remains constant over time (Snoeck et al. 2010). It may even increase in cocoa agroforestry systems set up in the savannah zones, by up to 3.1 % in plots that are more than 40 years old (Jagoret et al. 2012).

10.2.5.1.2 The Use of Macro- and Micro-Organisms for Biological Control

Some micro-organisms are inoculated directly on plants or habitats of pests, for example, entomopathogenic fungi (*Metharizium* spp. or *Beauveria* spp.) or bacteria (*Bacillius thuringiensis*). Their effectiveness, however, depends on the survival of these micro-organisms in the environment. Their use is particularly beneficial for post-harvest infections, when environmental conditions are more easily controlled (Bastiaanse et al. 2010). Finally, mycorrhization (on-site or before planting) can protect plants against plant-parasitic nematodes. Some parasitoids are also introduced for the biological control of pests.

10.2.5.1.3 The Use of Chemical Mediators

Some chemicals (pheromones) are used to lure pests into traps, for example, sordidin, the aggregation pheromone of the banana weevil.

10.2.5.1.4 The Use of Natural Substances

Plant extracts, algae, or mineral natural preparations can either have a direct effect on pests and diseases, or an indirect one by stimulating natural defence mechanisms. Research is being undertaken actively to find such solutions but certain hurdles remain. Approval processes are usually long and the results obtained cannot match those of conventional pesticides.

10.2.5.1.5 Other Agricultural Practices

Some agricultural practices inhibit the life cycle of pests, without the help of any natural mechanisms:

- prophylactic practices that try to destroy the sources of the inoculum (for example, sanitary deleafing in case of the black Sigatoka disease of banana) or its habitats;
- laying protective covers on the stands or on parts of plants, such as safety nets against insects, plastic sleeves on bunches of bananas;
- stand densities that can affect life cycles;
- pruning green leaves of sugarcane crops, which provides abundant mulching while limiting weeds in the succeeding crop;
- mechanical weeding in cotton plantations to decrease weed pressure.

10.2.5.2 Reducing Crop Damage Caused by Pests

Understanding the sequence of events between damage and crop losses offers new entry points (Savary et al. 2006) and allows farmers to strengthen plant tolerance through appropriate cropping practices. Thus, crop health is enhanced by a balanced and regular supply of water and minerals that boosts physiological functioning. In the case of foliar diseases of plants that continuously produce leaves (for example, the black Sigatoka disease of banana), rapid leaf sprouting offsets the reduced leaf area and maximizes photosynthesis in the upper foliage that is most exposed to solar radiation. Moreover, optimization of the flow of assimilates through specific removal practices (source-sink ratios) can enhance the flow to useful parts of the plant (for example, fruits or seeds). Such practices can increase the susceptibility of plants to pests and diseases, as in the case of the susceptibility of banana to crown rot (Lassois et al. 2010). Finally, in the case of this disease, leaf symptoms have a direct effect on the storage life of fruits that cannot be exported in the event of major attacks. Thus, the regular removal of necrotic stages limits the effects of the black Sigatoka disease on fruit quality, even in the case of heavy defoliation (de Lapeyre de Bellaire et al. 2013).

10.3 Consequences for Design Methods

New systems designed around ecological intensification practices with an objective of mastering all the ecological processes solicited are complex and need to incorporate multiple species. They have to be closely managed in order to be effective. Their implementation calls for specific and in-depth knowledge and know-how, and requires significant time for producers and local actors to become familiar with them. Thus, the three stages of the design process (diagnosis, *a priori* proposition of solutions, and assessment and adaptation of the proposed systems) have to be modified for a better integration of the dimensions of actors, space, and time.

10.3.1 Taking Local Knowledge into Consideration

Local knowledge and know-how on the use and management of biodiversity in cropping systems can almost always be found in all agricultural regions of the planet. Such knowledge is generally preserved in a more sustainable manner in traditional agricultural practices of the South. In West Africa, for example, cotton farmers manually prune the top of cotton plants, which then respond by sending chemical messages to prevent egg-laying by females of different species of predatory insects that feed on fruiting organs of cotton. This autochthon technique should be the first to be used, as it is an apt example of the mobilization of local biological resources adapted to their environments. Such knowledge can be combined with the technical expertise gained by research teams from other experiences in other parts of the world with similar ecological conditions and settings.

10.3.2 Ensuring a More Comprehensive Involvement of Producers During the Entirety of the Design Processes

Producers are the primary actors involved in the implementation of a complex change, with a significant risk of failure given the long stages required for the adaptation of and mastery over the systems, and the stabilization of the stimulated ecological processes. Solutions must be tailored to different types of producers, with each charting his own trajectory of change and fine-tuning his own cropping systems. New systems must be designed in a participatory manner with a systematic involvement of producers at every stage. Changes have to be designed and planned with their help, based on their resources, objectives, and the risk level that is acceptable in the design/adaptation process.

10.3.3 Involving Other Local Actors Too

The implementation of technical modifications often depends on conditions that are external to the farms and beyond the sole control of the producers. Thus, different producers and users of rural areas have to manage different resources required for production (land, biodiversity, water, biomass, etc.), some of which influence the viability itself of the agroecological systems concerned. Similarly, actors involved in providing technical assistance, leasing, or credit for the purchase of equipment and specific inputs, and public policy actors who create the legal, economic, and social conditions that promote such change should be involved at an early stage, from the moment there is consensus on the need for change.

10.3.4 Taking into Account Different Scales in the Design of Cropping Systems

To address the conditions that are imposed on them (market volatility, climate change, land degradation, population growth, pressure on resources, etc.) and that create further instability – and sometimes also opportunities –, small producers in the South must not only adapt their cropping systems, but also their production systems. To do so, they have to modify the balance between their different activities (new production, specialization, intensification, and diversification) and thus the allocation of the means and factors of production amongst these activities. They also have to negotiate with other actors (producers or non-producers) in matters concerning land development (land-use planning, provision of shelters for natural enemies, or protective hedges against pests, etc.) or access to certain collective resources (irrigation water, land, biomass, etc.).

10.3.5 Taking into Account the Temporal Aspect of the Agroecological Transition

The process of designing and adapting new agroecological systems usually requires several rounds of assessments and readjustments before deciding on the relevant systems for each type of producer. These systems require that producers not only clearly understand the underlying processes, but also master the technical management of these processes and learn to exercise proper control over them. Many of these processes take time to stabilize, with the benefits of different ecosystem services becoming evident only after several years of application. Finally, negotiating with the public actors and all local stakeholders to create conditions favourable to these systems takes time.

10.3.6 Favouring the Trajectories of Change Over Ready-to-Use Technological Packages

The deployment of an agroecological transition, simultaneously at different scales, and over the relatively long duration required for it to stabilize, shows that it is not suitable for distribution as a ready-to-use system to be applied according to recommendations. It is instead more useful to equip producers so that they can create their own path to adaptation and undertake the transition in a way that is most suitable to their context. Such an adaptation is often progressive. This trajectory must, nevertheless, conform to a certain technical consistency and must, at every stage, be technically feasible and exhibit productive efficiency.

10.3.7 Implementing a New Concept of Technical Support and Assistance for Actors

This process also raises the question of who provides support and how. The technician becomes less a prescriber of normative technical recommendations and more an adviser who points out different directions for reflection that allow various types of producers to build their own alternative systems. He thus becomes a facilitator for a co-construction process. Similarly, facilitators and managers are required for co-innovation platforms. Depending on the case, they can originate from local communities, associations, or agro-industrial entities involved in the agri-chains.

10.4 Conclusion

Research is underway to find new technical solutions, including in particular the incorporation of a certain functional plant biodiversity to support the process of designing new agroecological and sustainable systems within agri-chains. This leads to the system and its management becoming ever more complex. The aim of this increased complexity, however, is to improve the performance of various production processes. The new systems have not only to meet expectations of increased production but also of improved production, especially with regard to environmental and health impacts. The assessment that accompanies these design processes must therefore have a multi-criteria approach in order to incorporate the concerns of producers as well as those from other members of society.

In these design processes, the various technical levers mobilized have to be implemented over time; these processes thus become gradual and progressive, not instantaneous or too sudden. They should be able to improve the more deficient functions of the cropping systems gradually. Various types of trade-offs and decisions have to be made. These concern, on the one hand, the hierarchy and combination of functions which need to be improved on a priority basis and, on the other, the organization over time of the technical modifications that have to be made to the local systems to achieve this improvement.

References

Bancal MO, Ney B, Soussana JF (2006) La parcelle, lieu d'intégrations entre différentes populations. In: Doré T, Le Bail M, Martin P, Ney B, Roger-Estrade J (eds) L'agronomie aujourd'hui. Éditions Quæ, Versailles, pp 137–150

Bastiaanse H, de Lapeyre de Bellaire L, Lassois L, Misson C, Jijakli MH (2010) Integrated control of crown rot of banana with *Candida oleophila* strain O, calcium chloride and modified atmosphere packaging. Biol Control 53(1):100–107

Boudreau MA (2013) Diseases in intercropping systems. Annu Rev Phytopathol 51:499–519

Charbonnier E, Ronceux A, Carpentier AS, Soubelet H, Barriuso E (2015) Pesticides : des impacts aux changements de pratiques. Versailles, Éditions Quæ 400 p

de Lapeyre de Bellaire L, Essoh Ngando J, Abadie C, Chabrier C, Blanco R, Lescot T, Carlier J, Côte F (2009) Is chemical control of *Mycosphaerella foliar* diseases of bananas sustainable? Acta Hortic 828:161–170

de Lapeyre de Bellaire L, Guillermet C, Le Guen R, Fouré E, Dorel M, Lescot T (2013) Concepción y evaluación de sistemas de cultivo para crecer variedades de Cavendish sin control químico de Sigatoka. Scientific workshop on controlling black Sigatoka of banana crops in the Caribbean region. Havana, Cuba, 19–21 March 2013

Dorel M (1994) Mécanisation et tassement des sols en culture bananière. Proceedings XI Acorbat Meeting, San Jose, Costa Rica, pp 667–677

Dorel M, Roger-Estrade J, Manichon H, Delvaux B (2000) Porosity and soil water properties of Caribbean volcanic ash soils. Soil Use Manag 16(2):133–140

Dorel M, Achard R, Tixier P (2008) SIMBA-N: Modeling nitrogen dynamics in banana populations in wet tropical climate. Application to fertilization management in the Caribbean. Eur J Agron 29(1):38–45

Dorel M, Lakhia S, Pététin C, Bouamer S, Risède JM (2010) No till banana planting on crop residue mulch. Effect on soil quality and crop functioning. Fruits 65(2):55–68

Husson O, Séguy L, Charpentier H, Rakotondramanana (eds), Michellon P, Raharison T, Naudin K, Enjalric F, Moussa N, Razanamparany C, Rasolomanjaka J, Bouzinac S, Chabanne A, Boulakia S, Tivet F, Chabierski S, Razafintsalama H, Rakotoarinivo C, Andrianasolo HM, Chabaud FX, Rakotondralambo T, Rakotondralambo P, Ramaroson I (2013) Manuel pratique du semis direct sur couverture végétale permanente (SCV). Application à Madagascar. GSDM/CIRAD, Antananarivo. Volume 1 Chapitre 1: 1–24

Jagoret P, Michel-Dounias I, Malézieux E (2011) Long-term dynamics of cocoa agroforests: a case study in central Cameroon. Agrofor Syst 81:267–278

Jagoret P, Michel-Dounias I, Snoeck D, Todem NH, Malézieux E (2012) Afforestation of savannah with cocoa agroforestry systems: a small-farmer innovation in central Cameroon. Agrofor Syst 86(3):493–504

Lassois L, Bastiaanse H, Chillet M, Jullien A, Jijakli MH, de Lapeyre de Bellaire L (2010) Hand position on the bunch and source-sink ratio influence the level of banana fruit susceptibility to crown rot disease. Ann Appl Biol 156(2):221–229

Lescot T (2012) Banana Sector. Commonwealth of Dominica. Thematic Support for the Identification and Formulation of the Support Framework under the EU Funded Banana Accompanying Measures (BAM), Final Evaluation Report. European Union, Brussels, p. 26

Milgroom MG, Levin SA, Fry WE (1989) Population genetics theory and fungicide resistance. In: Leonard KJ, Fry WE (eds) Plant disease epidemiology. 2. Genetics, resistance and management. McGraw Hill, New York, pp 340–367

Mundt CC (2002) Use of multiline cultivars and cultivars mixtures for disease management. Annu Rev Phytopathol 40:381–410

Savary S, Teng PS, Willocquet L, Nutter FW (2006) Quantification and modeling of crop losses: a review of purposes. Annu Rev Phytopathol 44:89–112

Snoeck D, Abolo D, Jagoret P (2010) Temporal changes in VAM fungi in the cocoa agroforestry systems of central Cameroon. Agrofor Syst 78(3):323–328

Torquebiau EF (2000) A renewed perspective on agroforestry concepts and classification. Comptes rendus de l'Académie des sciences, Série III, Sciences de la vie, 323(11):1009–1017

zur Wiesch PA, Kouyos R, Engelstadter J, Regoes RR, Bonhoeffer S (2011) Population biological principles of drug-resistance evolution in infectious diseases. Lancet Infect Dis 11(3):236–247

Chapter 11
Evolution and Challenges of Varietal Improvement Strategies

Robert Domaingue, Sylvie Lewicki, Patrice This, Frédéric Bakry, Jean-Pierre Horry, Serge Braconnier, David Pot, and Gilles Trouche

Agricultural production and supply chains are currently facing profound and rapid changes, and are being subjected to additional and diverse constraints. The constantly increasing world population has resulted in a growing demand for food products in a context where the pressure on resources caused by this growth in population and urbanization, changes in food habits, as well as competition between different uses of food products (fresh and processed) and non-food products (energy, chemicals, construction) increases the diversity of situations to be addressed. Production methods must, for example, adapt to these changes and comply with increasingly stringent new requirements and product quality standards. Reducing the use of phytosanitary products is an important issue for the development of an ecologically intensive agriculture. World agricultural production must also confront other global phenomena such as climate change, which is manifesting mainly as a general increase in temperatures and atmospheric CO_2, or the erosion of genetic resources (Albajes et al. 2013). Genetic improvement is today being increasingly called upon to help meet the consequences of actual or expected climate change, and their impacts on agriculture (Jaramillo et al. 2011).

Breeders use varietal improvement techniques to create new races or varieties that help address the challenges of production systems and agri-chains, especially those that these will face in the future. A more diversified use of products within these agri-chains (Abécassis and Rousset 2012) requires a greater varietal diversification, leading to new breeding criteria. This changing context calls for a

R. Domaingue (✉) • S. Lewicki • P. This • F. Bakry • J.-P. Horry • S. Braconnier • D. Pot
G. Trouche
CIRAD – UMR Agap, Montpellier, France
e-mail: robert.domaingue@cirad.fr; sylvie.lewicki_dhainaut@cirad.fr; patrice.this@cirad.fr; frederic.bakry@cirad.fr; jean-pierre.horry@cirad.fr; serge.braconnier@cirad.fr; david.pot@cirad.fr; gilles.trouche@cirad.fr

E. Biénabe et al. (eds.), *Sustainable Development and Tropical Agri-chains*,
DOI 10.1007/978-94-024-1016-7_11

strengthening of links between different actors of an agri-chain, and requires interdisciplinary research that must be innovative, flexible, and responsive, as illustrated below in cases of improvement of banana and sorghum.

11.1 The Ideotypes of Tomorrow Have to Be Defined Today

The goal of varietal improvement is to incorporate a set of traits in ideotypes that will be used for cultivation in local or regional farms, traits for productivity, adaptation to cultivation conditions, resistance to pests and diseases, and quality and acceptability of the end products. Thus, successful plant breeding relies on a multidisciplinary approach bringing together genetics, plant pathology, entomology, ecophysiology, agronomy, processing of products, and the social sciences (Desclaux et al. 2013).

Varietal improvement is a long process: 10–25 years are needed, depending on the species (especially for perennials and forest species), to create a variety – a distinct, homogeneous, and stable product – that can be used by farmers, consumers, and other end users. Hence, the process requires that future scenarios be considered when breeders make their choices for hybridizations in order to define the varietal range which will be of interest at the end of this selection process. This projection into the future is especially important considering that, once the new varieties of crops are planted, they will be grown for periods ranging from one to several seasons in the case of annual crops to long cycles of up to 30 years for perennial crops (Bunn et al. 2014), and even a hundred years for trees.

Taking climate change into account requires reliable predictions of the climate and the taking into account – before the beginning of the breeding programme – of biotic, abiotic, and social factors, which will be categorized by key traits to be selected. For example, modelling approaches and methods help in studying the response of plants to drought in a better manner. Then, by identifying individual genes or gene families associated with the expression of adaptive traits, the pyramiding of favourable alleles and marker-assisted selection can help create better adapted varieties. Furthermore, climate change goes hand in hand with the uncertainties resulting from the emergence of new pathogens, such as the diseases of banana (Daniells 2009) or coffee leaf rust (Bunn et al. 2014).

In addition to this climatic challenge, the increase in and diversification of the demand for food call for the development of new varieties that are productive, diversified, and environmentally friendly (ecological intensification), with an increased focus on acceptance and appropriation (Hellin et al. 2014).

Breeding strategies must incorporate ecological intensification efforts that improve the sustainability of production systems (Ahmadi et al. 2013). These efforts call for a critical look at the various ecosystem services (carbon sequestration, fuelwood production, maintenance of biodiversity, etc.) that should be considered and require a rethinking of the improvement and breeding approaches by exploring other types of cropping systems. These systems challenge the approaches themselves, which is a prerequisite for varietal improvement *sensu-stricto*. The work of Litrico and Violle (2015) illustrate these concerns well.

In addition, the emergence of certified products or those from specific locations (protected geographical indications) require the availability of specific varietal types and profiles, such as organic farming, which is developing rapidly in some agri-chains (van Bueren et al. 2011).

11.2 Agri-Chains Shape Improvement Strategies

11.2.1 Varietal Improvement of Banana in an Export-Oriented Industrial Supply Chain

Global banana production, currently at 130 million tonnes/year, keeps on growing steadily. The banana chains, however, are lumbering giants with feet of clay, since almost 95 % of this production is based on a small number of different genetic combinations (7 to 14 at the most), corresponding to eating habits prevalent in production and consumption areas (plantains in West and Central Africa, highland bananas in East Africa, apple banana in India, Popoulou and Maoli in Oceania, Prata in Brazil). Besides, half of the global output, 97 % of which is destined for export, is represented by a few clonal varieties of the Cavendish group (Grande Nain, Williams, Robusta, Petite Naine).

Such a low genetic diversity for such a major crop, which forms the basis of a subsistence crop for millions of people worldwide, is exceptional, even in comparison with other plants propagated via clones (for example, potato or strawberry). It makes banana production particularly susceptible to emerging or existing diseases, whose spread is now facilitated by international trade, as shown by the recurrent appearance of new fungicide resistant races of the fungus *Mycosphaerella fijiensis*, the causal agent of black Sigatoka disease. Diseases play a major role in the low yields currently being recorded, which are far below the varieties' potential. The yields of plantain in West Africa and highland bananas in East Africa, for example, barely reach 2 to 4 T/ha, whereas their potential is 20 and 35–45 T/ha respectively under favourable ecological conditions.

11.2.1.1 Strengths and Weaknesses of the Cavendish Group

The Cavendish group is a good illustration of the very narrow genetic base of cultivated bananas. With 60 million tonnes produced annually, this group clearly dominates global banana production. Local markets absorb 45 million tonnes, with the remaining 15 million tonnes being destined for export markets. Several reasons explain the dominance of these varieties and the success of the production and export supply chains based on Cavendish. The fruit is pleasant to the taste with good post-harvest qualities of the green fruit. It has cold-storage times suitable for the long transportation periods needed for getting it from production to consumption areas (20–30 days) and a good tolerance to shocks and bruises. Finally, agricultural performance under intensive cropping conditions is satisfactory, with average yields touching 45 T/ha. However, the Cavendish varieties are very susceptible to foliar fungal diseases and require 20–50 fungicide applications per year, depending on the production region, with the use of alternating active molecules to prevent the development of resistance. Banana plantations are threatened by a new, highly pathogenic strain of the fungus *Fusarium oxysporum* (causative agent of Panama disease), while, until recently, the Cavendish varieties were resistant to other strains of this fungus. This new strain (TR4) was first identified in Taiwan, and has led to the decline in production throughout Southeast Asia (Philippines and China). It has recently been detected in Australia, Mozambique, Jordan, and Pakistan, and presents an imminent threat to West and Central African production and, eventually, to Latin American and West Indian production (Ploetz and Churchill 2011). So far, there are no effective means to control this infection.

11.2.1.2 The Banana Chain in the French West Indies (Martinique and Guadeloupe)

The banana chain in Martinique and Guadeloupe, which supplies bananas to the French market, is solely based on the monoculture cultivation of the Cavendish group and the beneficial traits it offers. Nevertheless, this agri-chain is subject to extreme parasitic pressure as well as other constraints related to island agricultural production (competition between urbanization and agriculture), environmental pressures that result in the tightening of phytosanitary regulations and a capping of pesticide use, high production costs, a strong structuring of marketing and export networks, and competition from non-European sources (Temple et al. 2008). Together, these factors reinforce the need to broaden the genetic base of the bananas grown, via the development and distribution of new hybrids that incorporate the benefits of the Cavendish while being more tolerant or resistant to disease. It is a matter of leveraging the great natural genetic diversity that remains untapped, particularly that of wild species.

11.2.1.3 Improvement of the Banana by Reconstruction

Most cultivated bananas are triploid varieties that are highly heterozygous, having a monospecific (*Musa acuminata*) or interspecific (*M. acuminata*/*M. balbisiana*) constitution. These varieties are vegetatively propagated and have a high degree of gametic sterility associated with a greatly reduced capacity of fertilization. The current triploid varieties hold little interest for improving export bananas. They are 'evolutionary dead-ends' with a very narrow genetic base. Moreover, Cavendish clones are sterile and produce practically no seeds. Genetic improvement must first be seen as a process that is constrained by recombination and seed production. Faced with such difficulties, the process of obtaining triploid hybrids directly from diploid resources was found to present some flexibility. Based on this observation, CIRAD initiated a breeding programme, over 20 years ago, that aimed to mobilize diploid genetic resources to create triploid varieties (Bakry et al. 2009).

Diploids present a vast and complex diversity that can be mobilized, both from cultivated varieties as well as from wild species. The crossing strategy, known as 'reconstruction', relies on data from phylogenetics (Perrier et al. 2011), phenotyping, molecular marking, and new genomic resources obtained from the sequencing of banana (D'Hont et al. 2012) in order to choose the best parents to cross, orient diploid-level recombination, and facilitate the actual breeding of seedless hybrids at the triploid level, according to the expectations of the export supply chain.

The aim of this pragmatic approach is to combine favourable traits from the two selected parents, based on our knowledge of evolution, and maximize heterozygosity in triploids progenies (Sanford 1983).

11.2.1.4 A Breeding Platform

In 2008, the banana chain of Guadeloupe and Martinique and CIRAD launched, with the support of both regions and departments and the Ministry of Agriculture, an ambitious project called the Sustainable Banana Plan (Chap. 9). This plan established two collaborative technical platforms between CIRAD and the banana chain of the two islands, one of which was a breeding platform for new disease-resistant banana varieties. A remarkable outcome of this plan was the direct involvement of the profession in the activities of the breeding platform, including the creation of the Tropical Technical Institute as an initiative of the Union of Banana Producer Groups of Guadeloupe and Martinique (UGPBAN) and CIRAD. This institute has filled a void in the organization of the research, validation, and transfer mechanism.

This participatory approach has allowed:

- the definition and recognition of the multiplicity of breeding criteria that determine a relatively constrained objective for the ideotype for the export agri-chain;
- the establishment of a plant breeding committee which includes producers who are responsible for decisions regarding the breeding of new hybrids;
- the establishment of a network of test plots to assess new varieties on the farms of pioneering producers in Guadeloupe and Martinique;
- the involvement of the banana chain's downstream transportation and ripening network in the evaluation and development of varieties;

- the creation of a joint team consisting of members from CIRAD and the Tropical Technical Institute.

An important outcome of the Sustainable Banana Plan was the strengthening of research and development capabilities to help support agri-chains. The agri-chain was, in this way, endowed with an ability for innovation and foresight. This was made possible by adopting breeding criteria suitable for each of the many stages of the agri-chain, from cultivation in the West Indies to marketing of the fruit in mainland France. The objectives of varietal improvement include: developing shorter plants to facilitate care while on the bunch and post-harvest, obtaining high bunch yields, increasing resistance to black Sigatoka and Panama diseases, imparting good adaptation to agronomic constraints, and developing taste quality and good post-harvest performance of the fruit, especially all through the export and marketing chains. A major constraint among these criteria is the cost of production, which includes the high labour costs in the French West Indies.

The breeding platform has steadily increased its capacity, going from 400 hybrid varieties produced per year in 2007 to a thousand now. Several among these hybrids have cleared the initial breeding stages but, so far, only one of them (Cirad 925) seems to meet most of the set selection criteria. Cirad 925 is currently at the testing stage in the farms of pioneer producers in Guadeloupe and Martinique, and downstream of the banana chain to assess its performance during transportation and at ripening facilities. Several hybrids tolerant to Sigatoka diseases, but which failed to meet strict export criteria, have been evaluated by some Caribbean partners (Lesser Antilles, Dominican Republic, etc.) within the framework of Interreg projects, involving the Tropical Technical Institute and CIRAD, with the objective of cultivating them for local or regional markets. The selected hybrids were registered in the plant variety catalogues (PVC), thus ensuring due recognition and protection.

11.2.2 Sorghum: From a Basic Subsistence Crop to a Multi-use Plant

Sorghum (*Sorghum bicolor* [L.] Moench) is a cereal crop very well adapted to semi-arid tropical regions due to its hardiness and moderate water needs. It is widely cultivated throughout Sahelian Africa and in India, mainly as a staple grain for hundreds of millions of people. Plant breeding efforts have enabled a gradual acclimatization of sorghum, allowing it to spread to temperate and subtropical regions, including to the United States, Australia, Mexico, Argentina, and, to a lesser extent, Europe. In these latter countries, sorghum is mainly grown as animal feed in the form of flour for the grain varieties and as green forage or silage for fodder varieties (multi-cut and single-cut).

11.2.2.1 An Agri-Chain for Multiple Products

As long as sorghum was mainly grown for its grain, whether for human consumption or animal feed, breeders focused their improvement programmes almost exclusively on a single objective: increasing grain yield, while taking into account certain quality criteria. As in the case of other cereals, the main strategy was to shorten the stems, resulting in a larger panicle, and thus improve the harvest index.[1] However, other production objectives gradually emerged, either in tropical countries, the regions of CIRAD's main involvement, or in France, which urged breeders to modify, and sometimes even completely redefine, their breeding strategies.

For example, with the intensification of livestock systems in Sahelian Africa, increased feed requirements led breeders at CIRAD to implement specific programmes to breed dual-use sorghum capable of simultaneously producing grain and easily digestible straw.

More recently, the preoccupation with the environmental impacts of our lifestyles, the sustainability of our production systems, and the emergence of new uses within the bioeconomy[2] framework have greatly expanded the potential uses of sorghum and, consequently, suitably reoriented the breeding objectives, especially for temperate countries.

In the context of global warming and depletion of fossil energy resources, it is now accepted that the lignocellulosic biomass will represent a renewable supply of energy, biomaterials, and chemicals (IEA Bioenergy 2009). The bulk of biomass will primarily be supplied by sustainable agricultural systems based on non-food grasses or by 'green' waste or by-products of agriculture, agro-forestry, and industry, as well as by household rubbish.

Sorghum enjoys a number of advantages that allows it to be grown as a dedicated biomass crop (Salas Fernandez et al. 2009). First, it is a type C4 plant with a high biomass production potential (current hybrid varieties produce 25–30 tonnes of dry matter per hectare under normal conditions). Second, its water needs are moderate, allowing it to be grown under water deficient conditions. Third, its agricultural practices are well established and, finally, there is still considerable scope for genetic improvement. Moreover, if sweet sorghum varieties are considered, it is the only plant capable of producing starch-rich grains and, at the same time, accumulating soluble sugars in the stalks with limited competition effects (Gutjahr et al. 2013).

---

[1] Ratio of usable plant mass to the total biomass.

[2] Biological sciences are adding value to a great number of products and services that are generically grouped under the term bioeconomy (OECD 2009).

11.2.2.2 Uses Dictate the Ideotypes

Researchers at CIRAD have been working for several years on these new objectives of sorghum production, seeking to better exploit its energy and biomaterials production potential, in association, or not, with grain producing ability, in accordance with the needs of the target country or production systems. This diversification of production objectives necessitated the adoption of new breeding criteria and development of new varietal ideotypes. It also led breeders to widen their working genetic base, which led to changes in their breeding strategies. Consequently, several sorghum genetic diversity panels, most notably the mini core collection[3] put together by CIRAD and ICRISAT, have been characterized for their sugar content and syrup production, as well as for their biochemical composition and stem digestibility. This enabled the identification of populations and lineages beneficial for improving each of these traits (Trouche et al. 2014) (Table 11.1).

11.2.2.3 The Example of Biomass Sorghums in France: Variable Ideotypes Based on Different Value Chains

Building agri-chains to produce and derive value from lignocellulosic biomass for purposes of energy and biomaterials production from cultivated sorghum and miscanthus is one of the primary goals of the research projects Biomass for the Future (http://www.biomassforthefuture.org/) and Biosorg. In both these projects, CIRAD, in collaboration with private French seed companies, INRA's public laboratories, MINES Paristech Graduate School at Sophia Antipolis, and some industrial partners, has carried out research on sorghum varieties suitable for the three target uses: methanization, production of construction blocks, and production of composite polymer materials for the automobile industry. Indeed, lignocellulosic biomass from sorghum represents a potentially useful feedstock for methanization to produce second generation biofuels and biomaterials for construction or other industries.

To this end, research on improving sorghum have mobilized multidisciplinary skills to characterize the biomass produced (biochemistry, histology), understand the growth of sorghum (with and without water constraints), and identify the genetic bases of useful traits. This multidisciplinary effort has been used to develop improved planting materials (definition of breeding schemes, mobilization of phenotyping and genotyping tools) (Pot et al. 2015).

The Biomass for the Future project has chosen to recommend the cultivation of specific sorghum varieties with high biomass productivity and whose biochemical composition and technological characteristics meet the specifications defined for the three value chains that are the main focus of this project. It is therefore not a

[3] A collection of 210 traditional sorghum varieties chosen to best represent the global genetic diversity of cultivated sorghums.

Table 11.1 Types of sorghum grown based on usage

Types	Zones	Production objectives	Countries concerned (projects involving CIRAD)
Single-use sugar (with high fibre digestibility)	Temperate zone	Silage for dairy cows or for methanization	France/Europe
Dual-use sugar	Tropical zones in emerging countries	Fuel ethanol + bagasse for cogeneration	Colombia, Brazil, Mexico
Multi-use sugar	Tropical zones in developing countries	Grains + soluble sugar (syrup or food grade alcohol) + bagasse	Haiti, Mali
Fibre biomass	Temperate zone	2G ethanol – Biomaterials and bioplastics	France/Europe

matter of identifying one single varietal type but rather a group of varieties representing several ideotypes, with each of them complying with the appropriate industrial specifications, being able to be grown in a sustainable manner, and being acceptable to the farmers.

11.2.2.4 Breeding Strategies

Of the three sorghum biomass-based value chains targeted in these projects, the methanization chain is the one that has best defined the technological and processing processes, the production potential, and the characteristics of the desired biomass (high digestibility of the whole plant which is strongly correlated with the methanogenic potential). Nevertheless, the key aspects of this value chain, such as the choice of the production system (dedicated crop or catch crop, irrigated or unirrigated) or the organization of sorghum biomass production at the territorial level, are yet to be defined. For the other two value chains under consideration, the breeding programmes are being conducted in parallel with the *de novo* establishment of these value chains, which presents an unusual situation for sorghum breeders. In this case, it is essential that these programmes incorporate, create, retain, and adequately characterize the genetic diversity at an early stage of the breeding process, also known as the pre-breeding stage. Thus, even as the specifications of biomass production systems, processing facilities, and collection-transportation and storage systems are defined for a given value chain, the conservation of this genetic diversity will ensure the preservation of useful biochemical or physical properties that could serve to meet the demands of a developing production system. In this perspective, factorial crosses based on a broad genetic base, or back-crosses, to produce BCNAM[4] populations are appropriate because these

[4] Backcross-Nested Association Mapping: this method is being applied in the BFF project to broaden the genetic base of female parents of hybrids.

mechanisms can provide genetic material that is not only useful for varietal improvement but also in the understanding of the genetic determinism of traits of interest.

Questions currently confronting breeders (Which production and quality traits to favour for a particular value chain and a particular production situation?) are illustrated by the following two examples. Material science researchers have established that only stalks of sugar-free sorghum varieties can be used to make composite polymers, which means that this criterion must be incorporated in hybrid development programmes for the value chain. No doubt, an alternative solution would be to develop methods of extracting soluble sugars with minimal effect on fibre structure and format: this is what the Biosorg project is intending to test. If this extraction of soluble sugars from the stalks is technically and economically feasible, the idea would be to move towards a cascading extraction of value by maximizing the use of the different fractions. However, the format of the fibres must be verified following the extraction of soluble sugars, as it greatly influences the end use of these fibres. To be used as feedstock in methanization systems, sorghum must not only have a good potential for biomass production in catch crop systems (3-month cycle) but the entire plant must also present good digestibility. However, some antagonism was observed between these two criteria; for example, low lignin genotypes have more digestible stems but are less robust in the initial growth stages and more susceptible to lodging. In this case, the breeders will have to choose the best trade-offs between production traits and quality traits of the biomass.

11.3 Conclusion

Varietal improvement, with its objective of creating new varieties to help develop production systems, must now address new challenges, which include taking environmental issues into account, anticipating risks and global changes that may result from climate change, and more diversified uses of plant material and products. It has to incorporate all these multiple objectives in its strategy.

Our two examples of banana and sorghum illustrate the challenges confronting varietal improvement when it comes to changes in production systems. Banana breeding efforts face severe constraints, not only because of the plant's genetic structure and its reproductive characteristics but also because of the way the export agri-chain is structured. These constraints have compelled researchers to come up with a wider varietal range. The improvement of sorghum has to accommodate new opportunities in terms of multiple uses of the plant and its products. In both cases, in order to help improve the sustainability of the agri-chain, plant breeding requires a more complete knowledge of domestication events, of the diversity of useful genetic resources, and of target traits. In order to provide the flexibility necessary to cope with new demands for biomass, in particular from new value chains, researchers are relying upon – and improving – current approaches and breeding tools and methods, especially those pertaining to pre-breeding. This new reality requires a very close working collaboration with end users and the development of innovative forms of partnership. The aim is to achieve a co-development between the needs expressed or sought and the varieties developed. Other ongoing projects at CIRAD are addressing issues of ecological intensification by studying mixtures of species, through agroforestry or mixtures of varieties, in particular to improve the durability of resistances to diseases and pests. Other projects are focusing on incorporating interactions between plants and other organisms living in the soil, mainly beneficial bacteria, into breeding programmes.

References

Abécassis J, Rousset M (2012) Quelles évolutions pour les filières céréalières ? Innov Agronomiques 19:1–11

Ahmadi N, Bertrand B, Glaszmann JC (2013) Rethinking plant breeding. In: Hainzelin E (ed) Cultivating biodiversity to transform agriculture. London, Springer, pp 91–140

Albajes R, Cantero-Martínez C, Capell T, Christou P, Farre A, Galceran J, Loópez-Gatius F, Marin S, Martín-Belloso O, Motilva M-J, Nogareda C, Peman J, Puy J, Recasens J, Romagosa I, Romero MP, Sanchis V, Savin R, Slafer GA, Soliva-Fortuny R, Viñnas I, Voltas J (2013) Building bridges: an integrated strategy for sustainable food production throughout the value chain. Mol Breed 32(4):743–770

Bakry F, Carreel F, Jenny C, Horry JP (2009) Genetic improvement of banana. In: Jain SM, Priyadarshan PM (eds) Breeding plantation tree crops: tropical species. New York, Springer, pp 3–50

Bunn C, Läderach P, Ovalle RO, Kirschke D (2014) A bitter cup: climate change profile of global production of Arabica and Robusta coffee. Clim Chang 129(1):89–101

D'Hont A et al (2012) The banana (*Musa acuminata*) genome and the evolution of monocotyledonous plants. Nature 488(7410):213–219

Daniells JW (2009) Global banana disease management. Getting serious with sustainability and food security. Acta Hortic 828:411–416

Desclaux D, Chiffoleau Y, Nolot JM (2013) From ideotype to realtype concept: Dynamic management of plant breeding by a multidisciplinary and participatory way. Example of organic durum wheat. Innovations Agronomiques 32:455–466

Gutjahr S, Clément-Vidal A, Soutiras A, Sonderegger N, Braconnier S, Dingkuhn M, Luquet D (2013) Grain, sugar and biomass accumulation in photoperiod-sensitive sorghums. II. Biochemical processes at internode level and interaction with phenology. Funct Plant Biol 40(4):355–368

Hellin J, Beuchelt T, Camacho C, Govaerts B, Donnet L, Riis-Jacobsen J (2014) An innovation systems approach to enhanced farmer adoption of climate-ready germplasm and agronomic practices, Capri Working Paper No. 116. International Food Policy Research Institute, Washington, DC, p 30

IEA Bioenergy (2009) Annual report 2009. IEA Bioenergy, p 136

Jaramillo J, Muchugu J, Vega FE, Davis A, Borgemeister C, Chabi-Olaye A (2011) Some like it hot: the influence and implications of climate change on coffee berry borer (*Hypothenemus hampei*) and coffee production in East Africa. PLoS One 6(9):e24528

Litrico I, Violle C (2015) Diversity in plant breeding: a new conceptual framework. Trends Plant Sci 20(10):604–613

OECD (2009) The bioeconomy to 2030: designing a policy agenda, OECD International Futures Programme. OECD Publications, Paris 322 p

Perrier X, De Langhe E, Donohue M, Lentfer C, Vrydaghs L, Bakry F, Carreel F, Hippolyte I, Horry JP, Jenny C, Lebot V, Risterucci AM, Tomekpé K, Doutrelepont H, Ball T, Manwaring J, De Maret P, Denham T (2011) Multidisciplinary perspectives on banana (Musa spp.) domestication. Proc Natl Acad Sci U S A 108(28):11311–11318

Ploetz RC, Churchill ACL (2011) Fusarium wilt: the banana disease that refuses to go away. In: Proceedings International ISHS-ProMusa Symposium on Global Perspectives on Asian Challenges, vol 897. Acta Horticulturae ISHS, Guangzhou, pp 519–526

Pot D, Trouche G, Le Moigne N, Carrère H, Luquet D, Soccalingame L, Vilmus I, Roques S, Berger A, Clément-Vidal A, Soutiras A, Jaffuel S, Verdeil JL, Gatineau F, Bastianelli D, Bonnal D, Vaksmann M, Rami JF, Jeanson P, Clamens S, Ventelon Debout M, Alcouffe J, Chapus M, Fabre F, Navard P, Vo L, Chupin L, Thera K, Témé N, Dufayard JF, Barrière Y, Höfte H (2015) Combining genetic analysis and breeding for multipurpose sorghum: from the development of a biomass phenotyping tool kit to the development of dedicated breeding schemes. XXIIIrd EUCARPIA Maize and Sorghum Conference, 10–11 June, Montpellier, France

Salas Fernandez MG, Becraft PW, Yin Y, Lübberstedt T (2009) From dwarves to giants? Plant height manipulation for biomass yield. Trends Plant Sci 14(8):454–461

Sanford JC (1983) Ploidy manipulations. In: Moore JN, Janick J (eds) Methods in fruit breeding. Purdue University Press, West Lafayette, pp 100–123

Temple L, Marie P, Bakry F (2008) Les déterminants de la compétitivité des filières bananes de Martinique et de Guadeloupe. Économie Rurale 308:35–53

Trouche G, Bastianelli D, Cao-Hamadou TV, Chantereau J, Rami JF, Pot D (2014) Exploring the variability of a photoperiod-insensitive sorghum genetic panel for stem composition and related traits in temperate environments. Field Crop Res 166:72–81

van Bueren EL, Jones SS, Tamm L, Murphy KM, Myers JR, Leifert C, Messmer MM (2011) The need to breed crop varieties suitable for organic farming, using wheat, tomato and broccoli as examples: a review. NJAS-Wagening J Life Sci 58(3–4):193–205

Chapter 12
Sustainable Processing Systems: What New Strategies for Tropical Agri-Chains?

Dominique Pallet and Jérôme Sainte-Beuve

Biomass processing systems are highly constrained by the variability and complexity of the products' life cycles (heterogeneity and biological activity of raw materials, diversity of processes and processing contexts). In addition, countries in the South are facing increasing constraints concerning access to water, energy, and raw materials; pollution control; the manufacture and maintenance of equipment; and the deriving of value from the use of co-products. Another constraint represents, by itself, a global challenge: reduction of losses – of raw materials, inputs, energy, finished products –, which are often substantial all along the agri-chain, especially at the processing level (Esnouf et al. 2011).

It is essential to design innovative and more efficient processing systems to take these issues of sustainability into consideration. Such systems must be capable, on the one hand, of reducing post-harvest losses, add value to local production, and generate employment and income, and, on the other, of managing environmental impacts, especially through energy efficiency and judicious water use (FAO 1997, 2004; Tran et al. 2015). The objective of CIRAD's approach is to propose methods to evaluate the diversity of existing processing systems, implement methods to design more innovative and sustainable new ones, and support them in real-world situations in association with the other actors of the tropical agri-chains. Several examples taken from food and non-food supply chains illustrate this approach and are presented in this chapter.

D. Pallet (✉)
CIRAD – UMR Qualisud, Montpellier, France
e-mail: dominique.pallet@cirad.fr

J. Sainte-Beuve
CIRAD – UMR IATE, Montpellier, France
e-mail: jerome.sainte-beuve@cirad.fr

E. Biénabe et al. (eds.), *Sustainable Development and Tropical Agri-chains*,
DOI 10.1007/978-94-024-1016-7_12

12.1 Imparting Quality and Sustainability to Processing Systems in View of Social and Economic Challenges

The approach to sustainability at the scale of the processing system takes into account technical aspects and know-how, the components of quality – nutrition, taste, and health in the case of food, and technological and mechanical characteristics for biomaterials (cotton, rubber) –, and the expectations of actors (processors, operators, beneficiaries, and consumers).

12.1.1 Organizational Forms Must Be Taken into Account While Designing Processes

Technical proposals must take organizational factors into account to find sustainable solutions within the agri-chains. One risk is to address technical issues at the expense of social ones (Box 12.1).

Box 12.1 Technical Solutions Proposed for the Natural Rubber Chain Affect the Modes of Organization in Its Upstream Part

Dominique Pallet and Jérôme Sainte-Beuve

Recent developments in the downstream sector of the natural rubber chain and the volatility of prices on the international market have encouraged manufacturers to adopt new strategies in the form of new requirements concerning product quality, a move that has profoundly affected the upstream sector, which is mainly composed of smallholders (80 %). The new organizations that were set up to respond to this requirement do not entirely satisfy manufacturers, who still complain about the wide variability of certain technological indicators. This represents a major challenge for the rubber chain. On their part, the primary processing factories (remillers) also seek to derive the maximum value from their products by negotiating long-term contracts with manufacturers, who make them commit to quantity and quality requirements. The recent Sustainable Natural Rubber initiative proposed by the IRSG[1] also stipulates quality control conditions that represent a challenge for the rubber chain.

[1]http://www.snr-i.org/Voluntary_Guidelines_and_Criteria_13_1.htm, retrieved 1 June 2016.

Box 12.1 (continued)

These dynamics force changes on the upstream sector, especially as concerns primary processing. New industrial production processes that produce grades suitable for export – *Technically Specified Rubber* (TSR 20), currently the most popular grade on the international market – are energy and water intensive, but use little labour. The older methods demand less investment but are labour intensive; they are used to produce ribbed smoked sheets grades (RSS3). The village sector, which still uses a part of this old process, mainly in Thailand, albeit without completely drying the rubber, cannot export the product. It can, however, sell it to primary processing facilities. This smallholding process is disappearing in the only country that had preserved it up to now. We are thus witnessing the deliberate disappearance of a type of usage that is adapted to smallholders (sheet type), now being replaced by a shorter supply chain that may lead to lower added value for the smallholder.

12.1.2 Contributing to the Sustainability of Small-Scale Processing

In a context of growing urbanization and changes in food consumption patterns, a part of the cultivated and gathered products is no longer processed at the family level but at an artisanal level or by small agrifood enterprises, especially in Africa. These structures – small-scale farms and agricultural enterprises – contribute to local food security through their diversity, not only by providing healthy and diverse food items but also by creating jobs and generating income. Two of the factors that constrain traditional post-harvest processes located close to cultivation areas are the excessively hard labour required and low value addition. These constraints have been overcome in the case of fonio with the help of a specially adapted huller (Box 12.2).

Box 12.2 The Mechanization of Fonio Hulling Has Reduced the Severity of Labour Required and Enhanced the Agri-Chain's Sustainability

Jean-François Cruz, Thierry Goli, Thierry Ferré, and Patrice Thaunay

In the late 1980s, the production of fonio, ancestral cereal cultivated in West Africa, was in sharp decline throughout its production area that stretched from Senegal to Lake Chad. This decline was mainly due to the difficulties faced by African women in hulling the tiny grains using a mortar and pestle, requiring exhausting and barely productive work (2 kg/h).

(continued)

Box 12.2 (continued)

The mechanization of post-harvest operations in the early 2000s, especially of hulling-milling and cleaning, helped increase the availability and improve the quality of processed products (milled fonio, pre-cooked fonio, etc.). The growing interest of urban consumers in this cereal, and the development of efficient machines, including a hulling machine called the 'GMBF fonio huller' (where GMBF stands for Guinea-Mali-Burkina-France) with a capacity of 100 kg/h, prompted the emergence of numerous processing units (small enterprises, women's groups) that today supply to urban and export markets.

The sustainability of this fonio chain relies *inter alia* on the cost of using the huller: it drops below that of manual processing when 50 or more tonnes of grain are processed in a year, which improves the economic viability of small processing facilities. The most significant benefit from the mechanization of the hulling-milling operations has been to reduce substantially the drudgery of women. It has also helped promote the revival of this long-forgotten tiny cereal (Cruz et al. 2011) (Figs. 12.1 and 12.2).

Fig. 12.1 Demonstrating the operation of the GMBF fonio huller in a village in Guinea (© Jean-François Cruz, CIRAD)

(continued)

Fig. 12.2 GMBF fonio hullers in an enterprise in Burkina Faso (© Thierry Ferré, CIRAD)

12.1.3 Taking the Evolution of Consumer Demands into Account in Traditional Processing Methods

Urbanization not only brings about dietary changes based on new expectations (nutritional, organoleptic, health, etc.) of consumers in the South, but also a nutritional transition of the concerned populations (Maire and Delpeuch 2004) towards a high-energy diet coupled with an increasingly sedentary lifestyle. This latter point may even jeopardize the consumers' health. In such a context, traditional foods exhibit a good potential for sustainability and must retain their place in the diets of the populations of the South (Box 12.3).

Box 12.3 Fermented Foods, Mainstays of Traditional Food in Africa, Already Meet Several Sustainability Criteria

Sylvie Lortal and Christian Mestres

Fermentation of products has enhanced food security and diversity for over 10,000 years. A large number of raw materials of animal or vegetable origin, depending on the region of the world, can be fermented – including milk, meat, fish, eggs, cereals (rice, maize, wheat, rye, oats, sorghum, millet, etc.),

(continued)

Box 12.3 (continued)

fruits and vegetables (cabbage, olives, pickles, soya bean, beans, carrots, cassava, etc.), cocoa or tea – to produce food items (cheese, ham, sauerkraut, etc.), alcoholic beverages (wine, beer, etc.), and non-alcoholic beverages (kefir, *nuoc-mâm*, etc.).

These foods meet many sustainability criteria: traditional fermentation processes are very simple – they do not require sophisticated equipment – and have historically relied solely on local resources. They require little or no external energy and significantly lengthen the storage life of raw materials without the use of a cold chain. The fermentation process also improves the nutritional value of the finished product considerably (micronutrients, reduction of anti-nutritional factors, etc.) and can dramatically diversify the organoleptic qualities (taste, aroma, texture) of the same raw material. Thus, when natural resources lack diversity or are produced seasonally, fermentation creates relatively varied diets that helps survival in extreme habitats. This represents another form of sustainability. Prolonging the storage life and providing a diversity of different tastes that are appreciated by local consumers amount to value addition by the processor, who is often a woman in a small-scale operation (Obodai et al. 2015).

Despite playing a key role in the food security in countries of the South and in the sustainability of food systems, traditional fermentation processes are yet to be explored in terms of issues pertaining to the health and survival of populations and the strengthening of value chains. CIRAD is studying, characterizing, and helping improve numerous fermentation processes, ranging from the cereal porridge used as baby food to the fermentation of cocoa, which is exported and represents an important source of value addition. Research is also focused on nutritional value and health security with the aim of disseminating good processing practices that use fermentation, especially in the context of the European project After (After 2014).

Another challenge for the new engineering approaches to traditional processes is to ensure consumer satisfaction by making sure that the quality of the original product remains unaltered, as in the case of cassava products (Box 12.4).

Box 12.4 Applying Effective Industrial Solutions to Traditional Methods of Cassava Processing: Downscaling

Dominique Pallet and Jérôme Sainte-Beuve

The future holds great potential for large cassava processing industries due mainly to a growing population and urban centres. Existing small- and

(continued)

Box 12.4 (continued)

medium-scale technologies are inefficient in dealing with this growing demand (excessive use of energy and water, low yields), resulting in high production costs and environmental impacts. There is, however, substantial scope for improving their efficiency through (Tran et al. 2015):

- increased yields;
- increased energy efficiency;
- deriving of value from co-products and waste;
- economies of scale.

The production of biogas from factory wastewater can help reduce methane emissions (greenhouse gases), replace fossil fuels (coal, oil) with renewable energy, and reduce the factory's energy bill.

The aim is to downscale industrial knowledge and processes to benefit small-scale processing operations. To this end, the approach includes:

- assessing existing small-scale technologies in order to identify the critical stages of processing methods, especially the ones that are most energy- or water-intensive and least efficient;
- identifying effective and sustainable industrial technological solutions and assessing their potential for downscaling;
- improving performance by applying tools and methods of the design process.

In this approach, multi-criteria tools for optimization-modelling were used to take the technical, economic, social, and environmental aspects into account. The aim is to ensure that the improved method fits well in the local socio-economic context and that it is not abandoned soon after adoption. The quality of the finished product is crucial to this approach. It is particularly important to ensure that the product obtained from the modified method has the same organoleptic qualities as the original product so that consumers accept the product. It is indeed a necessary criterion for the sustainability of the improved process.

12.1.4 Ensuring Food Health Safety for Consumers

The notion of food health safety can be crucial for the sustainability of an agri-chain. This is true for fish, a highly perishable product, whose processing by freezing, fermenting, or smoking must comply with very strict and standardized practices. The sustainability of these agri-chains in an international context is based on designing new processes in association with the processors.

Global consumption of fish has more than doubled over the last five decades, driven mainly by population growth, rising incomes, and urbanization. The increased availability of this vital source of protein is due to an improved efficiency of fish distribution channels and a growth in production, linked increasingly to the expansion of aquaculture, an activity that now accounts for close to half of the global production of fish destined for human consumption (FAO 2014).

Although fish is not the primary source of dietary protein in developing countries, it contributes significantly to meeting human protein needs, especially in coastal areas and in the neighbourhood of lakes and ponds. Fish is sold in processed forms (smoked, fermented, salted, or dried) not only to overcome its high perishability but also, increasingly, to satisfy hedonistic needs. There is a growing demand for these specialized products, both locally and on export markets, and value chains based on these products are able to generate good income to benefit local food security. However, several obstacles continue to hamper the further expansion of these activities, particularly as regards quality management and adherence to standards in production facilities, often artisanal ones, due to the lack of suitable equipment.

As part of the European project After (After 2014), the Senegalese smoked fish (*kong*) and the Beninise fermented fish (*lanhouin*), representative of their respective product families, have been the focus of extensive studies to improve the sustainability of these activities. The agri-chains were analyzed in terms of the health safety of the products, efficiency of the processes (including a life cycle assessment), and the acceptability of the improved products on the local and export markets. Recommendations were made and demonstrations organized for professionals. A process of multi-criteria analysis has shown the way forward for designing equipment to smoke *kong* in Senegal that complies with health practices and standards.

12.2 Integrating the Environmental Dimension of Processing

A global vision of the environmental sustainability of processing activities involves, on the one hand, reducing the use of energy (Boxes 12.5 and 12.6) and of water in the processes and, on the other, deriving value from co-products and managing and controlling pollution (Box 12.5).

12.2.1 Reducing Energy Consumption

Reducing energy consumption is a key factor in imparting sustainability to food and non-food processing operations in the countries of the South. Using a detailed assessment of consumption (Box 12.5), research and design tools can help improve the energy efficiency of processes (Box 12.6).

Box 12.5 The Processing of Rubber, a Highly Energy-Intensive Sector

Dominique Pallet and Jérôme Sainte-Beuve

The primary processing of natural rubber consumes a significant amount of energy, mainly for mechanical and thermal processes. The breakup of the installed power requirement for each sub-process is shown in Fig. 12.3 for a plant processing 10 T/h of natural rubber. Electrical energy is used for crumbling, homogenization, and part of the drying. This energy can be obtained from the national grid, from generators that run on fuel in very remote regions, or, in some few rare cases, from cogeneration fuelled by biomass collected from rubber plantations. A comparative study (Energy Efficiency Index in Rubber Industry 2007) conducted in Thailand on TSR20 grade rubber highlights significant differences in specific energy consumptions (SEC): a plant with an output of 5 T/h has an installed requirement of about 2.8 MJ/kg. Another study carried out in 2010 in Indonesia (Utomo et al. 2010) recorded a slightly higher consumption for the TSR20 manufacturing process (3.05 MJ/kg), but the distribution between requirements for mechanical operations (60 %) and thermal operations (40 %) remains similar.

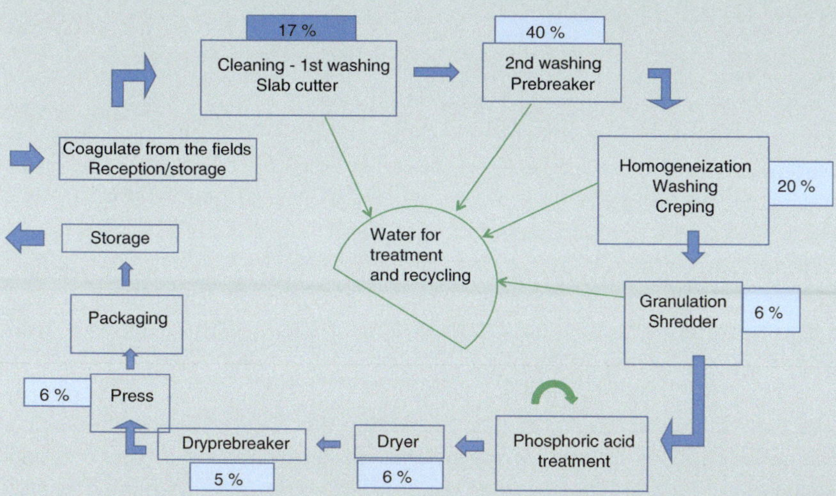

Fig. 12.3 A flowchart of the processing of TSR20 rubber and a breakup of the installed power requirement for each sub-process (*grey boxes*)

(continued)

Box 12.5 (continued)

As far as raw dry natural rubber is concerned, the last 20 years have seen the emergence of a number of more efficient oil-fired dryers (mainly improved insulation) which, on average, have reduced consumption by 30 %. Similarly, the use of high-performance burners, subject to local availability of LPG (liquefied petroleum gas) as fuel, has helped bring down the average consumption to 28 l per tonne of dry rubber (i.e., by 35 %). This innovation, which is now spreading to many countries, is a good alternative and has the added benefit of low CO_2 emissions. Electricity consumption is directly dependent on the power of motors in the pre-processing and processing sections of the production line.

However, these are only partial solutions to an energy issue that is present at all the levels of the rubber chain, and current trends do not indicate any satisfactory answer soon. The quality requirements of the automotive industry demand a fundamental rethinking of the primary processing stage in a perspective of substantial energy reduction.

Box 12.6 Diagnosis and Prospects of Sustainability of the Dried Mango Chain in Burkina Faso

Michel Rivier and Jean-Michel Méot

Drying is a traditional method, widely practiced in Africa, to preserve food products and derive value from them. However, it requires a large amount of energy, which is a major obstacle to its development and wider adoption. In Burkina Faso, 350 Atesta gas dryers dry 8000 tonnes of mangoes, which are then exported to Europe over a period of three to 4 months. The Burkinabe units together employ 1500–2000 people. Controlling production costs depends on optimizing energy use of drying technologies that currently consume a large amount of gas (Rivier et al. 2009). Finding alternative energy sources for drying equipment and improving efficiency will go a long way in improving the sustainability of agri-chains in the South.

CIRAD has thus undertaken thermal and air flow studies of the drying process which have helped reduce the drying time from 20 to 12 h, enabled the production of homogeneous batches in terms of water content, and ensured a reduction in energy consumption of about 40 % by recycling part of the air flow. Butane gas, currently utilized in dryers in Burkina Faso, is easy to use and requires little investment. However, its cost is becoming an increasingly large portion of the total cost, and the subsidy on gas imports is a heavy burden for this country's economy.

(continued)

Box 12.6 (continued)

These units have power requirements of 20–400 kW, which is impossible to obtain solely from solar power, an energy source that is intermittent and with low energy density. The use of bioenergy from waste (sawmill, agrifood) and/or biomass can constitute a reliable alternative. The mango stones (kernels) can be used as fuel in conjunction with butane gas, thus contributing almost 30 % of the total energy consumed.

12.2.2 Issues of Water Consumption and Pollution Control

Water is often a limiting factor in processing operations in many countries of the South. In addition, some processing operations use chemical inputs and/or discharge polluting effluents in contexts where societal and normative constraints create an unavoidable requirement for managing environmental impacts, especially through possible redesigns of the processes themselves. Some agri-chains (for example, of cassava, wet coffee, rubber, etc.) consume large amounts of water and there is a huge scope for saving water through more efficient processes and through recycling (Box 12.7).

Box 12.7 Recycling Water and Reducing Emissions: New Environmental Challenges to Primary Processing of Natural Rubber

Dominique Pallet and Jérôme Sainte-Beuve

The processing of natural rubber requires large amounts of water, mainly for cleaning. This water is mainly drawn from underground aquifers, lakes, canals, or rivers. There is a 20 % loss (by weight) of raw material in liquid form during the first processing step, which must also be treated in addition to the wash water. Processing methods around the world vary widely, often due to the need to adapt processes to the non-consistency and contamination levels of the local raw material. For example, the manufacturing process in Africa for the TSR20 grade of rubber can have two or three cleaning tanks, each with a capacity of 10 m^3, whereas these tanks are twice as big in Asia (Indonesia and Malaysia). Similarly, creping machines, which allow efficient washing and homogenization of the rubber, use large amounts of water (about 50 m^3 per day per machine). In addition, bringing the level of impurities in the rubber down to permissible levels (< 0.2 % w/w) results in high water consumption, of the order of 30–40 m^3/T of rubber (TSR20). On the whole, the water consumption by plants processing 100–200 tonnes of dry rubber per day is estimated to be between 3000 and 8000 m^3/day, or an average of 23 m^3/T of dry rubber (Hien and Tao 2012).

(continued)

Box 12.7 (continued)

The actors of the sector upstream of the agri-chain, including the equipment manufacturers, have recognized the need to economize on water use, despite the fact that most processing plants are located in tropical areas with high rainfall (greater than 1500 mm annually). Projected improvements for the next decade target a more widespread recycling of wash water, to up to 30 % of the requirement without compromising product quality. All plants currently have water treatment units to reduce BOD (Biochemical Oxygen Demand), COD (Chemical Oxygen Demand), and suspended solids in wastewater effluents in accordance with national recommendations and standards. These systems are based mainly on conventional biological processes (lagooning, anaerobic digestion, and activated sludge) (Leong et al. 2003). Several studies have been undertaken to improve the efficiency of treatments by lagooning, making them more robust and less dependent on weather conditions. They propose systems to speed up the process (Latex Industry and Block Rubber 2001), which seems increasingly necessary in order to reuse some of the liquid effluents. Although systems using microorganisms have also been tried out to treat liquid effluents (Jai Shanker and Girish 2014) that are mainly composed of organic matter, they are still at an experimental stage.

The processing of natural rubber by manufacturers uses a large amount of chemicals. Through its Reach programme (2006) (*Registration, Evaluation, Authorization and restriction of CHemicals*), the European Union has, since 2007, implemented an integrated system to manage all chemicals that may pose a risk to human health or the environment. This constraint has impacted the entire manufacturing industry. Another challenge is reducing CO_2 generated by passenger cars, which has to fall to 95 g CO_2/km by 2020, compared to 130 g CO_2/km in 2015.[2] In order to be able to conform to this requirement, the weight of automobile tires has to be reduced, which would consequently result in a reduced demand for natural rubber.

[2]Regulation (EU) No. 333/2014 of the European Parliament and of the council of 11 March 2014 amending Regulation No. 443/2009 to define the modalities for reaching the 2020 target to reduce CO_2 emissions from new passengers cars. *Official Journal of the European Union* – L 103/15.

12.3 Steps to Incorporate Sustainability Criteria in the Design of Processing Systems

12.3.1 Key Issues for Developing New Multi-criteria Approaches

New approaches in designing sustainable processing systems must, as discussed above, incorporate several relevant constraints and criteria pertaining to raw materials, the finished product, and the local environment. To develop and propose such approaches is in itself a matter for research, whose main issues include:

- How to characterize the performance of a complex agrifood process with its technical, social, and environmental aspects taken into account?
- What are the functions expected (fermentation, drying, etc.) to be applied to food items, especially traditional ones, and which simple processing operations (unitary operations) can be used to accomplish them?
- What strategy to use to define an efficient and sustainable technological itinerary?
- What technological alternatives to use to make better use of raw materials? Should existing processes be modified or new ones be introduced? Should a new study be undertaken to suggest a sequencing of unitary processing operations?
- What equipment to use in order to derive more value from raw material in the challenging context of countries of the South?

These issues were at the heart of approaches used in cassava processing systems and the dried mango chain in Burkina Faso (Box 12.6).

12.3.2 Modelling the Design of Sustainable Processing Systems

Food processing methods involve several unitary operations. The choice of operations and their sequential organization are crucial to the quality of the finished product and the performance of the process. The number of unitary operations, the multiplicity of possible arrangements, and the strong interlinking of operations render the study and optimization of the process as a whole more complex. The modelling approach formalizes and streamlines the selection of technological solutions that comply with sustainability conditions established upstream (Briffaz et al. 2014a, b; Reach 2006).

A first step is to develop an observation model that describes – by using input variables and parameters (size, initial properties, intrinsic properties, etc.) – the mechanics of and the coupling between the various phenomena involved in the processing system (material transport, heat transfer, chemical kinetics, etc.) and having an impact on the desired properties (taste, texture, compactness, energy efficiency, etc.). Particular attention has to be paid to the product-process interaction. A second step takes a relook, using an optimization process, at the range of solutions provided by the model in order to select the one that meets the required sustainability criteria. To this end, it is necessary to specify the sustainability constraints so that they can be cross-checked with the output data provided by the model. An example of this type of approach is shown in Fig. 12.4. In this example of designing rice cookers, the criteria of consumer preferences were linked to water and energy consumption (Box 12.8).

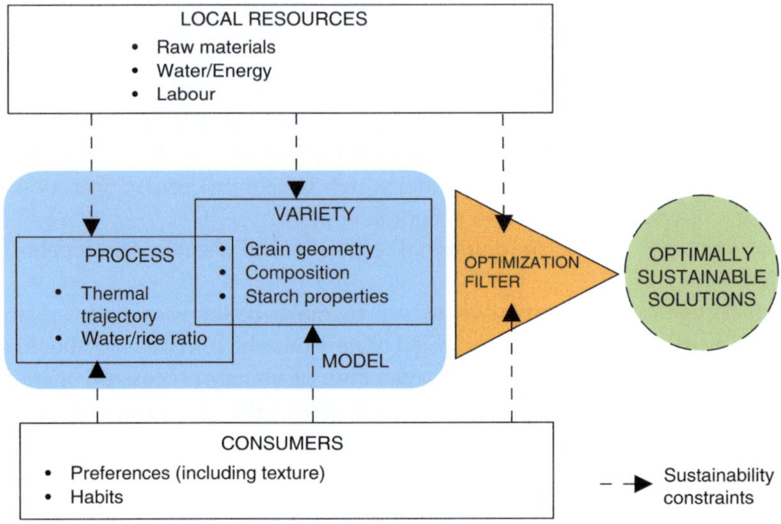

Fig. 12.4 Consideration of sustainability criteria in a model for rice cooking

Box 12.8 Modelling Rice Cooking Allows the Incorporation of Elements of Sustainability in the Design of Equipment

Aurélien Briffaz and Christian Mestres

Rice is the most common cereal in the human diet, even more so than wheat and maize (source: International Grains Council). In the countries of the South, local constraints (for example, access to energy) and specificities (for example, consumer preferences, traditions) should be taken into account in the design of food processing equipment such as grain cookers. Sustainability can be assessed in several ways:

- economic: proposing a technical solution that entails reasonable manufacturing costs, which allows the generation or continuation of local economic activities, and helps make better use of local agronomic varieties;
- environmental: proposing low-energy solutions with reasonable levels of water consumption;
- through food security: proposing a process that preserves the product's nutritional qualities;
- social: proposing a solution in line with consumer expectations (taste, texture, convenience, etc.), particularly with regard to the grain's firmness and stickiness.

(continued)

Box 12.8 (continued)

In order to assess the sustainability of the technical solutions without having to undertake an extended experimental phase, a mathematical model simulating all the physico-chemical phenomena involved in rice soaking-steaming operations was developed and validated, both in terms of process parameters (time, temperature, water/rice ratio, etc.) and the rice variety selected (Briffaz et al. 2014a, b). An optimization phase allows the selection of possible solutions maximizing the sustainability criteria that were established initially. Modelling thus helps understand and assess the sustainability of technical solutions. A set of soaking-steaming itineraries meeting the sustainability criteria defined upstream has been identified. These itineraries allow the texture of cooked rice to be adjusted to the preferences of consumers while limiting the amount of water and energy used for cooking.

12.4 Conclusion

Processing system for agricultural and food products are complex because they consist of operations that must meet different – and sometimes contradictory – needs and specifications (material, food, energy, water), depending on the spatial and temporal scale under consideration. In the South, the agrifood economy is often based on artisanal processing units that can be categorized as small enterprises. Through its research, CIRAD supports these actors and promotes their diversity which, in a constrained economic context, can be considered a measure of resilience.

Even though we are specifically interested here in the processing part of the agri-chain with its own scales – unitary operation and process –, it is also a matter of integrating the constraints of other scales: production basins and supply and marketing areas. To deal with this complexity, which encompasses a diversity of multi-disciplinary knowledge, our analytical approaches must take into account criteria such as the quality of the finished product in its various forms, the use of water and energy, efficient use of the raw material, ecological balances, etc. The incorporation of the dimension of sustainability in the design of food and non-food processing systems therefore requires a multi-dimensional approach. To this end, new design processes for sustainable processing systems rely on multi-criteria analysis methods.

References

After (2014) Technical Guidelines. http://www.after-fp7.eu/en/results/results-dissemination/technical-guidelines. Retrieved 2 June 2016

Briffaz A, Bohuon P, Méot JM, Dornier M, Mestres C (2014a) Modelling of water transport and swelling associated with starch gelatinization during rice cooking. J Food Eng 121:143–151

Briffaz A, Bohuon P, Méot JM, Pons B, Matencio F, Dornier M, Mestres C (2014b) Modelling of brown rice and limited-water cooking modes and its potential use for texture prediction. J Food Eng 141:99–106

Cruz JF, Béavogui F, Dramé D, (2011) Le fonio, une céréale africaine. Versailles/Wageningen (The Netherlands)/Gembloux (Belgium), Éditions Quæ/CTA/Presses agronomiques de Gembloux, coll. Agricultures tropicales en poche, 160 p

Esnouf C, Russel M, Bricas N (dir.) (2011) Pour une alimentation durable. Réflexion stratégique duALIne. Versailles, Éditions Quæ, 288 p

FAO (1997) The state of food and agriculture: the agroprocessing industry and economic development. Rome, FAO. http://www.fao.org/docrep/017/w5800e/w5800e.pdf. Retrieved 2 June 2016

FAO (2004) The state of food insecurity in the world. FAO, Rome ftp://ftp.fao.org/docrep/fao/007/y5650e/y5650e00.pdf. Retrieved 2 June 2016

FAO (2014) The state of world fisheries and aquaculture 2014. FAO, Rome. http://www.fao.org/3/a-i3720e.pdf. Retrieved 2 June 2016

Hien NN, Tao LT (2012) Situation of wastewater treatment of natural rubber latex processing in the Southeastern region Vietnam. J Vietnam Environ 2(2):58–64

Jai Shanker HP, Girish K (2014) Rubber processing industry effluent treatment using a bacterial consortium. Int J Curr Microbiol App Sci 3(10):775–782

Latex Industry and Block Rubber (2001) Industrial sector code of practice for pollution prevention (Cleaner Technology). Latex Industry and Block Rubber, Thailand 47 p

Leong ST, Muttamara S, Laortanakul P (2003) Reutilization of wastewater in a rubber-based processing factory: a case study in Southern Thailand. Resour Conserv Recycl 37(2):159–172

Maire B, Delpeuch F (2004) La transition nutritionnelle, l'alimentation et les villes dans les pays en développement. Cahiers Agricultures 13(1):23–30

Ministry of Energy (2007) Energy efficiency index in rubber industry, department of alternative energy development and efficiency. Ministry of Energy, Thailand, 73 p

Obodai M, Oduro-Yeboah C, Amoa-Awua W, Anyebuno G, Ofori H, Annan T, Mestres C, Pallet D (2015) Kenkey production, vending, and consumption practices in Ghana. Food Chain 4 (3):275–288

Reach (2006) European regulation (regulation no. 1907/2006)

Rivier M, Méot JM, Ferré T, Briard M (2009) Le séchage des mangues. Versailles, Éditions Quæ, coll. Guide pratique, 112 p

Tran T, Da G, Moreno Santander MA, Vélez-Hernández GA, Giraldo-Toro A, Piyachomkwan K, Sriroth K, Dufour D (2015) A comparison of energy use, water use and carbon footprint of cassava starch production in Thailand, Vietnam and Colombia. Resour Conserv Recycl 100:31–40

Utomo TP, Hasanudin U, Suroso E (2010) Comparative study of low and high-grade crumb rubber processing energy. In: Proceedings World Congress on Engineering 2010, vol III, 30 June-2 July, London

Chapter 13
A Step Towards Food Security

Nadine Zakhia-Rozis, Éric Scopel, and Dominique Pallet

Feeding the 9–10 billion people that the UN foresees will inhabit the planet in 2050 will be a truly major global issue. Agricultural research and agri-chains will have to rise to the challenge of providing these billions with food that is healthy in a sustainable and equitable manner. To do so, they will have to, first, take into account global changes (climate, environmental, nutritional, and policy) and, second, construct a real transversality and interdisciplinarity.

In order to contribute to global food security according to the four pillars defined by the FAO in 1996, transformative action will be required at the level of agri-chains. This will mean ensuring:

- food availability (related to production capacity);
- access to food (related to poverty and social inequalities);
- stability of prices (related to market variability and various hazards);
- food utilization (related to food and nutritional transitions in the North and the South).

The governance and public policy dimension is inseparable from these four pillars to achieve the goals of providing healthy food in a sustainable and equitable manner.

This chapter focuses on one of the pillars: food availability.

N. Zakhia-Rozis (✉)
CIRAD – Department Persyst, Montpellier, France
e-mail: nadine.zakhia-rozis@cirad.fr

É. Scopel
CIRAD – UPR Aida, Montpellier, France
e-mail: eric.scopel@cirad.fr

D. Pallet
CIRAD – UMR Qualisud, Montpellier, France
e-mail: dominique.pallet@cirad.fr

E. Biénabe et al. (eds.), *Sustainable Development and Tropical Agri-chains*,
DOI 10.1007/978-94-024-1016-7_13

13.1 Several Approaches to Consider

Different approaches can improve food availability at the global scale but none of them is exclusive. Indeed, capacity to improve food availability and meet the needs and demands of the Earth's populations will likely be the result of implementing a combination of these approaches.

The first and obvious approach is to extend the area of agricultural and cultivated land. But this approach brings with it risks of deterioration of soil quality, damage to natural ecosystems, and land tensions, especially because of competition between food and energy uses of agricultural production and between urban development and agricultural development.

A second approach aims to increase agricultural yields, either through the use of various inputs, such as more productive varieties, fertilizers, or pesticides, or by improving the efficiency of use of available resources. Despite considerable advances in this field, it is clear that a gap still exists between the theoretical or potential yields and actual yields obtained. Furthermore, negative environmental impacts are also to be considered when the intensification practices adopted are not ecological.

A third way is to reduce post-harvest losses (products lost after leaving the field and during transport or storage) and wastage of food (products lost by the consumer after processing and distribution). This issue, which was a priority in the international research agenda and in those of donor agencies in the 1970s and 1980s, is now occupying centre stage once again given the high percentage of current post-harvest losses, especially in tropical regions. This approach of reducing losses amounts to optimizing the different links of the agri-chain (production, processing, distribution, consumption), while taking into account the technical, social, economic, organizational, and political aspects of the value chain's context. In other words, this third way comes down to working on the interactions between agricultural practices, the efficiency and sustainability of food processes, and the organization of actors and value chains in order to reduce losses and wastage of food. This is a complex topic of research involving multiple stakeholders and disciplines.

13.2 An Agenda for Research

When seen through this prism, the contribution of agri-chains to food security can be divided into three themes, each of which, in turn, can be broken down into research issues that involve agricultural research in its disciplinary diversity and which form the research agenda presented here.

13.2.1 Theme 1: Increasing the Quantity and Improving the Quality of Food Products Through Upstream Developments (Fields, Livestock)

- Understanding the influence of agroecological practices on the increase in quantity and the improvement in quality of upstream production, through efficient use of resources and/or the reduction of potentially contaminating products.
- Developing relevant quality indicators that are shared by all the actors (from producers to consumers) in order to ensure better communication and ease of trade all through the agri-chain. These quality indicators should incorporate the health, nutritional, and organoleptic determinants of food quality.
- Better consideration and management of the variability and diversity of agricultural raw materials produced.
- Studying the impact of standards (health, nutritional, economic, social, security) on the organization of agri-chains and actors.
- Understanding the mechanisms by which the needs and expectations of consumers and civil society orient practices of food production and processing.
- Managing trade-offs between product quantity and quality (pesticide reduction versus lowered yields) or between environmental impacts of production and product quality (use of agricultural effluents versus health quality of harvested products, for example) in order to optimize the supply chain's functioning. To be able to make these trade-offs in an informed manner requires multicriteria modelling and assessment tools, which then lead to the development of decision-making tools for use by actors and policymakers.

13.2.2 Theme 2: Tighter Integration Between Production, Processing, and Consumption

- Analyzing actor networks and their functioning in order to identify possible room for manoeuvre to promote efficient and sustainable agriculture and food systems.
- Studying the concepts of proximity and integration, which can be geographical (for example, undertaking the first processing on the production site itself, with the aim of a quick stabilization of the product and value addition) or concern intra-agri-chain levels (between different stages, in different locations, or by different actors).
- Making more efficient and sustainable food processing systems by supporting the creation of mobile processing units (which can be positioned close to production sites) that are flexible (which can adapt to the variability and diversity of agricultural raw materials) or by sharing processing equipment (like the cooperative use of agricultural equipment), in pursuit of an improved

distribution of wealth, responsibilities, and tasks between the agri-chain's actors. These changes require an integrated approach and new skills (in logistics, design, infrastructure, or biomass exploitation).
- Modelling the functioning of agricultural and food systems at the scale of demarcated territories to find balances in the use of natural resources (soil, climate, water) and human resources (actors, governance). Indeed, the concept of the territory emerges as a relevant unit of sufficiently large operational size for an integrative approach aiming to optimize the functioning of and interactions between agri-chains in pursuit of food security.

13.2.3 Theme 3: Limiting or Reducing Post-Harvest Losses and Recycling, Reusing, and Extracting Value from Co-products

- Using a systemic approach to research and identifying the factors underlying pre-harvest and post-harvest losses to establish the hierarchy and interactions of causes.
- Identifying possible room for manoeuvre in terms of agronomic and technological itineraries to limit these losses. Different scenarios can coexist, such as: prevention (designing itineraries that produce less waste), reduction (identifying new commercial outlets for waste products of agri-chains currently considered as losses, such as the agri-chains of 'ugly' fruits and vegetables, rejected by traditional distribution networks because of their heterogeneous appearance but sold via different commercial routes), and recycling (for example, cascade reuse of agricultural co-products and by-products).

Limiting losses requires managing trade-offs between, for example: recycling of agricultural effluents and managing problems of quality of health of products; cascading use of various co-products and by-products such as the use of agricultural residues for soil fertilization and amendment versus for livestock farming versus for processing into biofuels; or the use of additional energy (refrigeration, for example) to reduce food perishability.

In conclusion, it is a matter of combining options for improving production (quantity and quality) through agroecology, optimizing proximities and integrations within agri-chains, and minimizing post-harvest losses while promoting recycling. The challenges facing agricultural research in its efforts to contribute to sustainable development, food security, and global changes require an unprecedented scientific openness and a truly interdisciplinary participation. If one had to define a single clear priority, it would have to be the need to design, at the territorial scale and with the involvement of the agri-chains' actors, efficient and innovative agricultural and food systems capable of producing sufficient quantities in a sustainable manner.

Part III
Diversity of Uses of Biomass and Inter-Agri-Chain Dynamics

Chapter 14
Agricultural and Forestry Chains in the Countries of the South in the Age of Bioenergy

Laurent Gazull

Global production of total primary energy reached 510 exajoules[1] (EJ) in 2010, and increased by 2 % every year for the subsequent 5 years (IEA 2010). Asia, and in particular China, accounted for more than half of this increase in production. The contribution of biomass energy to this global primary production is around 50 EJ, i.e., about 10 %, and has remained relatively steady over the past two decades. This biomass originates mainly from wood (45 EJ) and, to a much lesser extent, from agricultural products and residues (5 EJ).

The various prospective studies carried out by the Intergovernmental Panel on Climate Change (IPCC) indicate that biomass energy used as a substitute for fossil fuels would have to double to around 100 EJ by 2050 and quadruple by 2100 to achieve targets for reduction in atmospheric CO_2 concentrations (Bruckner et al. 2014). This increased use of biomass energy is already noticeable. The International Energy Agency's (IEA) trend scenarios for 2030 show that biomass used for electricity generation should increase by 5 % every year, for traditional cooking and heating by 6 %, and for the biofuel sector also by 5 %. Many developed countries – and increasingly developing countries too – are adopting incentive policies to promote the use of biomass for energy and its cultivation for energy generation. Such incentives make use of various technical, legal, and fiscal tools: direct investments, mandatory incorporation of biofuels, targets for renewable energy mix, feed-in tariffs for energy, differential taxation on agricultural products based on end-use, loan mechanisms for the installation of production infrastructure, etc.

Several agri-chains already produce energy in a big way in the form of electricity, heat, and solid and liquid fuels. Indeed, on a global scale, 12 % of the vegetable oils are converted to biodiesel, 15 % of the sugarcane is converted to bioethanol,

---

[1] One exajoule (10^{18} J) is equal to one billion gigajoules.

L. Gazull (✉)
CIRAD – UPR BSEF, Montpellier, France
e-mail: laurent.gazull@cirad.fr

E. Biénabe et al. (eds.), *Sustainable Development and Tropical Agri-chains*,
DOI 10.1007/978-94-024-1016-7_14

12 % of the coarse grains (maize, sorghum) are converted to bioethanol, and, finally, over 50 % of the world's timber is used for energy production (OECD/FAO 2014).

The energy market is thus becoming increasingly important in the organization and management of agri-chains. The objectives of these agri-chains can differ: orienting production towards a new mass market at the national or international level, diversifying outputs to mitigate the effects of the vagaries of agricultural markets, reducing their own energy consumption and carbon footprint, optimizing material flow by reducing greenhouse gas emissions or by optimising nutrient cycles (C, N, P, K), etc.

Biomass production must adhere to several sustainability criteria in order to meet a growing global demand, one that is constrained by targets to reduce greenhouse gas emissions and limited availability of land and water resources. These criteria include carbon neutrality, positive energy balance, low environmental impact, and production models that contribute to rural development. The sustainability of bioenergy production systems is a major challenge and a necessary requirement for the growth of these systems. Since bioenergy is often associated with mass production systems, questions on competition with food cultivation, land grabbing, and the role of industry in agricultural and forestry production are ever present, often accompanied by a range of concerns and controversies.

The objective of this chapter is to present an overview of the place of energy production in four major agricultural and forestry chains in the countries of the South: oil palm, sugarcane, cassava, and timber. For each of these agri-chains, we will briefly recall the current state of production dedicated to energy, the interest of the agri-chain's actors in energy production, the integration of sustainability criteria, and the technical and organisational changes that have resulted from these new energy uses.

14.1 The Oil Palm Chain

Three sources of biomass can contribute to energy production in the oil palm chain. The first is palm oil, which can be converted into biodiesel via a chemical process called transesterification. The second source, far less significant, consists of solid biomass – stalks, fibres, shells – used in cogeneration units to produce steam, heat, and electricity. A third source is oil mill effluents which, through methanization, can produce a synthesis gas that can also be use to generate heat and electricity.

14.1.1 Production of Biodiesel

An estimated 5–10 % of global palm oil production is currently converted to biodiesel (FAO 2013; OECD 2013). Indonesia, Malaysia, Thailand, and Colombia are the leading producers of biodiesel. With the exception of Ghana, no African country yet produces biodiesel from palm oil. Based on government objectives for blending into common diesel fuel, OECD and FAO predict a doubling of the global production of biodiesel between 2010 and 2025, a significant portion of which will be derived from palm oil (OECD/FAO 2014). Apart from Indonesia and Malaysia, which have oriented their production mainly towards exports, most producer countries use the biodiesel they produce primarily to meet national requirements. Thus, 40 % of the palm oil produced in Thailand and nearly 50 % of Colombia's production are consumed domestically as biodiesel, while this share is only 8 % and 2 % respectively in Indonesia and Malaysia. Among smaller producers, Nigeria and the Philippines export their entire production.

From the processing point of view, the dominant technical model is a high capacity industrial one. For example, biodiesel plants in Colombia have, on average, a processing capacity of 50,000–100,000 T/year. This corresponds to supply basins of 10,000–25,000 ha respectively (Fedecombustibles 2014). Outputs from artisanal supply chains (micro oil mills) are not included in these figures, despite ongoing efforts at transesterification by micro units, a fact that largely explains the absence of African countries – where the supply chains are artisanal for the most part– from the biodiesel market.

Biodiesel production has little impact on farming practices. On the one hand, the final conversion of oil into biodiesel requires no special adaptation of planting or harvesting methods and, on the other, since production is primarily oriented for the domestic market, there is little or no need for conformance with international sustainability standards (ISO). The need for certification arises mainly for export, especially to the EU, currently the main importer of biodiesel (1.6 million tonnes in 2013). The EU requires, as part of the RED directive, producers and processors to be certified under EU-approved voluntary schemes. However, these schemes, the most widely used of which are 2BSvs for imports into France and ISCC for imports into Germany (455 certified companies), are not overly exacting. The RSPO certification (Roundtable on Sustainable Palm Oil) should be more restrictive, but it is only in the process of being approved by the EU and it is based on the voluntary involvement of producers (Paoli et al. 2010). However, biodiesel is mired in controversy, even in the limited context of production for domestic use, regarding its environmental and social impacts. In Latin America, especially in Colombia and Brazil, the promotion of palm biodiesel is accompanied by communication campaigns, incentives for good agricultural practices with regard to the environment, and integration of smallholders (RSPO certification programmes in Colombia, *Selo Combustível Social* in Brazil). Thus, it appears that the biodiesel market encourages agricultural practices that are at least as virtuous as those observed in the absence of this market.

14.1.2 Producing Energy from Residual Biomass

Furthermore, outside the biodiesel market, environmental concerns (waste management, reduction of CO_2 and greenhouse gases emissions) and economic concerns (increase in energy prices) impel processors to make increasing use of solid and liquid waste residues to produce electricity.

Cogeneration using fibres and shells is an old practice, generally providing enough heat and electricity to meet the needs of factories, and sometimes even more. This production is expected to increase in the coming years with all possible sources of solid biomass being used as feedstock: granulation projects for oil palm stalks (Empty Fruit Bunches), leaves, trunks, and sawdust are already operational, in particular in Malaysia. This extensive utilization of solid waste puts it in competition with the use of residual biomass to maintain the soil fertility structure and organic balance. Indeed, EFBs and leaves are now largely left on the fields, either raw, or in a semi-composted or composted state, along with liquid effluents (Palm Oil Mill Effluent), leading to a significant reduction – by up to 20 % – in the use of fossil or synthetic fertilizers.

14.1.3 Methanization of Effluents

In a similar way, the practice of methanization of palm oil mill effluents is spreading rapidly. Since 2010, Malaysia has implemented a programme to install methanization units in 500 oil mills in order to generate energy for factories and local communities, and even feed the excess energy into the national grid (MPOB 2014). With the objective of reducing greenhouse gas emissions now forming part of RSPO's Principles and Criteria, the number of methanization units is steadily increasing.

14.2 The Sugarcane Chain

In the sugarcane chain, three sources of biomass contribute to the production of a large amount of energy: the first is the cane itself which, when fermented and distilled, produces ethanol that can be directly blended into gasoline. The second, and equally important, is bagasse – the fibrous residue that remains after crushing sugarcane stalks – which, when used in cogeneration units, produces steam, heat, and electricity. And finally, a third source consists of solid residues left behind in the field: sugarcane tops and leaves which, if harvested, can be used in cogeneration units.

14.2.1 Ethanol Production

Ethanol has thus become a major product of sugarcane chains around the world. Currently, an estimated 45 % of the sugarcane produced is used for ethanol in Latin America, 4 % in sub-Saharan Africa and only 1 % in Asia; this figure goes up to 4 % in Pakistan and Thailand (FAO 2013; OECD 2013). Brazil is the undisputed leader in ethanol production from sugarcane, accounting for 65 % of world production in 2013. (In the same year, its global share of ethanol produced from all sources was 35 %.) Outside Brazil, the largest producers of ethanol from sugarcane are India (5 %), Thailand (3 %), Argentina (1.6 %) and Colombia (1 %). Mozambique, Tanzania, and South Africa are major producers in sub-Saharan Africa, making up about 1 % of the world's production.

OECD and FAO predict a 60 % increase in ethanol production by 2025. This growth will be concentrated mainly in Brazil, driven by domestic demand and exports to the United States. At the same time, however, global trade in ethanol is only expected to show a small increase with increased production primarily used to meet the requirements of domestic users.

As is the case for biodiesel from palm oil, the dominant technical process for ethanol production remains a high capacity industrial model: ethanol distilleries are generally attached to large sugar refineries. Their capacities range between 100,000 T/year to 300,000 T/year (Fedecombustibles 2014; Droulers and Carrizo 2010). Artisanal supply chains are not part of this energy production, even though a few efforts by small distilleries are underway. Thus, in Colombia, for example, where the Panela artisanal sector is as large as that of industrial sugarcane, small independent producers do not make any ethanol.

14.2.2 Production of Electricity by Cogeneration

From the production point of view, the energy market is having an increasing impact on agricultural practices and plant breeding. In the last ten years, sugar refineries have become major electricity producers by using their existing cogeneration units and by making new investments in state-of-the-art high pressure turbine generators. In this way, several countries have developed significant sugarcane-based electricity generation capacities: Brazil (12,000 MW), India (4500 MW), Thailand (2500 MW), and Mauritius and South Africa (250 MW each). Power generating units running on bagasse have an average capacity of about 10 MW (Irena 2015).

This new market encourages cultivators to increase their production of solid biomass in the field. Several sources can contribute to this increase:

- breeding sugarcane plants that combine high sugar content with high biomass;
- increased collection of leafs and sugarcane tops;

- cultivation devoted to production of solid biomass by growing high-fibre sugarcane varieties or by introducing other plants, including woody species (acacia, eucalyptus). These plantations are usually grown on fields considered unsuitable for growing sugarcane for sugar.

The collection of leaves and tops is already a widespread practice, but it poses numerous storage and logistics problems which explain the increase in the number of timber plantations which offer significant yields of solid biomass that can be harvested all year round. Moreover, the long term effect of removing plant matter on the organo-mineral balance of soils is still poorly understood and numerous studies are underway to determine the maximum amounts of biomass removable from a plot.

As is the case for biodiesel, since the ethanol produced is primarily meant for domestic markets, its production systems are subjected to little or no requirements of conforming to sustainability standards (Bonsucro or others). However, ethanol that is produced for export to the European Union is subject to the same rules as those governing biodiesel: producers and processors must be certified under voluntary schemes approved by the European Commission (Pacini and Assunção 2011). The Bonsucro standard is approved by the EU, and most Brazilian and Colombian exporters have obtained certification under this system. The guiding principles of this benchmark, much like those of the RSPO (palm oil) or FSC (Forest Stewardship Council for timber) are as follows: protection of forests and wetlands, soil protection, ratification of International Labour Organization (ILO) conventions, and avoidance of activities that could jeopardize local food security.

14.3 The Cassava Chain

The high starch content of the cassava tuber makes it an excellent choice for ethanol production, which can be obtained directly from either the root or from the pulp left over after the starch extraction process.

This production is still in its infancy. Indeed, except in Thailand where an ambitious development programme was launched in 2008 (Alternative Energy Development Plan), the production of biofuel from cassava is, in most cases, still either at the project testing stage or limited only to low-volume facilities. High production costs and small-scale production facilities, when compared to sugarcane, preclude any significant expansion of this agri-chain at present. Large scale production of biofuel from cassava is, *a priori*, not possible, except in a few countries like Nigeria (largest producer in the world) or Ghana, where cassava cultivation is well established.

Nevertheless, OECD and FAO suggest that cassava cultivation for bioethanol production could hold significant potential for developing countries (OECD/FAO 2014), and could amount to 15 % of the global ethanol production (excluding

Brazil's output) by 2020. However, if Brazil's output and sugarcane production is included, the percentage of cassava-derived ethanol will remain low (<4 %).

Cassava is traditionally grown by small farmers on relatively poor soils (as compared to cereal crops). Many governments regard ethanol production from cassava as an opportunity to attract private investors to a small-scale sector and to integrate minor producers into a sector dominated by large industrial sugarcane producers. Two case studies in Thailand and Colombia illustrate these objectives and the challenges that stand in their way.

Thailand is the largest producer of cassava in Asia. Considered a cash crop, it is grown by small farmers, primarily in the north-eastern region of the country. It is cultivated mainly for export to China as cassava chips for animal feed and as starch. In 2010, the government began promoting cassava as a raw material for ethanol production (Salvatore and Damen 2010). It adopted strategic objectives in developing the market of ethanol derived from cassava: decreased dependence on China, diversification of markets, and furthering of national objectives of increasing biofuel production. The production model developed will be rolled out over a relatively constant area (1.5 Mha) through increased yields, abandoning of less productive land, and a gradual conversion of upland rainfed rice fields to cassava when conditions are favourable. However, the cassava-ethanol chain is struggling to take off as it has to contend with strong competition from sugarcane. The sugarcane chain is more entrenched and better organized, and even undertakes campaigns to convert cassava producers in order to increase production for large sugar refineries and distilleries.

Colombia has traditionally been a cassava producing country. But this crop is primarily cultivated as a subsistence crop, which makes its agri-chains fairly unstructured. In Colombia, the International Centre for Tropical Agriculture (CIAT) and the Latin American and Caribbean Consortium to Support Research and Development of Cassava (CLAYUCA) are currently working in areas not connected to the national electricity grid on generating energy from biomass of sugar crops, including cassava, sweet sorghum, and sweet potato. The Rural Social Bio-refineries (RUSBI) model, based on the production of biomass for local use, aims to produce 400–500 litres of anhydrous ethanol per day and per production platform (CIAT 2009). However, such small bio-refineries are yet to be established. Other industrial-scale projects have been attempted, but strong competition from sugarcane-derived ethanol, lower yields, and the impossibility of cogeneration (fibre too rich in cellulose) are deterrents for investors.

14.4 Woodfuel Chains

Forests and their timber have been used since time immemorial to produce energy. By definition, woodfuel encompasses all kinds of solid, liquid, or gaseous combustible material derived from wood. Its most common forms include traditional logs for heating or cooking; domestic charcoal that is extensively used in cities in the countries of the South; woodchips or pellets used mainly for thermal power stations, as also in modern domestic wood stoves; and, finally, second generation biofuels (ethanol and biodiesel).

The IEA estimates the current global demand for woodfuel to be around 3700 million cubic metres, 85 % of which is consumed in developing countries. In comparison, according to the FAO, global consumption of roundwood (wood for construction or for manufacturing paper) is estimated to be 1600 million cubic metres. The amount of woodfuel consumed is thus more than twice the amount of wood consumed in any other way. Woodfuel is, by far, the main way wood is used in the world. Currently, woodfuel consumption in households in developing countries (for cooking and heating) varies between 60 % to 90 % of these households' overall energy consumption, and 76 % of the population depends on woodfuel as the primary source of energy (IEA 2006).

The use of woodfuel seemed to decline in the 1980s in the countries of the North and the South. In Africa, for example, most governments introduced or had already rolled out proactive policies to reduce woodfuel consumption in favour of gas or kerosene. However, since 2000, climate change hazards and the unpredictability of hydrocarbon markets have led governments and wood and energy industries to look at this resource with renewed interest (Gazull and Gautier 2014). Demand for woodfuel in the countries of the North, as well as in developing countries, is increasing mainly due to the installation of collective boilers, the conversion of large coal-powered electricity generating facilities to woodfuel based ones, and the establishment of projects to produce second-generation ethanol. Demand is expected to continue increasing in the future in the countries of the South, particularly in Africa, due to population growth and lack of alternative energy sources. Woodfuel consumption forecasts for 2030 suggest it will account for around 80 % of the domestic consumption in Africa, 70 % in Southeast Asia and 40 % in Andean America. Rural electricity needs also contribute to increased demand, leading to the establishment of large wood-fuelled power plants in Kenya, Tanzania, and South Africa. Small-scale power plants are also coming up, mainly in India, but also in the Brazilian Amazon (remote sites) and in Africa (South Africa, Benin, Madagascar, Uganda).

The woodfuel chains in the countries of the South are largely informal and are becoming increasingly liberalized: access to the supply chain is free or weakly regulated at all levels. It is estimated that over 13 million FTE (full-time equivalent) individuals work directly in bioenergy supply chains in Africa and over 30 million across the world (Openshaw 2010). In comparison, FAO estimates that there are 14 million FTE employees in the formal, industrial, and artisanal forestry chains.

Woodfuel is mainly obtained from forests and natural woodlands (savannah, Cerrado, Caatinga, etc.), with timber plantations managing to meet only about 5 % of woodfuel requirements. While it is undeniable that the availability of timber as a resource remains high and that woodfuel extraction is not the main cause of deforestation, it is equally undeniable that tree resources are becoming scarce in many areas, most notably around major African and Latin American cities. Despite the awareness of the threat this demand puts on tree resources and the implementation of sustainable management policies, especially around cities (strategies for domestic energy in West Africa), the immense majority of timber is extracted without adherence to any rules of sustainability. Unlike for other types of biofuels, few standards have been established specifically for woodfuel, even in the context of international trade. For example, Europe does not impose any standards on pellets of non-EU origin.

The countries of the South, which still retain abundant tree resources, are thus now faced with a rising demand from two sources: their own burgeoning needs and a rapidly growing international market. As in the case of biodiesel, the European Union is currently the main importer of woodfuel, with Japan and South Korea also notable for their imports. There already exists widespread international trade in wood pellets for producing electricity and large energy industries are not only sourcing immense quantities of this resource from developing countries but are also establishing woodfuel plantations in Africa, Southeast Asia, and Latin America mainly to supply the European market. The demand for woodfuel is so huge that the first voices are being raised in civil society to condemn its potential impact on forests in the countries of the South, to warn against the dangers of extensive land grabbing for energy plantations, and to caution about the carbon footprint of this neoteric removal of timber (Smolker 2013).

Confronted by this expected increase in demand and issues of sustainability at different levels, a first response is to develop ‘fast wood’ plantations with high productivity. ‘Very short rotation’ or ‘very, very short rotation’ plantations are particularly suitable for woodfuel production. Although the technical itineraries of such plantations are widely known and proven, further research is still needed for improvements in planting material, reduction of soil acidity, maintenance of soil mineral balance, compensation of biodiversity loss and, above all, improvement in social acceptability.

A second possible solution is the intensification of extraction from existing forests and the integration of the production of woodfuel with that of lumber and pulpwood. In the present systems of forest resource management, the bulk of the woodfuel resources come from logging residues, including stumps; small-diameter trees removed during clearing or first thinning; wood unsuitable for sawing or peeling; resources that are not exploited because of their relative inaccessibility; and residues from processing industries. Although large logging companies are testing different production methods, an intensification in the extraction of forest produce (tree tops, stumps, small wood) may pose environmental problems (soil compaction, biodiversity loss, nutrient loss) as well as social tensions generated through competition with local supply chains for woodfuel destined for domestic use.

Thus, it is clear that, given the present situation, sustainable models of woodfuel production are yet to be invented or redesigned based on numerous experiments already carried out across the world.

14.5 Conclusion

These four examples clearly demonstrate how bioenergy is now a major component of many agricultural and forestry chains in the countries of the South. Energy has evolved from merely being a by-product to a co-product, one whose importance will continue to rise in the coming decades.

This new bioenergy production is already influencing agricultural practices, the adaptation of planting material, changes in land use patterns, and, in general, the management of material flows (solid and liquid biomass) throughout the production and processing stages. Bioenergy is also encouraging large oil multinationals (Total, Petrobras, Shell, BP, Exxon) and major electricity companies (EDF, Engie, EON, etc.) to recognize and invest in the agricultural and forest sectors. Finally, even though the bulk of the energy produced is destined for local consumption, international trade in biofuels (raw oil, biodiesel, ethanol, wood pellets and chips) is on the rise. In a world characterized by a growing demand for energy and the emergence of transcontinental energy actors, bioenergy production is currently conceived and organized in mass production systems that have the potential of destabilizing markets and existing agri-chains.

In this context, the sustainability of bioenergy production models in agri-chains warrants special attention. First and foremost, biofuels are at the centre of controversies and public debates pertaining to North-South relations; the energy independence of States; the very role of agriculture; the carbon market; industry's place in agriculture; etc. Although these debates sometimes do not seem to be, strictly speaking, related to issues of sustainability of agricultural production systems, they do concern producers, especially in the context of responding to the apprehensions of civil society by implementing new sustainability standards (RSPO, Bonsucro, RSB, etc.). Second, the agri-chains, in general, continue to take advantage of the complementarities between the markets for food, energy, and materials. They do take care to see that practices imposed by one market do not permanently compromise the possibilities of transferring from one market to another. This is particularly true in the case of sugarcane, where the collection of leaves and the cultivation of sugarcane with a higher content of lignocellulose (high-fibre sugarcane) are still not fully accepted, especially because of doubts about their sustainability. This is even truer in the forestry domain, where woodfuel production is encountering strong resistance from traditional forest users.

Finally, despite ongoing changes and the broad-based nature of bioenergy production, few innovations have been made in the agricultural and forestry production systems to incorporate energy as a separate and independent production. With the exception of sugarcane systems in Latin American, energy has only resulted in marginal changes and a few innovations that are mainly confined to processing systems. Some models have emerged to optimize energy, forestry, livestock, and grain production systems, and although these agro-silvo-pastoral models deserve special encouragement, a great deal of research still needs to be carried out to confirm their technical feasibility and sustainability.

The lack of clarity on energy resources of the future (uncertainties pertaining to shale gas, progress in the use of hydrogen fuel cells, revival of hydroelectric programmes) is undoubtedly one of the reasons why investments into bioenergy-related research continue to be low. Another reason is that many countries have yet to clearly define the framework of intervention of State policies on bioenergy. Indeed, these policies often impose, or promote, cultivation methods that consign bioenergy to a secondary production. Such is the case, for example, in environmental laws that allow the growing of energy crops only as intermediate crops in France, in incentives to plant *Jatropha* as an intercrop in West Africa, or in subsidies offered in Brazil to grow castor beans in association.

Such a context is not favourable for the emergence of sustainable solutions, which would require a coherent vision and policy messages on the part of States to indicate the part they wish to devote to biofuels in their energy mix and their agriculture.

References

Bruckner T., Bashmakov I.A., Mulugetta Y., Chum H., de la Vega Navarro A., Edmonds J., Faaij A., Fungtammasan B., Garg A., Hertwich E., Honnery D., Infield D., Kainuma M., Khennas S., Kim S., Nimir H.B., Riahi K., Strachan N., Wiser R., Zhang X., 2014. Energy systems. In: Climate change 2014: mitigation of climate change. Contribution of working group III to the fifth assessment report of the intergovernmental panel on climate change (Intergovernmental panel on climate change, ed), Cambridge University Press, Cambridge/New York, 88 p

CIAT (2009) Inauguran Bio-Refinería Social Para Comunidades Pobres. http://ciat-library.ciat.cgiar.org:8080/jspui/bitstream/123456789/6681/6/boletin_73.pdf. Retrieved 18 August 2016

Droulers M, Carrizo S (2010) Stratégies agro-industrielles autour de la filière des biocarburants au Brésil et en Argentine. Revue géographique de l'Est, 50 (3–4). Available online at http://rge.revues.org/3107. Retrieved 19 April 2016

FAO (2013) FAOSTAT. Food and Agriculture Organization of the United Nations. Available online at http://faostat3.fao.org/home/E. Retrieved 19 April 2016

Fedebiocombustibles (2014) Fedebiocombustibles, informacion General. Available online at http://www.fedebiocombustibles.com. Retrieved 19 April 2016

Gazull L, Gautier D (2014) Woodfuel in a global change context. Wiley Interdiscip Rev Energy Environ 4(2):156–170

IEA (2006) Energy for cooking in developing countries. In: World Energy Outlook 2006 (International Energy Agency). OECD/IEA, Paris, pp 419–446

IEA (2010) World Energy Outlook 2010. OECD/IEA, International Energy Agency, Paris, 736 p

Irena (2015) Irena renewable energy capacity statistics 2015. International Renewable Energy Agency, Madar City (Abu Dhabi), 44 p

MPOB (2014) Chapter 9: deepening Malaysia's palm oil advantage, 35 p. Available online at http://www.mpob.gov.my/home/153-demo-content/992-nkea-national-biogas-implementation-epp5. Retrieved 19 April 2016

OECD (2013) OECD statistics, available online at http://stats.oecd.org. Retrieved 19 April 2016

OECD/FAO (2014) Biofuels. In: OECD/FAO Agricultural Outlook 2014–2023. OECD Publishing, Paris, pp 109–126

Openshaw K (2010) Biomass energy: employment generation and its contribution to poverty alleviation. Biomass Bioenergy 34(3):365–378

Pacini H, Assunção L (2011) Sustainable biofuels in the EU: the costs of certification and impacts on new producers. Biofuels 2(6):595–598

Paoli GD, Yaap B, Wells PL, Sileuw A (2010) CSR, oil palm and the RSPO: translating boardroom philosophy into conservation action on the ground. Trop Conserv Sci 3(4):438–446

Salvatore M, Damen B (2010) Bioenergy and food security: the BEFS analysis for Thailand, environment and natural resources working paper No. 42. FAO, Rome 107 p

Smolker R (2013) Wood bioenergy: green land grabs for dirty "Renewable" energy. Global Forest Coalition/Biofuelwatch, 32 p

Chapter 15
Diversifying Biomass Uses Through New Cropping Systems

François-Régis Goebel, Jean-Louis Chopart, Christophe Poser, Serge Braconnier, Jean-François Martiné, and Edward Gérardeaux

As discussed in Chap. 14, the global demand for renewable energy from plant biomass is growing as a result of policy choices – whose aim is to reduce dependence on other energy sources, especially fossil fuels, and to shrink CO_2 emissions – population growth, and rising per capita energy needs. This energy requirement is a major obstacle to economic development in tropical islands and remote landlocked areas. The possibility of using plant biomass from fast-growing crops holds great potential in areas that enjoy favourable agro-pedo-climatic conditions.

In a world where feeding the planet's population and securing energy resources are major challenges for humanity, we must redouble our ingenuity to produce sufficient plant resources on increasingly valuable land, while fulfilling the imperative of protecting the environment. The utilization of the whole plant for energy

F.-R. Goebel (✉) • J.-F. Martiné • E. Gérardeaux
CIRAD – UPR Aida, Montpellier, France
e-mail: francois-regis.goebel@cirad.fr; jean-francois.martine@cirad.fr; edward.gerardeaux@cirad.fr

J.-L. Chopart
CIRAD – UPR Aida – Roujol, Petit-Bourg, Guadeloupe, France
e-mail: jean-louis.chopart@cirad.fr

C. Poser
CIRAD – UPR Aida, La Bretagne, Réunion, France
e-mail: christophe.poser@cirad.fr

S. Braconnier
CIRAD – UMR Agap, Montpellier, France
e-mail: serge.braconnier@cirad.fr

E. Biénabe et al. (eds.), *Sustainable Development and Tropical Agri-chains*, DOI 10.1007/978-94-024-1016-7_15

use, which is part of the envisaged plan, involves upstream policy decisions to address the question of competition between food and energy crops on arable land that is getting dramatically scarcer (Fischer et al. 2009; Botha 2010). This chapter covers the technical and organizational issues that arise at the scale of cropping systems when they are used for energy production. Indeed, producing a resource that will be used in a local energy supply chain will involve the adaptation of production systems, both for family farming and large scale farms. Such farming and organizational changes represent a major challenge. Research efforts presented below will help better understand the consequences of changing the end uses of the productions, design better agricultural practices, and select suitable varieties.

The present chapter focuses on sugarcane and sorghum in particular. At present, sugarcane – grown in more than 120 tropical countries and accounting for 80 % of the world's sugar – and maize, cultivated mainly in temperate zones, are the main crops used to produce ethanol for use as a biofuel. In this context, many farmers, from family farming and large farms alike, are keen on taking advantage of opportunities offered by the diversification of this agri-chain's products. The abolition of European sugar quotas will affect the profitability of the sugar chain in African, Caribbean, and Pacific countries. Farmers in countries from these regions and in French overseas territories benefiting from these quotas will have to adapt quickly to a new global market. Agricultural and industrial stakeholders are exhibiting a growing interest in other plants such as sorghum, a member of the grass family related to sugarcane. It reflects the desire to contribute to the food security of populations who grow these plants and to produce abundant biomass for alternative energy. With the help of a series of innovative projects that draw useful value from these plants' biomass, we demonstrate how the goal of producing biomass for non-food purposes involves developing and adopting new farming practices.

15.1 Sugarcane, a Multi-resource Plant

Sugarcane is a multi-purpose plant. Products derived from sugarcane not only include sugar and rum, but also electricity, biofuel, and recently, bioplastics. A number of products can be obtained from it, with uses in almost all economic sectors: pharmaceutical, agrifood, aerospace, construction, building, etc. Studies on improving sugarcane quality (2008–2012) took into account the potential use of the total biomass, or parts of it, for the production of sugar, energy, and molecules that are useful for refineries using sugarcane and other mixed sources (Martiné et al. 2012; Roussel et al. 2013; Sabatier et al. 2015). Following these initial results, it became necessary to expand on these studies on types of sugarcane suitable for the energy sector in order to broaden the range of biomass sources and their uses.

15.1.1 Using the Total Biomass of Sugarcane to Generate Electricity: Towards an Adaptation of Technical Itineraries?

We draw lessons from two research projects undertaken recently in Guadeloupe (*Rebecca*, Research on sugarcane biomass-energy in Capesterre) and in Réunion (*Sypecar*, Sugarcane-based energy production systems in Réunion). The ultimate goal of these programmes is to remove identified scientific, technical, and organizational barriers in order to allow the emergence of agro-industrial supply chains for generating electricity through the combustion of biomass grown locally for this purpose. These programmes depend on sugarcane to provide plant biomass by selecting fibre-rich ancestral varieties to meet the new goals of combustion. Their main aim is to determine the technical and economic feasibility of the sugarcane-energy supply chain and to develop cropping systems dedicated to energy production from high-fibre sugarcane or other energy crops.

15.1.2 Performance of Planting Material

In Guadeloupe, the project specifically targets sites polluted by the pesticide chlordecone in order to replace food crops whose consumption, and therefore marketing, are prohibited. In Réunion, the project targets high-altitude plots (>600 m) where temperature and inaccessibility make sugarcane cultivation unprofitable. As a result, it was decided to use the entire sugarcane biomass as fuel instead of extracting sugar from it. To meet these objectives, it was first necessary to identify sugarcane varieties most suited to different environments and to use as fuel. Once high performing varieties were identified in Guadeloupe, agronomic studies were conducted to test several new cropping systems adapted to crop requirements, in terms of environmental standards and industry requirements (Chopart and Bachelier 2012). In all, about 50 varieties were tested, six of which are common to both islands. Such planting material has a very variable ratio of fibres and sugar (multipurpose varieties) (Fig. 15.1). It originates mainly from CIRAD's collection in Guadeloupe, from eRcane's programme which was designed to broaden the sugarcane genetic base and from introductions from the *West Indies Central Sugar Cane Breeding Station* (WICSBS, Barbados), and from varieties that have been selected for their high biomass. In addition to the sugarcane varieties selected in Guadeloupe, two varieties of *Erianthus* (NG28007 and IK 76048) and a *Saccharum spontaneum* (Moentai) cultivar were tested but they were found to be not as suitable as the sugarcane varieties.

Since energy performance, expressed as lower calorific value (LCV) of the dry product, shows a strong correlation with the dry biomass produced, regardless of the proportions of fibre and sugar or stems and leaves (Chopart et al. 2013), we focused on the dry biomass.

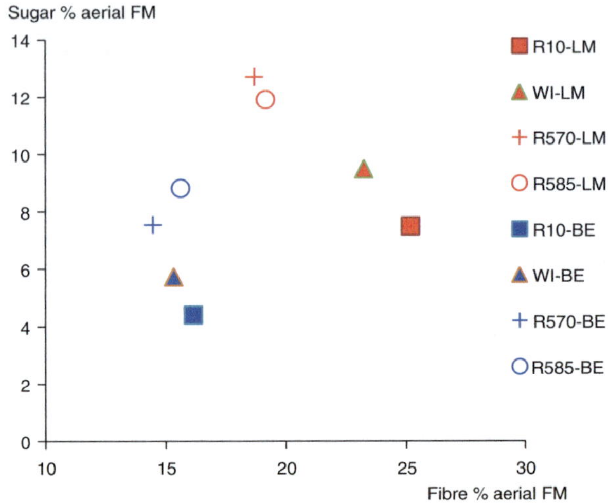

Fig. 15.1 Fibre (%) and sugar (%) of different types of sugarcane after 12 months of growth in Réunion's coastal areas (*LM* La Mare – altitude 50 m) and in the highlands (*BE* Berive – altitude 680 m). High-fibre clones from eRcane (R10), multipurpose clones from Barbados (WI), and commercial sugarcane varieties (R570, R585). *FM* fresh matter

Figure 15.2 shows the relative performances of traditional sugarcane and high-fibre sugarcane varieties as compared to local controls, based on results obtained from two sites. The maximum dry biomass potentials observed are 84.9 T/ha in Guadeloupe from the WI81/456 variety over three 12-month cycles and 67.2 T/ha in Réunion from a 14-month planting cane using the R585 variety.

In Guadeloupe, after 3 years of cultivation two varieties were found to be particularly high-yielding and well-adapted. They originate from WICSBS in Barbados. Figure 15.3 shows the performance of these two superior high-fibre sugarcane varieties when tested against the local R579 variety. The average total dry matter during a three-year cultivation period for these two varieties is amazing, with yields at more than 80 T/ha/year. This yield places them among the best annual productions of plant biomass – from both herbaceous and tree species – grown in temperate or tropical regions.

The first lesson learnt from these experiments: the potential of the gene pool from Barbados is 20–70 % more productive than the R579 control in Guadeloupe and the R570 control in Réunion (Fig. 15.2) in the first year of cultivation (Fig. 15.4). The second take-away message: we note that the R585 variety, recently selected by eRcane to produce sugar in high-altitude regions, yields 60 % more dry biomass than the control. This is a proof of the effectiveness of eRcane's work on selection and suggests that crosses, especially with varieties from Barbados, hold promise.

The moisture content of the material at the time of harvesting is a crucial determinant of its usefulness as fuel. This figure is between 60 and 65 % of fresh weight in this humid environment, implying that drying is required prior to its use as fuel down to a minimum of 45 % (moisture content of the bagasse currently used as fuel). This need for pre-combustion drying thus reduces the plant material's energy performance. Preliminary tests carried out in Réunion to dry it on the field indicate a possible weight loss of around 4 % per day for crushed cane in the first

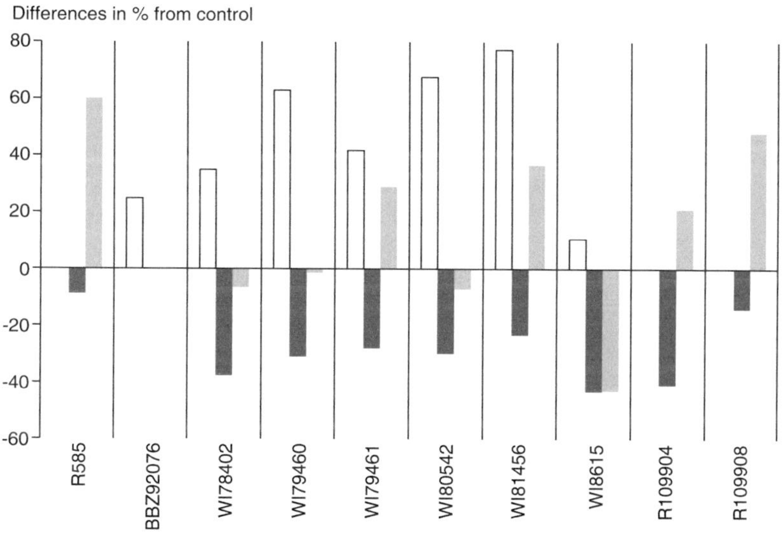

Fig. 15.2 Differences in outputs (in %) in dry biomass yield for sugarcane between different varieties and the control plot. Sites: Guadeloupe, Fromager, 3 years, 12 months (*white rectangle*); Réunion, La Mare, plating cane, 14 months (*dark grey*); Réunion, Berive, first crop, 14 months (*light grey*). Control crop: Guadeloupe, R579 (47.9 T/ha); Réunion, R570 (73 T/ha in La Mare and 42 T/ha in Berive)

5 days, under favourable conditions (hot and dry), leading to a moisture content equivalent to that of bagasse fuel used as reference. It seems possible to use an industrial process to dry high-fibre sugarcane to about the same level while removing 30 % of the moisture. These results give an idea of the possible losses in energy performance of the supply chain in relation to post-harvest conditions and the industrial choices opted for.

The same varieties were tested on a different site with a drier climate, where they were subject to drought. Under such water deficient conditions, the sugarcane varieties that were bred in humid conditions were found to be very resistant (Fig. 15.4). Although the yield of the best variety, recorded at 49 T/ha of dry matter, was lower than those at the Capesterre site (Fig. 15.3), it was considered satisfactory.

15.1.3 Adaptations of 'High-fibre Sugarcane' Cropping Systems with Respect to Those for 'Traditional Sugarcane': Crop Cycle and Harvest Period

The crop cycle time must be as long as possible in order to meet the yearlong requirements of industrial units. The sugarcane planting period and crop cycle time

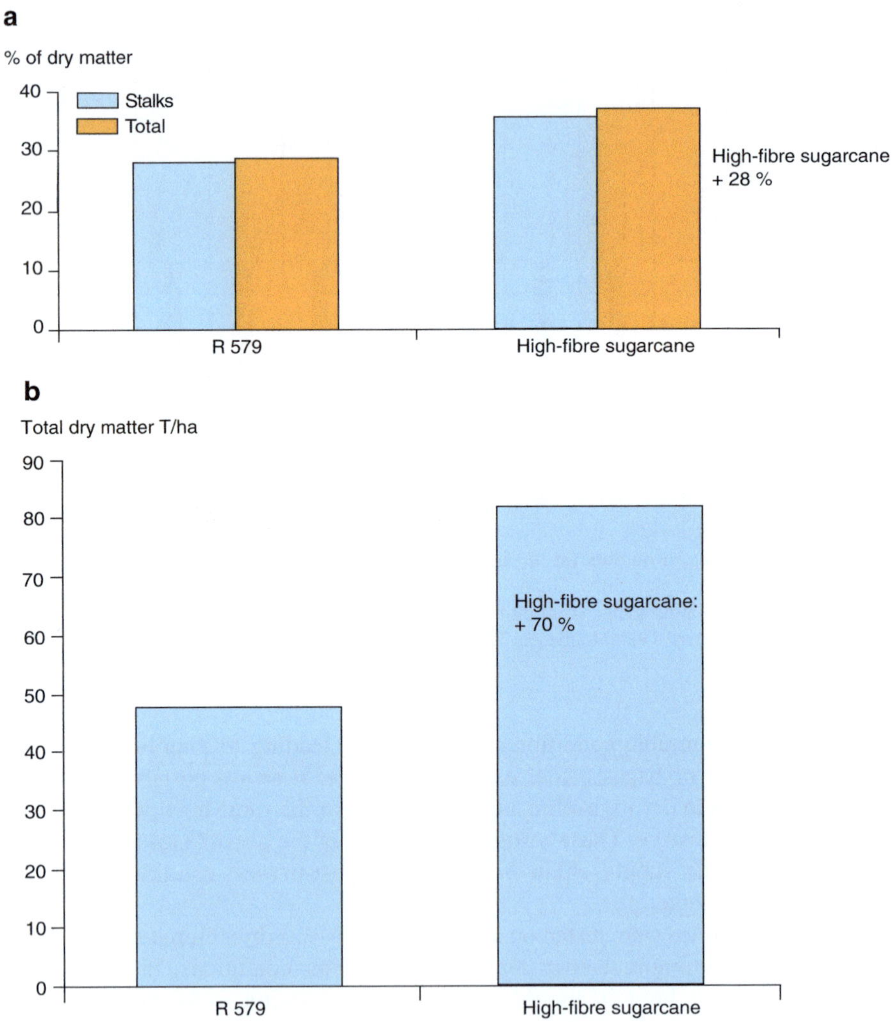

Fig. 15.3 Average yield over 3 years for the two best varieties of sugarcane (WI81456, WI79460) after 12 months of cultivation in the Capesterre wetlands (altitude 100 m), annual rainfall of 3000 mm (Source: Chopart et al. 2015)

should therefore be extended. Studies in Guadeloupe show that it is possible to plant a high-fibre sugarcane variety over an extended period between June and November (Fig. 15.5).

The most suitable crop cycle time for biomass production from traditional sugarcane is 12 months (Chopart et al. 2015). Harvesting is possible between 8 and 14 months after planting or from the time of ratooning, with little decrease in yield. A steady supply of plant biomass could be assured for the industries throughout the year if planting dates on farms and mill supply areas are staggered

Fig. 15.4 Experimental system in Capesterre (Fromager), aerial view (©J.-L. Chopart, Cirad)

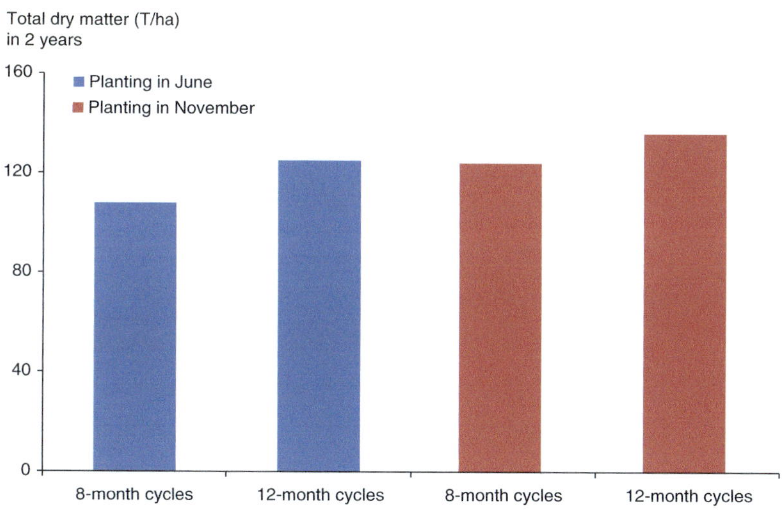

Fig. 15.5 Guadeloupe project. Cumulative production over 2 years (in T/ha of total dry matter) of one of the three best varieties of sugarcane for combustion (BBZ 92076) as a function of the planting season and crop cycle time (3 cycles of 8 months or 2 cycles of 12 months, based on Chopart et al. 2015)

between June and November, and if crop cycles range between 8 and 14 months. In contrast, to obtain an acceptable biomass yield without too much moisture in the highlands of Réunion, a crop cycle time of 12 months – or even 14 – will be required. MOSICAS, a sugarcane cultivation model, was calibrated and validated on both sites for dry biomass production and the level of dry matter. It predicts the interaction between the genotype and the environment, and consequently varietal classifications according to sites (Fig. 15.6). It can therefore also help estimate the quantity of supply to industries, while optimizing cropping systems according to the concerned production site.

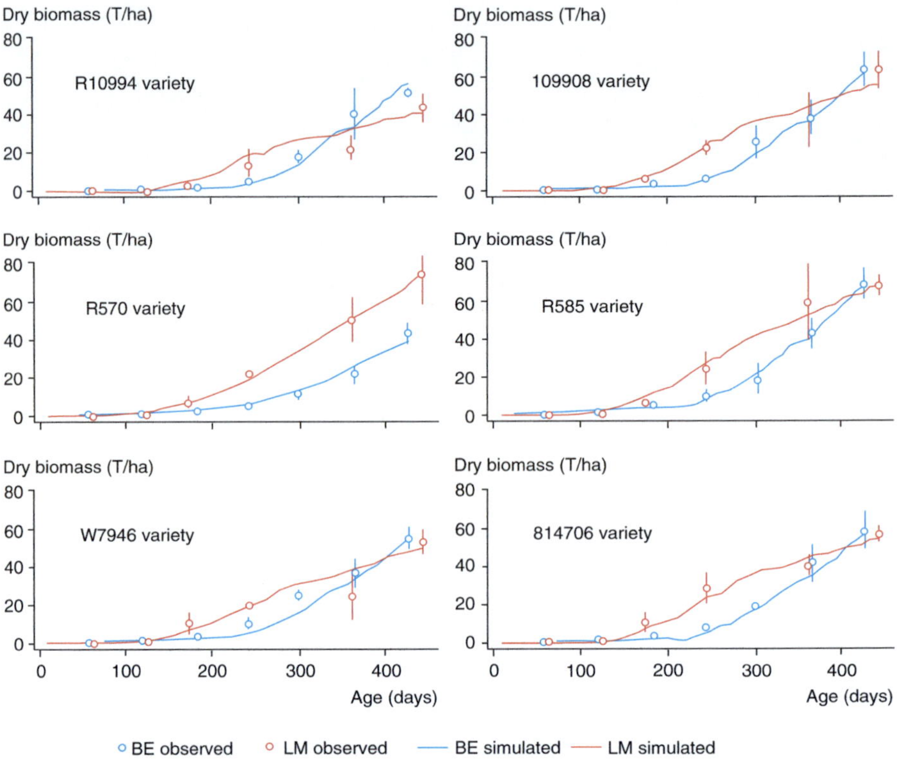

Fig. 15.6 A comparison of changes in biomass dry matter observed and simulated by the model in Réunion, on two sites with contrasting climates: *LM* La Mare (altitude 70 m) and *BE* Berive (altitude 550 m)

15.1.4 Farming Practices

Different types of sugarcane function differently at the physiological level. Sugarcane grown for sugar must be exposed to some degree of stress (reduced availability of water and nitrogen, and low temperatures) at the end of the cropping cycle to maximize sugar content. High-fibre sugarcane varieties, on the other hand, do not need this. Moreover, sugarcane grown for sugar flowers little compared to high-fibre sugarcane varieties, which bloom profusely.

Despite these differences and changes in planting dates and crop cycle times, varietal and agronomic experiments conducted in Guadeloupe over 4 years indicate that tilling, planting, fertilizing and harvesting techniques for high-fibre varieties should be the same as for traditional sugarcane production. Since high-fibre sugarcane varieties show higher vegetative growth than traditional sugarcane, it should become easier to control weeds since limited light penetrates to the ground level. This should also lead to a reduction in herbicide use as compared to traditional sugarcane and permit compliance with agri-environmental rules governing agroecological production. Indeed, high-fibre sugarcane was grown in Guadeloupe for 4 years without any application of insecticides or fungicides. Under such conditions, fewer *Diatraea saccharalis* (*Lepidoptera: Crambidae*) sugarcane borers were found in the stalks of high-fibre varieties than in traditional sugarcane varieties (Chopart et al. 2015).

15.1.5 Maintenance of Soil Organic Matter

A major concern pertaining to the cultivation of varieties for uses other than extraction of sugar is the management of soil organic matter. Indeed, only stems are harvested in the case of sugarcane grown for sugar. The leaves and the top portions of the plant are left behind in the field. Experiments carried out as part of the project showed that this residual organic matter represents a large quantity of biomass, about 12 tonnes of dry matter per hectare, to which must be added about 2 more tonnes of dry matter to account for leaves that fall during the crop cycle. In contrast, when cultivating high-fibre varieties, the top portions and leaves attached to stems can theoretically be harvested for use as fuel. However, this is not possible if we want to maintain the same level of contribution of annual biomass by the plant and thus the same level of organic matter in the soil. The biomass from leaves that fall naturally to the ground during the crop cycle is lower for high-fibre sugarcane varieties (about 1 T/ha in Guadeloupe). Although this is partially offset by a slightly higher root biomass for these varieties than the local variety R579–3.2 and 2.6 T/ha respectively (Chopart and Sergent 2015a, b, c) – a portion of the aboveground biomass of high-fibre sugarcane varieties has to be intentionally left behind on the field. It is suggested that the plant's top portion (8.5 T/ha) could be left behind. The rest of the requirement (about 3.5 T/ha) could also be met from the biomass produced on site, but this reduces useful production. It could also come in the form of an application of compost. Ultimately, such decisions will have to be taken by the farmer, based on relative prices of high-fibre sugarcane and compost.

15.2 Impact of Diversification of Sorghum Uses on the Cropping System: Results from the Sweetfuel Project

The goal of the Sweetfuel project was to produce bioethanol from sorghum and, in the process, recommend planting material that is better suited to temperate and tropical semi-arid areas, as well as optimized technical itineraries and harvesting and storage methods (Chap. 11). The project brought together various partners from seven countries on four continents, including private players (research institutes, seed companies), government bodies (national agricultural research centres,

Table 15.1 Characteristics of different sorghum ideotypes sought for in the Sweetfuel project

Use	Type of sorghum	Ideotype
2G ethanol or biogas	Biomass sorghum: grain production is not essential	High biomass (height > 3 m)
		Good tolerance to low temperatures
		Sensitivity to photoperiodism suitable for late flowering, good biomass quality (low lignin content to increase digestibility)
		Good lodging resistance
		Good tolerance to water scarcity
1G ethanol + grain + fodder	Sweet sorghum: grain production is essential	High biomass (20–30 T/ha of dry matter)
		Production of grain at 1.5–3 T/ha
		High soluble sugar content of stalks, Brix° of 15–20 with 80 % sucrose
		Juicy stalks
		High fodder value of bagasse (good digestibility after incorporating the Brown Midrib (BMR) trait)
		Suitable for cropping seasons in India
		Good drought tolerance
1G ethanol + cogeneration	Sweet sorghum: grain production is not desired	High biomass (40–50 T/ha of dry matter)
		High soluble sugar content of stalks, Brix° of 15–20 with 80 % sucrose
		Juicy stalks
		Bagasse with high calorific value for cogeneration (high lignin content)
		Adaptation to marginal soils (acidic, aluminium toxicity, deficient in P)
		Adaptation of crop cycles (good complementarity with sugarcane)

universities, advanced research institutes) and a CGIAR (Consultative Group on International Agricultural Research) centre (http://www.sweetfuel-project.eu/).

Breeding objectives were defined according to the selected areas and the biomass transformation processes. Thus, in semi-arid areas, the ideotype in focus was a sweet sorghum, with different characteristics based on biomass use (Table 15.1). In Europe (temperate zone), considering the availability of methods to produce second-generation ethanol on an industrial scale, the identified ideotype was a biomass sorghum with improved tolerance to low temperatures and lodging resistance, and low lignin content. Such a planting material would also be suitable for biogas production.

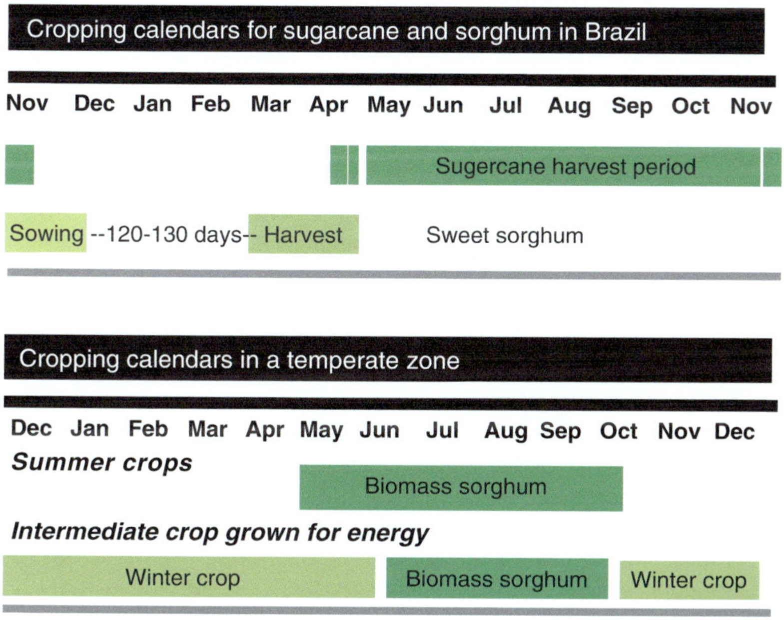

Fig. 15.7 Crop calendars. Complementarity of sweet sorghum cycles and sugarcane cycles in Brazil, complementarity of sorghum and a winter crop in temperate areas

15.2.1 The Sugarcane and Sorghum Chains are Complementary for Bioethanol Production

One of the project's major findings is that the sorghum chain for producing ethanol has a much better chance of being successful in areas where an industrial sugarcane chain is already present. In fact, it is very difficult for an ethanol production unit from sweet sorghum to be profitable when it is the only activity because factories would work only for 3 months in a year due to the crop's seasonality and the need for rapid processing of the stalks immediately after the harvest (Eggleston et al. 2015). The complementarity between the sugarcane chain and the sorghum chain arises due to their different crop cycles (Fig. 15.7).

Brazil currently has about 10 million hectares under sugarcane, and presents a typical example of how sorghum can be grown along with sugarcane for the production of ethanol (Schaffert et al. 2011). Indeed, farms that produce ethanol from sugarcane characteristically follow two main processes: they replant sugarcane every 5 years, normally allowing a legume crop cycle between two 5-year sugarcane periods and sugar refineries must usually wait until the sugar content of the stalks is sufficiently high to begin extraction, which means that they do not operate from December to April. The idea is to replace the legume cycle by a sweet sorghum cycle. The harvest of sweet sorghum stalks would then allow refineries to extend their operating period to March and April, with an attendant increase in production (by up to 25 %) without additional investment (except for the seedlings, since sorghum is grown and processed with the same equipment) or increase in cultivated area (every year, 20 % of the area under sugarcane is made available for sorghum). It is, however, important to maintain soil fertility once the legume crop cycle is discontinued.

At the technical level, this concept can theoretically be recommended for all sugarcane growing areas, especially in India which has the largest area under sugarcane cultivation. But in this particular case of a mixed sugarcane and sorghum production system, and discounting administrative problems (for example, licensing for alcohol distillation, etc.), two key barriers still need to be overcome. Unlike in Brazil, sugar mills in India are not yet equipped with distillation units and therefore would require significant investments to produce alcohol. In addition, with production originating from small farmers (average area of 1–2 ha), there would be a need to organize sugarcane growing areas comprising of tens of thousands of farmers to ensure a continuous supply of feedstock to extraction plants, since sweet sorghum cannot be stored.

In Haiti, the 'S3F for Haiti' project (http://s3f-haiti.cirad.fr/) showed that several uses could be developed for sweet sorghum: production of potable alcohol, fodder for the beef industry, beans to revive the poultry industry, etc. However, this kind of development requires efficient logistics involving organized producer groups, as well as policy support through hard decisions, especially for the revival of the poultry industry (Leclerc et al. 2014).

Two cultivation systems can be adopted to produce 2G ethanol (or biogas) from sorghum biomass in temperate climates. In the first, sorghum is grown as a spring/summer crop and planted in April/May and a biomass productivity of 30–40 T/ha/year can be expected. In the second, it is treated as an intermediate crop for energy purposes and sown after a winter cereal in late June, in which case there is a lower biomass productivity, of 20–25 T/ha/year (Fig. 15.7).

The second option is increasingly being preferred, since it does not compete with a food crop, which is planted during the previous cycle.

15.3 Conclusion

Developments within the agri-chains have resulted in the remodelling of cropping systems for both the plants presented here, depending on the end-use of the production: non-food (the focus being on high-biomass production) and/or food. The generation of income through the bioeconomy, from the multiple products originating from these two crops can form the basis of future perspectives that could be extended, depending on policy options for ensuring energy and food transitions, to all cultivated grasses. This chapter shows the benefits of undertaking research to better integrate these new uses in the objective of cropping systems and thus develop suitable agricultural practices. Research efforts can help extend the range of options and overcome some barriers. In Guadeloupe, the cultivation of high-fibre sugarcane does not compete with that of traditional sugarcane, since the former can be grown on fallow land or on soil contaminated by the pesticide chlordecone. High-fibre sugarcane cultivation can also be integrated into innovative cropping systems, using cover crops that provide a set of useful ecological services (fewer weeds, pest control, increased soil fertility, etc.). Moreover, such a high-fibre sugarcane chain, which requires new agricultural practices, would diversify the income sources of farmers in the context of difficulties created by the liberalization of the sugar market, which is planned for 2017.

In the case of ethanol production in Brazil, the development of sorghum to complement the production of sugarcane will have a marked impact on the sugarcane cultivation system. This latter system will gradually shift from a crop rotation system whose structure consists of a five-year sugarcane cycle, followed by a legume crop cycle, followed by a five-year sugarcane cycle, etc. to an essentially similar system where the legume crop is replaced by sorghum. However, although this modification will increase the production of alcohol, it will lead to problems of soil fertility due to the removal of the legume cycle. This is an issue that will have to be addressed to preserve the sustainability of the production system.

In Europe, the desire to migrate from a fossil-fuel based economy to an economy based on renewable resources demands an increased production of biomass for non-food uses. If we want to retain production for food purposes so as not to jeopardize food security, one solution is to produce biomass during intermediate cycles. It must be borne in mind that this solution would subject land to more intensive farming, making the maintenance of soil fertility an increasingly important issue for the sustainability of the system.

References

Botha FC (2010) Future prospects. In: RJ H, Kole C (eds) Genetics, genomics and breeding of sugarcane. Science Publishers, Enfield, pp 249–264

Chopart JL, Bachelier B (2012) Propriétés et performances comparées de 16 cultivars de Poacées (*Saccharum sp.* et *Erianthus*) en vue d'un usage énergétique. In: Proceedings Congrès sucrier. Afcas/Artas, La Réunion, September, 9 p

Chopart JL, Sergent G (2015a) Root biomass quantification of sugar and multipurpose cane varieties for sustainable production. In: Proceedings of ISSCT Agronomy Workshop, Durban, 24–28 August, 1 p

Chopart JL, Sergent G (2015b) Estimation de la biomasse à laisser au sol en culture de canne combustible dans le Sud de la Guadeloupe. Note scientifique Rebecca WP1 no. 2, 15 p

Chopart JL, Sergent G (2015c) Estimation de la faisabilité d'une coupe mécanique de cannes fibreuses au cours de l'année dans la zone cannière de Capesterre-B-E à Sainte-Rose (Guadeloupe). Note scientifique Rebecca WP1 no. 4, 11 p

Chopart JL, Bonnal L, Martiné JF, Sabatier D (2013) Functional relationships between dry aboveground biomass and the energy yield of sugarcane. In: Proceedings of XXVIII ISSCT (International Society of Sugar Cane Technologists) Congress, São Paulo, 14–27 June, vol 28, 11 p

Chopart JL, Sergent G, Goebel FR (2015) Cropping systems for energy cane grown on volcanic soil in a tropical climate: initial results on planting dates, cycle duration and pest pressure. In: Proceedings of ISSCT Agronomy Workshop, Durban, 24–28 August, 1 p

Eggleston G, DeLucca A, Sklanka S, Dalley C, St. Cyr E, Powell R (2015) Investigation of the stabilization and preservation of sweet sorghum juices. Ind Crop Prod 64:258–270

Fisher G, Teixeira E, Tothne Hizsnyik E, van Velthuis H (2009). Land use dynamics and sugarcane production. In: Sugarcane ethanol, contributions to climate change mitigation and the environment (P. Zuurbier, J. van de Vooren, dir.). Academic Publishers, Wageningen ,29–62

Leclerc E, Pressoir G, Braconnier S (2014) L'avenir prometteur du sorgho sucré en Haïti. Field Actions Science Reports, 9. Consultable sur. http://factsreports.revues.org/2801, Accessed 8 Apr 2016

Martiné JF, Roussel C, Sabatier D (2012) Influence de paramètres agro-climatiques sur la production de sucre et d'énergie. In: Proceedings Congrès sucrier, Afcas/Artas, La Réunion, September, 7 p

Roussel C, Martiné JF, Petit A, Sabatier D, Corcodel L (2013) Harvest schedule for multi-purpose cane production. In: Proceedings of XXVIII ISSCT (International Society of Sugar Cane Technologists) Congress, São Paulo, 14–27 June

Sabatier D, Martiné JF, Chiroleu F, Roussel C, Letourmy P, van Antwerpen R, Gabrielle B, Ney B (2015) Optimization of sugarcane farming as a multipurpose crop for energy and food production. GCB Bioenergy 7(1):40–56

Schaffert RE, Damasceno CMB, Parrella RAC (2011) Breeding strategies for sorghum as a feedstock for first and second generation technologies for production of bioenergy in Brazil. In: 6th Frontiers in bioenergy: United States-Brazil symposium on sustainable bioenergy, 15–18 May, West Lafayette (Indiana), Embrapa

Chapter 16
Inter-supply Chain Recycling of Residues

Tom Wassenaar, Jean-Marie Paillat, François Guerrin, Philippe Lecomte, Jean-Michel Médoc, Laurent Parrot, Jérôme Queste, Paulo Salgado, Emmanuel Tillard, and Jonathan Vayssières

As discussed elsewhere in this book, any expansion of agri-chains can result in competition for a territory's resources. The rapid growth of agri-chains, which have, over the past half century, become increasingly specialized, integrated, and globalized, has resulted in their disconnection from the territory. Yet, the extent of their sustainability is largely expressed in relation to local segments of the

T. Wassenaar (✉) • J.-M. Paillat • F. Guerrin
CIRAD – UPR Recycling and Risk, Montpellier, France
e-mail: tom.wassenaar@cirad.fr; jean-marie.paillat@cirad.fr; francois.guerrin@cirad.fr

P. Lecomte
CIRAD – UMR Selmet, Dakar, Senegal
e-mail: philippe.lecomte@cirad.fr

J.-M. Médoc
CIRAD – UPR Recycling and Risk – LEMSAT, Dakar, Senegal
e-mail: jean-michel.medoc@cirad.fr

L. Parrot
CIRAD – UPR Hortsys, Montpellier, France
e-mail: laurent.parrot@cirad.fr

J. Queste
CIRAD – UPR Green – FOFIFA-DRFP, Antananarivo, Madagascar
e-mail: jerome.queste@cirad.fr

P. Salgado
CIRAD – UMR Selmet, Antsirabe, Madagascar
e-mail: paulo.salgado@cirad.fr

E. Tillard
CIRAD - UMR Selmet, Ligne-Paradis, Réunion, France
e-mail: emmanuel.tillard@cirad.fr

J. Vayssières
CIRAD – UMR Selmet – LEMSAT, Dakar, Senegal
e-mail: jonathan.vayssieres@cirad.fr

E. Biénabe et al. (eds.), *Sustainable Development and Tropical Agri-chains*,
DOI 10.1007/978-94-024-1016-7_16

agri-chain, and is centred mainly on the local context (in terms of its environmental and, more importantly, social dimensions), a fact that necessitates a response that is specific and appropriate to this context.

The absence of relationships – or, worse, the presence of conflicting ones – between segments of various compartmentalized agri-chains that coexist in territories hinders the sustainable development of these agri-chains, and, more broadly, of the territories concerned. Creating links to the territory by identifying synergies between them could help overcome this limitation. A good example is the case of organic residues on Réunion Island (Wassenaar et al. 2015). More generally, this constraint concerns the consumption of limited local resources (for example, water, soil, agricultural land on which regulations allow waste to be spread, and landfill capacity, in the case of Réunion) or degradation of these resources by the dumping of various substances. In the context of a general and increasing scarcity, industrial ecology (the study of material and energy flows through industrial systems) intends using locally the gains obtained from increased energy and material efficiency via an inter-sectoral approach, known as 'industrial symbiosis'. This term is used in the broadest possible sense to include all production resulting from human effort and is also popularly known as the 'circular economy', which seeks to establish connections between activities, and thus between the links of the supply chains. The promise of the circular economy is to transform such waste into local resources by establishing inter-supply chain relationships that reduce the consumption and degradation of resources.

Local relationships between supply chains, especially between agricultural supply chains, have not all disappeared. Indeed, some of them are well-known and understood, and even promoted, by CIRAD. In the following sections, we will present a set of analyses and conceptions of relationships between local segments of different agri-chains (links of supply chains that transcend territories) using, respectively, socio-economic analyses and flow analyses.

However, where such relationships do not exist, the issue is how to create or re-create links between local segments of supply chains, whether agricultural or not. In material terms, these segments are hardly watertight: to pass on a product from an upstream segment to a downstream one within their supply chain, other incoming flows are generally used, which, in turn, generate still other outgoing flows within their territory. How to connect the lateral flows, which consist of orphan branches going out from a supply chain, to incoming orphan branches of other supply chains (Fig. 16.1), while allowing each segment to grow in step with the development of its own supply chain? This chapter presents the outcome of an initial attempt at industrial symbiosis involving agri-chains.

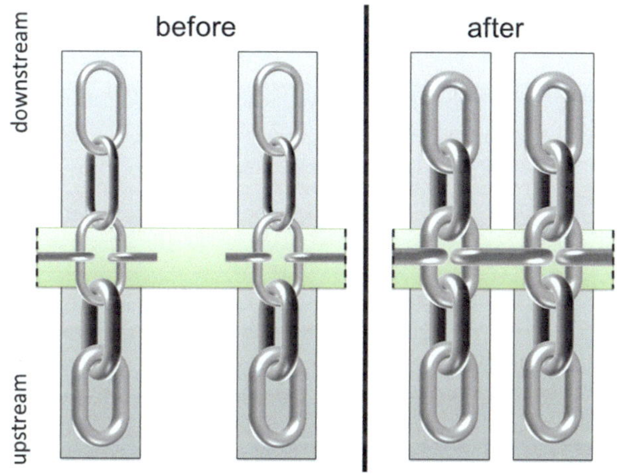

Fig. 16.1 Agri-chains (vertical links) can have segments (represented by links here) within the same territory (horizontal band) In case the development of the agri-chain is hindered by local incoming and/or outgoing flows of these segments, industrial symbiosis could seek to transform them into territorial links

16.1 Socio-economic Analyses of Recycling Situations

A common and age-old practice is for one agri-chain to recycle the organic residue generated by other supply chains. The dynamics of development continue to lead to the opportunistic emergence of new examples of such inter-supply chain links, which are sometimes documented by CIRAD (Box 16.1). Two cases of recycling are analyzed in detail here: the economic analysis of recycling urban waste by peri-urban agriculture in Cameroon, and the link between sugarcane cultivation and milk production in Réunion.

Box 16.1 Chicken manure from eastern Côte d'Ivoire used in cocoa plantations in the country's west

Tom Wassenaar and Jean-Marie Paillat

Cocoa farms in Côte d'Ivoire first began using fertilizers in the early 1980s, in that country's Soubré region. Some farmers, concerned about the early mortality of their cocoa trees due to poor soils in the area, turned to the application of chemical fertilizers, all on their own, without any public or private support. This development would eventually help the 'cocoa fertilizer' market to grow to nearly 80,000 tonnes by 2003.

Just when the fertilizer industry began marketing chemical fertilizers for cocoa farms, family farms initiated a diversification of their fertilization strategies, towards a greater use of organic inputs. The use of chicken manure (the product of a new value chain) as a fertilizer for cocoa began as an innovation around the 2000s, and found considerable widespread acceptance

(continued)

Box 16.1 (continued)

by the 2010s. It was only in 2011 that a section of industry began showing an increasing interest in organic fertilizers, but the inescapable fact remains that chocolate multinationals came to this innovation much later than the pioneering villagers.

This practice was adopted much earlier in the west of the country than in the east, due in part to the poorer soils in the west and a much higher migrant population there. The increase in cocoa prices and improvements in the road network created conditions that were propitious to the development of a relationship between chicken manure and cocoa farms. An entrepreneur identified a potential financial benefit in this interaction: transportation of a by-product from farms in the east, where it had no value, to plantations in the west, where it had substantial and growing value. Profit margins on chicken manure were attractive enough to help overcome the associated costs and risks.

The results of this activity, however, were not uniformly positive. The dosage, location of application, and period for using this input were decided very empirically; some farmers even recorded a higher cocoa tree mortality. (Based on Ruf et al. 2015)

16.1.1 From Collection of Municipal Waste to Preparation of Organic Fertilizers for the Horticultural Chain in Cameroon

In Cameroon, nearly 75 % of household waste generated in the capital city, Yaoundé (two million inhabitants), comprises an organic portion (Parrot et al. 2009a) that is collected along with other waste by a private city-based firm and dumped in a landfill, without it being put to any beneficial use. At the same time, 13.6 % of the households practice urban and peri-urban agriculture in the city's low lying areas where buildings cannot be constructed. No rules or standards govern such production systems, and the use of chemical products has contributed to the pollution of these areas. The proximity of household organic waste dumps to agricultural production areas is conducive to the development of relationships between the horticultural chain and units producing organic fertilizers consisting mainly of organic amendments (compost) (Sotamenou and Parrot 2013). Thus, one third of these producers use organic matter (chicken manure, compost) purchased from artisanal units that have organized themselves into associations (Parrot et al. 2009b) (Fig. 16.2).

Studies conducted in Cameroon have attempted to assess the economic sustainability of the waste recycling chain. They have focused on understanding the economics of the entities involved (agricultural households, artisanal production

Fig. 16.2 Compost being made from household waste in Bafoussam (© Serge Simon, CIRAD)

associations, municipal collection services) in these interactions (Parrot et al. 2009b); on identifying, using econometric studies, the determinants of the adoption of organic amendments and synthetic inputs (Parrot et al. 2009a; Sotamenou and Parrot 2013); and economic analyses to estimate the real value of organic fertilizers (defined by their contents of N, P_2O_5, K_2O and international reference prices) in relation to their market value (defined by their actual selling price on local markets).

The results confirmed the findings of previous studies on how the distance between plots and storage facilities (usually the residence) influenced the use of organic or synthetic inputs. Territorial anchoring, characterized by soil and climate conditions, geographical proximity, and short circuits, plays a decisive role in the use of products whose transportation costs are high. Land tenure security encourages the use of organic amendments, either solely or in combination with synthetic inputs (Sotamenou and Parrot 2013).

In contrast, an agroeconomic analysis has revealed that there is no link between the real and commercial values of organic amendments. In the case study in Cameroon, it was found that the commercial value of compost was overestimated (sold at five times its actual value) because the production process resulted in amendments with poor N, P_2O_5, and K_2O content and very low humic potential. At the same time, an analysis of broiler manure showed that this organic amendment was being sold for one-third its real value. This is an example of information asymmetry in which buyers are not made aware of the real value of products they are buying.

Determining a market value that allows an economic development over the long term depends on the quality of the product which, in our case, in turn depends on the training of these small-scale producers. But to ensure that municipal waste collection is planned in the right perspective, and that the official waste collection chain encompasses a broader territorial and municipal range, its market value must also incorporate the transportation costs between points of collection, processing, and distribution, and assess the agronomic impacts in different soil and climatic conditions. New business models based on local opportunities and constraints are called for.

16.1.2 Integration of Sugarcane Cultivation and Dairy Farming on Réunion Island

Crop cultivation and livestock farming are important economic activities on Réunion Island. At present, a fodder shortage in the dry season and high cost of inputs (feed concentrates and mineral fertilizers) jeopardize the socio-economic and environmental viability of farms. An option to improve the situation is to create a more self-sufficient cultivation system where livestock manure is used as organic fertilizer in sugarcane fields, and sugarcane by-products (bagasse, straw cane, and cane sprouts) are used as cattle feed.

In order to measure the impacts of a closer integration between crop cultivation and livestock farming, and to better understand the reasons which hinder this integration among some farmers, a mathematical programming model was created, using the General Algebraic Modelling System (GAMS) software. It takes into account the technical, socio-economic, and environmental impacts of sugarcane cultivation and dairy farming in Réunion (Randrianasolo 2012). The model uses a normative approach to identify the optimal solution. The model's objective function is to simultaneously maximize the incomes from sugarcane cultivation and dairy farming.

Three scenarios were studied to reflect the different degrees of integration:

1. minimal integration between the two activities;
2. partial integration;
3. total integration of crop cultivation and livestock farming activities.

Results from different scenarios confirm the hypothesis that an increased interaction between the two agricultural activities, through the adoption of alternative techniques, increases the overall efficiency of the larger production system encompassing both these activities. Gradually, as the degree of integration increases, the margin on direct costs of the combined agricultural activities tends to rise. However, the distance between the place of production and the place of delivery of co-products acts as a limiting factor for adopting this integration. The quantity of fossil fuel used to produce 100 l of milk drops by about 30 % in the third scenario, when compared with the baseline situation. On the one hand, the use of sugarcane co-products reduces the amount of concentrated feed consumed per animal while, on the other, the use of livestock manure reduces the need for mineral fertilizers and, consequently, the energy required to produce milk, making this system more independent and less reliant on external inputs.

The model's outcomes highlight the usefulness of this approach to assess the impacts of modifications in agricultural practices in a context of constantly changing economic and environmental constraints. For example, the economic impact of a tripling of the price of imported inputs (feed concentrates and fertilizers), as happened in 2015 on the world market, on the incomes of dairy farming and sugarcane cultivation is now estimated as a net loss of approximately EUR 32 million per year. The fact that livestock farmers can now rely on alternatives to better utilize their forage resources and local fertilizers increases the resilience of their farms, and consequently minimizes the negative economic and environmental impacts of a rise in input prices. These economically efficient and independent farming systems also fulfil the expectations of society in terms of respect for the environment, job creation, and economic viability.

16.2 Analysis of Recycling Situations Through the Management of Biomass and Nutrient Flows

Understanding the relevance and importance of recycling linkages between local segments of supply chains in terms of mutual benefits (for example, potential economic or environmental gains, or employment), helps conserve, develop, and stimulate them better. It is a matter of quantifying the flow of materials exchanged, and their effects within and around each segment at appropriate spatial and temporal scales. We present here three analyses of existing or proposed exchange systems between segments of two agri-chains: within mixed farms; between farms in urban and peri-urban areas; and between farms at the regional level.

16.2.1 An Analysis of Intra-farm Interactions Between Crop Cultivation and Livestock Farming

Systems that integrate crop cultivation and livestock farming now account for over 50 % of the global meat and food production, and still predominate, especially in the tropics. This integration can be analyzed in terms of biomass recycling: pastures or crops supply plant biomass which can be used as animal feed, while these animals provide organic manure that can fertilize the plants. In recent years, quantitative methods derived from analytical methods to study food webs, with a focus on nutrient and energy cycles, have been applied to quantify the degree of integration between crop cultivation and livestock farming in various regions of the world: in Ethiopia, Kenya, and Zimbabwe (Rufino et al. 2009); Réunion

(Vayssières et al. 2011); Madagascar (Alvarez et al. 2012); and, more recently, in dry areas of West Africa (Bénagabou et al. 2014).

In Madagascar, the quantification of nitrogen flows within farms with integrated systems, using the Network Analysis method (Alvarez et al. 2012), confirmed the vital role of animals in maintaining soil fertility. Modifications in the feeding practices of dairy cows (increased use of concentrates) and the management of livestock waste (storage and application of manure) were simulated and revealed changes in nitrogen flows and in environmental sustainability indicators (nitrogen balance and efficiency, degree of nitrogen recycling, degree of integration of nitrogen flows, etc.). Improved manure management practices lead to reduced nitrogen losses through leaching and volatilization and increased soil nitrogen storage to levels similar or greater than those obtained by the application of additional mineral fertilizers, and, finally, to improved nitrogen recycling on farms. An increase in complementation results in a small improvement in nitrogen efficiency on farms. It must be noted that the association of these complementation practices with improved manure management practices proved to be most efficient in terms of nitrogen utilization at the farm level. Thus, the role of livestock rearing in maintaining soil fertility, through improved nutrients recycling, can be ensured by adopting manure management practices that limit nitrogen losses and can be further increased through the quality of feed.

These studies carried out in the tropics show a wide diversity, as much in between regions as within an area with the same soil and climate conditions. This diversity shows a good scope for progress in increasing the economic and environmental efficiency of agricultural production systems by promoting exchanges between crop cultivation and livestock rearing.

16.2.2 An Analysis by Simulation of the Use of Livestock Wastes in Peri-Urban Market Gardening: The Case of Rufisque (Senegal) and Mahajanga (Madagascar)

CIRAD analyzed the recycling potential of livestock wastes in market gardening in two peri-urban areas, that of Rufisque, a suburb of Dakar (N'Diénor et al. 2013), and that of Mahajanga, a coastal city northwest of Madagascar (Ramahefarison et al. 2013). These analyses were based on a dynamic simulation of a territorial system using a computer model (Guerrin 2001) in order to assess and design suitable methods for using these organic residues.

In the Rufisque department, organic residues are in great demand by market gardeners for two purposes: as a fertilizer for crops, and as mulch to maintain soil moisture. The territory produces large quantities of various kinds of organic residues: dung from draught horses, effluents from livestock (poultry, cattle), waste from agrifood industries (peanut dust, fish meal, and slaughterhouse waste). Several carters operate here to transport the organic residue within the territory, albeit with limited capabilities and only over short distances. As a result, organic residue is not always available to garden farmers, nor is it evenly distributed to be able to meet everyone's needs. Thus, while a shortage of organic matter is a problem for market gardeners, the surplus organic residue that piles up in some local farms that produce it poses risks to the environment.

Market gardening in the suburbs of Mahajanga mainly involves growing leafy vegetables in very short cycles (up to 12 cycles/year). Production is very variable and is dependent on a good understanding of intensification. While there is substantial demand for organic fertilizers, supply is limited due to temporal or spatial unavailability of organic residues. Even then, the city produces 10,000 tonnes of dung and droppings in numerous small-scale urban farms, which is equivalent to about 87 T/year of nitrogen, whereas the nitrogen demand of market gardening is estimated at only 6 T/year. At the same time, manure from most urban farms is not used productively, and thus represents a source of pollution and nuisance.

To simulate the regional management of organic residues, we used the Magma model (Guerrin 2001). Based on the dynamics of systems, it represents the territory in the form of a set of stocks linked by flows. The simulation involves calculating the flow of organic residues between production and consumption units, and stock inventory levels resulting from transport activities and the spreading of organic residues.

Three principal scenarios were developed to help simulate and assess:

1. current practices: excessive fertilization in Rufisque; surpluses and scarcities in Mahajanga;
2. consequences of an environment-friendly fertilization on the supply of organic residues to market gardeners;
3. logistical facilities necessary to transport organic residues to areas distant from urban centres.

Simulations that mirror ground reality in Rufisque show that some consumption units suffer from a lack of organic residues, while others suffer from an excess of it. In scenario 2, adjustments were made to the fertilization practices and, as a result, to the demand for organic residues from consumption units. Transportation and storage were also adapted to improve the situation simulated in Scenario 1. This resulted in a better distribution of organic residues between crops, a reduced risk of nitrogen pollution, and a reduction in labour time. Scenario 3 involved the relocation of some of the consumer units a hundred kilometres from Rufisque, due to pressure from increased urbanization and water shortages. Simulations have helped understand the characteristics of transportation and storage required in this hypothetical situation.

In Mahajanga, simulations of scenario 1, based on current practices of repeated application of organic manure and chemical fertilizers, revealed excessive fertilization in consumption units. A significant amount of livestock manure (63 %) generated in urban areas was lost as it could not be used by market gardens located at a distance, on the periphery of Mahajanga. The shortage of manure during peak market gardening season was mainly caused by a lack of transportation between production units and consumption units. In the current situation, simulations helped quantify excessive fertilization and inefficient use of surplus manure, which poses obvious environmental pollution risks. Adjustments in the dosage and frequency of application of organic residues and chemical fertilizers (scenario 2) partly helped reduce these risks. The relocation of some farms or the transportation of waste to the town's outskirts (scenario 3) would result in a better utilization of organic residues and to minimize environmental risks. To be effective, this new strategy should be supported by improved transportation facilities between production and consumption units, and the use of new logistical means (for example, trucks) to transport the manure produced in the city to farmers located at a distance of 30–90 km from Mahajanga who need organic residues.

These examples show the importance of coordinating supply and demand for different types of organic residues over space and time to ensure proper distribution in territories with several hundred farms. Modelling allows us to simulate dynamically the functioning of these territorial systems and to attempt to improve it from an agronomic and environmental point of view.

16.2.3 Nutrient Balance at the Territorial Level in a Perspective of Exchanges Between Livestock Farms and Fish Farming Units in the Red River Delta, Vietnam

A boom in pig production in Vietnam since 1990 has resulted in an average annual growth of 10 %, which has been maintained at this high level for 25 years. Thai Binh Province is one of 18 provinces in the Red River Delta, southeast of Hanoi, in northern Vietnam. This delta supports 18 million people, with record population densities of more than 1200 inhabitants/km^2. A rapid and continuing increase in the demand for animal products is forcing traditional livestock farms to adopt more intensive industrial techniques.

CIRAD and the National Institute of Animal Husbandry (NIAH) developed a tool in 2005 and 2006 to calculate the nutrient balance at the provincial and local levels in Thai Binh Province. Assessments of the nutrient absorption capacity of crops and ponds were thus compared with the production of livestock wastes, in particular pig dung. Results showed that fish ponds can consume significant amounts of livestock wastes, since they can make use of both liquid and solid effluents. The surface area of market gardens is not large enough to absorb all the liquid pig effluent, resulting in surplus amounts, which are steadily increasing from the already high levels in 2004, posing a real danger of chronic pollution. During winter, there is a marked mismatch between the stock of effluents available in the rural communes and the ability of crops to absorb them.

Very conscious of the environmental dangers that livestock farming poses, provincial authorities have validated the established diagnosis. Calculating spatial balances has become a common practice within a framework such as of territorial ecology that encompasses all the territory's sectors and agri-chains. As far as organic matter is concerned, the balances of different elements over the entire territory, at various scales, constitute valuable information to define targeted and relevant public interventions.

16.2.4 Lessons Learnt from the Situations Studied

The analyses presented indicate that, in addition to exchanges within mixed farms, there is significant potential for value enhancement and exchange of products between specialized farms belonging to different agri-chains, and this at various geographical scales. Although these analyses are mainly based on nitrogen or phosphorus requirements – other elements being either in surplus or supplemented by chemical fertilizers –, they show the potential of deriving value from waste produced by different agricultural or agro-industrial units by using it in cropping systems and fish farming. The importance of organization and coordination also partly explains why exchanges at these supra-farm scales exist rarely without external intervention. It also shows us that, at the geographic scale of the territory, the latter should be treated like an entity, that is to say a network of actors sharing and regulating activities within an area. Techniques can then be used to create or strengthen links within this network between actors of distinct and relatively compartmentalized agri-chain segments, but with a potential linkage on the basis of common interests. We can, nevertheless, ask ourselves whether the partial apprehension of the territory that such a relationship constitutes represents the most suitable exchange for the entire territory from the agronomic and economic points of view. How easy it is to find an answer to this question depends on the complexity of the territory concerned.

16.3 Creating an Industrial Symbiosis Between Supply Chains: An Example of Integrated Management of Organic Residues on Réunion Island

This section presents an industrial symbiosis approach to ensure the promotion of relevant exchanges between actors of (multiple) segments of (multiple) supply chains that make up a complex territory.

Réunion Island is an apt example of a territory exhibiting high potential for industrial symbiosis centred on organic residues. On the one hand, the agricultural sector remains heavily dependent on fertilizer imports. On the other hand, the island also imports large amounts of nutrients, both to meet the needs of its inhabitants and of its livestock farms. Since economic outflows in terms of nutrient exports are very low, the resulting structural surplus has a substantial impact which is increasing in tandem with the population, which is growing at a rate of about 10,000 inhabitants/year on the island. The pressure this population growth puts on land resources further restricts the development of agricultural activities which, given the national and European regulatory frameworks, is already hard-pressed to achieve its goal of increasing the share of local production in the local market (products from market gardening and livestock rearing). At the same time, it is becoming increasingly difficult for the already overflowing landfills to handle organic residues from urban and agro-industrial activities. The desire to use and derive value from these new residues in agricultural activities has given rise to competition for agricultural land on which regulations allow waste to be spread.

While their motivations may vary, the concerned actors are willing to consider changes to overcome this impasse. Therein lies a prerequisite for industrial symbiosis: the benefits of participating must more than compensate the costs of mutual commitments. It follows that an integrated management of a territory's organic residues is perceived as beneficial since their accumulation will hinder the development of all the value chains concerned. However, other conditions are necessary, as experience has shown that initiatives rarely emerge spontaneously, even in such a situation. This is where the research community can play a dual role, which its status as a disinterested party without any direct interest in the value chains allows it to do: to catalyze a process of consensus by offering solutions; and also supporting this process with its knowledge, while facilitating consultations to arrive at a consolidated and acceptable solution. In order to fulfil this role of catalyst, the legitimacy of the research community's abilities must be recognized on the basis of its knowledge of the local situation.

16.3.1 The Approach and Its Implementation in Réunion

CIRAD has legitimate competency to initiate such a process in Réunion due to its extensive knowledge of local organic residues, of the benefits of their use under local conditions and the attendant risks, and, more generally, of the local agriculture. This skill set allowed it to bring together an initial group of stakeholders based on a framework proposal, and then to study it in its role of an actor involved in the implementation of the approach (Wassenaar et al. 2014). A comprehensive group of stakeholders' representatives was created to study the following assumptions:

- the management of organic material would benefit from being organized at the inter-communal scale;
- raw organic material is not suitable to be directly applied to crops, or to meet farmer needs;
- the production and marketing of standardized organic fertilizers from organic matter present in the territory, adapted to the needs of crops and farmers, would

go some way in solving local environmental problems in a sustainable manner and in reducing imports of synthetic fertilizers.

The approach adopted was based on several methodological proposals. Douthwaite et al. (2002), basing themselves on their method called 'Follow the Technology' (FTT), proposed involving stakeholders in an *in situ* assessment of workable technological solutions. The researcher, in this approach, becomes a supporter of farmers' innovations. Companion Modelling or Commod (Étienne 2011) is based on the same principle of extending support to actors for exploring organizational innovations. This approach uses modelling stages to study conditions and consequences of scenarios, without necessarily any practical implementation. The approach adopted for this study is based on both these methods by providing support to the stakeholders' representatives in exploring conditions and consequences linked to implementing a technology considered workable.

It was decided to study and understand the inter-communal level of Territoire de la Côte Ouest, a partner in the project with a population of 180,000 inhabitants, in a complex and dense institutional and regulatory context. The consultation system implemented endeavoured to support the reflection process within three arenas pertaining to different legitimacies (Queste and Wassenaar forthcoming; Wassenaar et al. 2015, for the co-construction stages):

- at the institutional level: a steering committee brought together representatives mandated by their institutions (governments, companies, consular chambers, etc.) to discuss and comment on the project guidelines. These steering committees have provided an institutional legitimacy to the project and the solutions studied;
- at the technical level: *ad hoc* working groups were established based on the co-optation principle. They brought together individuals identified by members of the project team as having knowledge, know-how, experience, or expertise relevant to discussions underway. The participation of these experts provided scientific and technical legitimacy to solutions co-constructed within the project framework;
- at the practical level: participatory workshops were organized that brought together members of any of the twelve target groups directly concerned by deriving agricultural value from organic matter (livestock farmers, green-areas maintenance services, planters, etc.). The participation of these representatives of target groups provided a twofold legitimacy to the project. First, it could claim to have taken into account, at least to some extent, the interest of stakeholders and, second, a regular consultation with these actors on solutions developed reinforced their social acceptability.

16.3.2 Proposals for Co-constructed Solutions

This process culminated in two prospective scenarios (detailed description in Wassenaar et al. 2015). These differ significantly from the original proposal, in terms of residues used (for example, the sludge from waste water treatment plants was finally excluded due to institutional reasons), treatment processes (for example, various regulatory constraints led to the abandoning of methanisation) and scale (for example, due to industrial design constraints, one of the planned recycling chains can only be implemented at an island-wide level, thus exceeding the inter-communal scope).

In the 'minimal' scenario, a value chain for a new fertilizer product will be introduced. A mixture of shredded green waste, poultry litter, and hog manure will be composted at a facility. The finished product would conform to the organic amendment standard NFU 44–051. This product will be of interest to market gardeners, as it would give them a complete fertilizer to be used once every two or three crop cycles. It could also be used as an exclusive fertilizer for sugarcane and pastures, but only when applied using mechanical means, which greatly reduces the extent of the area that can be benefited. The 'optimal' scenario extends the first scenario by developing a second value chain to produce both organic and organo-mineral fertilizers to cater to the needs of the main fertilizer market already identified: manual fertilization of the sugarcane ratoon crop. The challenge here is to produce a fertilizer that meets the annual needs of sugarcane without exceeding the maximum acceptable dose for manual application, estimated by farmers to be 2 T/ha. This value chain will be composed of many small co-composting plants whose products would form the organic base. These raw materials would subsequently be transported to a complementation and granulation factory that would develop two products meeting the NFU 42–001 standard.

Value can thus be derived from a huge part of the recoverable pool (which, in the case of pig manure, for example, remains well below the total quantity available). In the case of livestock wastes, this part goes up from 40 % for pig manure to 90 % for droppings from broiler chickens, and reaches 100 % for laying hens. The requirement of co-composting facilities for shredded green waste is half of the currently available quantity and thus will not jeopardize the needs of those who use green waste compost, including municipal green-areas maintenance services. The amount of waste which would finally have to be managed would decrease, resulting in economic, societal, and environmental benefits. Furthermore, the planned production of fertilizers would substitute a substantial amount of chemical fertilizer (about 1850 tonnes), which would fulfil an important objective collectively identified at the start of the project.

The creation of these value chains, which would be a practical application of the circular economy principle, requires, based on the conditions of the territory under study, considerable public financial support (Wassenaar et al. 2015). Even if such support is obtained, other obstacles will remain:

- long-term coordination between actors of a value chain will remain to be established;
- market constraints and price volatility of chemical fertilizers will limit profitability and rapid return on investment;
- the lack of certain technical references required to predict the agronomic effects of planned products, and the impossibility of producing them in the available time.

Following this participatory research, Qualitropic, a competitiveness cluster of industrial and agricultural actors in Réunion, is taking forward this project of industrial production of organic fertilizers and is working towards removing these obstacles.

16.4 Conclusion

In a world that is constantly evolving, links between different agri-chains or with other supply chains appear and develop with little contribution from the research community. While their very existence is a proof of their usefulness, these links are usually very short-term, opportunistic, and often unstable. This organizational instability is compounded by an inefficient utilization of agricultural products, resulting from an imperfect knowledge of materials and requirements, as illustrated in Box 16.1: the link between poultry litter and cocoa farms is very unstable, with raw chicken manure probably not very suitable for the requirements of cocoa trees. Other instances of bilateral exchanges are tried out in an opportunistic and *ad hoc* manner. All this justifies the usefulness of going beyond these tentative and unsystematic efforts.

Existing or desired relationships between local segments of supply chains are clearly very diverse in extent and nature, varying in complexity, magnitude, scale, involvement, etc. Through several examples, this chapter proposes different methods and tools for analyzing such systems of relationships, and for foresight and support, in both the technical and organizational dimensions. The diversity of situations described is largely explained by the composition of the territory and the dynamics of its development. A general rule seems to emerge: the more one endeavours to tailor these relationships to contribute to sustainable development – relationships that we would therefore want to be stable –, the more complicated they become and the more difficult to build. Direct and bilateral relationships are the ones that often form spontaneously. At the other extreme is the industrial symbiosis within territories in rapid development and often not sustainable. In this latter case, these relationships are recognized as new value chains: short-term, local, circular, transversal, which challenge the notions of the supply and value chain themselves. Similarly, the question of the scale of sustainable development arises: agri-chain, territory, transversal recycling chain, or a combination of all of them. The imperative of sustainable development compels us to find common ground between the point of view of the territory and that of the agri-chain. Thus, the territorial anchoring of agri-chains, through the development of inter-supply chain links, helps make these meeting points springboards for mutually beneficial relationships.

References

Alvarez S, Rufino MC, Vayssières J, Salgado P, Tittonell P, Tillard E, Bocquier F (2012) Whole-farm nitrogen cycling and profitability of rice-dairy systems in the highlands of Madagascar: an application of network analysis. Agric Syst 126:25–37

Bénagabou OI, Vayssières J, Vigne M, Vall E, Lecomte P, Bougouma V, Nacro HB, Blanchard M 2014 Recyclage des biomasses au sein d'exploitations de polyculture-élevage et efficience d'utilisation de l'énergie brute. In: 4e semaine scientifique Coraf, 16–20 June 2014, Niamey, Coraf, 1 pp

Douthwaite B, de Haan NC, Manyong V, Keatinge D (2002) Blending "hard" and "soft" science: the "follow-the-technology" approach to catalyzing and evaluating technology change. Conserv Ecol 5(2):13

Étienne M (2011) Companion modeling: a participatory approach to support sustainable development. Éditions Quæ, Versailles 368 p

Guerrin F (2001) Magma: a model to help manage animal wastes at the farm level. Comput Electron Agric 33(1):35–54

N'Diénor M, Guerrin F, Paillat JM (2013) Simulating the management of wastes at a territory level in the peri-urban market gardening systems of Dakar (Senegal): the case of the Rufisque department. In: 15e conférence internationale Ramiran : Recyclage des résidus organiques pour l'agriculture : de la gestion des déchets aux services écosystémiques, 3–5 June 2013, Versailles, Inra

Parrot L, Sotamenou J, Dia Kamgnia B, Nantchouang A (2009a) Determinants of domestic waste input use in urban agriculture lowland systems in Africa: the case of Yaoundé in Cameroon. Habitat Int 33(4):357–364

Parrot L, Sotamenou J, Dia Kamgnia B (2009b) Municipal solid waste management in Africa: strategies and livelihoods in Yaoundé, Cameroon. Waste Manag 29(2):986–995

Queste J, Wassenaar T (forthcoming) à paraître. A stakeholder matrix to co-design plausible solutions: a dialogue protocol for adaptive co-management. Ecol Soc

Ramahefarison H, Guerrin F, Paillat JM 2013 Simulation modeling of management scenarios of livestock manure in urban and periurban zones of Mahajanga, Madagascar. In: 15e conférence internationale Ramiran. Recyclage des résidus organiques pour l'agriculture : de la gestion des déchets aux services écosystémiques, 3–5 June 2013, Versailles, Inra

Randrianasolo J 2012 Modélisation économique des interactions entre systèmes de culture et d'élevage à l'échelle des exploitations et des territoires (Réunion et Madagascar). Doctoral thesis, University of Réunion Island, 132 p

Ruf F, Kla AG, Dja K, Kiendré J 2015 Chicken manure in the cocoa plantations of Côte d'Ivoire: an agro-ecological and social revolution, a 'frugal' village innovation, 15 p. Retrieved 18 May 2016, http://inter-reseaux.org/IMG/pdf/Ruf_Chicken_manure_in_the_cocoa_plantations_of_Cote_d_Ivoire.pdf

Rufino MC, Tittonell P, Reidsma P, Lopez-Ridaura S, Hengsdijk H, Giller KE, Verhagen A (2009) Network analysis of N flows and food self-sufficiency – a comparative study of crop-livestock systems of the highlands of East and Southern Africa. Nutr Cycl Agroecosyst 85(2):169–186

Sotamenou J, Parrot L (2013) Sustainable urban agriculture and the adoption of composts in Cameroon. Int J Agric Sustain 11(3):282–295

Vayssières J, Vigne M, Alary V, Lecomte P (2011) Integrated participatory modelling of actual farms to support policy making on sustainable intensification. Agric Syst 104(2):146–161

Wassenaar T, Doelsch E, Feder F, Guerrin F, Paillat JM, Thuriès L, Saint Macary H (2014) Returning organic residues to agricultural land (RORAL) – Fuelling the follow-the-technology approach. Agric Syst 124:60–69

Wassenaar T, Queste J, Paillat JM, Saint Macary H (2015) La co-construction de filières de recyclage de résidus organiques à La Réunion. Innov Agron 43:161–175

Chapter 17
A Counterpoint on Waste

Benoit Daviron

In view of the problem of sustainable development, the issue of waste is crucially important because our very existence is probably threatened as much by the amount of waste we generate as by the resources we deplete. While CO_2 may be the waste currently in the headlines, with its dangers being actively debated, others that are just as menacing also exist. For example, the various forms of active nitrogen in circulation have profound effects, both at the local level (nitrification of groundwater, green algae, air pollution, etc.) as well as at the global level (climate change, acidification of oceans, etc.).

According to the French environment code, waste is defined as: 'Any residue from a process of production, transformation, or utilization, any substance, material, product, or more generally, any item that is discarded, or that the owner intends to discard.' The fact that waste is 'discarded' implies that, one way or another, there exists a commons – a road, a river, the atmosphere, etc. – in which it will be disposed off. Thus, waste denotes 'products' with no value ascribed to them, but which affect the quality of the commons.

All processes for transforming matter generate waste. Production functions used by economists, where all inputs are transformed into outputs, are obviously not correct. There will always be some waste, even if it is only heat which, in one way or another, must be dissipated.

We know from history that human societies have experienced, from the eighteenth century onwards, major upheavals in their relationships with resources under the labels of development and industrialization. Anthony Wrigley, an English historian, makes a distinction in this regard between the concepts of 'organic economy' and of 'mineral economy' (Wrigley 2004, 2010). According to this author, industrialization must be interpreted as the transition from a material basis

B. Daviron (✉)
CIRAD – UMR Moisa, Montpellier, France
e-mail: benoit.daviron@cirad.fr

E. Biénabe et al. (eds.), *Sustainable Development and Tropical Agri-chains*,
DOI 10.1007/978-94-024-1016-7_17

almost solely consisting of biomass (for food, shelter, clothing, heating, travel, health care, ensuring soil fertility, etc.) to a material basis that is essentially dependent on minerals, of which fossil resources such as coal and petroleum are prominent examples.

This shift from the organic to the mineral has major implications for waste production, especially on the ability of living organisms to use it for their metabolism. Two problems can be distinguished here:

- that of waste that is already produced by organic societies, but in very small quantities, and which can be characterized today by phenomena of saturation. Such is the case with CO_2 or active nitrogen;
- that of waste that has no place in the natural processes of living organisms. This category consists primarily of synthetic products, such as plastics or pesticides, developed by chemical industries from the beginning of the twentieth century. Also obviously part of this category are nuclear waste and heavy metals; the latter do not naturally exist on the planet's surface and have to be mined from the ground.

The concept of waste, used earlier in this chapter with reference to resources, is also employed by authors such as Zygmunt Bauman to refer to populations that are displaced by the march of modernization and industrialization (Bauman 2009). The homeless population is becoming a defining image of many cities, even in the North. What transforms entire populations into 'waste' is the fact that they no longer have any space, other than public spaces, to use for basic existence.

One of the many questions we can ask ourselves is whether waste and its invasive presence are not an inevitable price to pay for generating development, prosperity, and wealth. However, the difficulty for our world, which bases its choices on economic calculations, is that waste, by its very nature, has no price, which makes it impossible to compare earnings and costs. The problem obviously becomes even more complicated if we include displaced populations in such waste.

There may be a way to reflect on this, at least as a metaphor, by relying on the laws of thermodynamics. The first law tells us that 'nothing is lost, nothing is created', i.e., although matter and energy can change forms, they are preserved at an overall level. The second, which is the one of interest to us, states that entropy always increases in a closed system, i.e., the amount of usable energy always diminishes or, more simply, disorder always increases. This means that, in a closed system, the creation of local order will always be at the cost of disorder somewhere in the system, at a magnitude greater than that of the order created. In short, we have to export disorder – a lot of it – in order to create order – a little of it.

The characteristic way for organic societies to function is as open and cyclic systems that use solar energy through plants and their photosynthesis. The main waste created in the process is helium, which results from fusion of hydrogen in the sun, a waste product we need not worry about given the distance that separates the earth from the sun. In contrast, the characteristic way for mineral-based societies, like the ones we live in, to function is as if the earth is a closed system, by drawing all energy required, or most of it, from resources here on earth. Although this ingenuity of the human race has allowed an unprecedented accumulation of wealth, it forces us today to deal with the question of determining at what price of disorder has this been achieved and whether this disorder is not ultimately greater than the order created.

The same question can be asked about agri-chains, which can be represented as successions of economic operators working in cooperation. It is this division of labour that leads to productivity gains and helps provide increasing amounts of agricultural products to consumers. From this point of view, agri-chains can likened to a form of order. But is this order really better than the disorder that inevitably accompanies it? This is the question that the world persistently asks of us today. To put it more starkly, is the pollution generated (pesticides, CO_2, active nitrogen, etc.) or the populations displaced (populations of tropical forests replaced by plantations, fishermen banned from fishing in coastal Guadeloupe because of chlordecone, etc.) worth more or less than the tonnes of tea, coffee, oil, or banana produced? Life cycle analyses will, no doubt, allow us to identify the technical itineraries that produce the least waste, but they will be unable to tell us whether the waste produced is lower in value than the wealth created, one that is measured solely by market prices.

References

Bauman Z (2009) Vies perdues: la modernité et ses exclus. Rivages poche, Paris, 254 p
Wrigley EA (2004) Two kinds of capitalism, two kinds of growth. In: EA W (ed) Poverty, progress and population. Cambridge University Press, Cambridge, pp 68–86
Wrigley EA (2010) Energy and the English industrial revolution. Cambridge University Press, Cambridge, 288 p

Part IV
Agri-Chains and Evaluation of Sustainability

Chapter 18
Why and How to Assess the Contribution of an Agri-Chain to Sustainable Development?

Denis Loeillet and Catherine Macombe

To ask if an agri-chain can be a vector of sustainable development is to determine if it contributes to global sustainable development. What then must be the basis of the assessment tools? To evaluate the real contribution of an action to sustainable development, we must determine the sum total of its consequences on the entire planet, and this over a very long time scale. Given that such an assessment is impossible, we rely on assigning a value (evaluation) to certain criteria that supposedly reflect the sustainability of actions undertaken within the agri-chain. This value is assigned to either the agri-chain's characteristics at a given time (performance assessment) or the consequences of such a change occurring in the agri-chain (impact assessment) (Table 18.1). In this chapter's first section, we explore the question of why we should address the environmental and social impacts of agri-chains. We then examine the issue of selecting the boundaries and the nature of impacts to consider. We then highlight the usefulness of life cycle assessment methods in studying agri-chains. Finally, we discuss the different roles attributed to assessment methods by decision makers.

D. Loeillet (✉)
CIRAD – UPR Geco, F-34398 Montpellier, France
e-mail: denis.loeillet@cirad.fr

C. Macombe
UMR ITAP, Irstea – Centre de Montpellier, 361, rue Jean-François Breton, BP-5095 34196 Montpellier, Cedex 5
e-mail: catherine.macombe@irstea.fr

E. Biénabe et al. (eds.), *Sustainable Development and Tropical Agri-chains*,
DOI 10.1007/978-94-024-1016-7_18

Table 18.1 Difference between performance, effect, and impact

	Performance	Effect	Impact
What is evaluated	Current situation of a company in an agri-chain	Social phenomena resulting from a change (with potential impacts) caused by a company in an agri-chain	Effect of a change caused by a company in an agri-chain that is directly felt by individuals
Examples	Gender, child labour, trade-union freedom	Loss/creation of jobs, formation of new networks	Change in health status, improved confidence

Source: Macombe et al. (2013)

18.1 Why Focus on the Agri-Chains' Environmental and Social Effects?

Tools to assess the efficiency of production systems, i.e., yields and different productivity criteria, as well as to gauge their profitability, are widely known and available. But evaluation has now progressed beyond just the economic and financial aspects; it incorporates all three dimensions of development: economic, social, and environmental (Box 18.1).

Box 18.1 Using Network Analysis to Evaluate the Systemic Sustainability of Agri-Chains

Paule Moustier

A systemic approach to sustainability is one that is comprehensive (focused on the entire entity under study) as it combines the study of the normative aspects of sustainability (economic, social, and environmental performances) with that of its functional aspects (method of organization and configuration of components of the entity under study). Actors contribute to sustainability when they anticipate the long-term consequences of their decisions and actions, and try to find the right balance between the three dimensions of sustainability. This vision, in terms of the system, has been successfully used to study the management of natural resources, health, and human resources. We applied it to two food chains as part of the 'Sustainable, local or localized, innovative food value chains project' (ANR Illiad project): Haute Provence small spelt (protected geographical indication, PEHP) and organic rice from Camargue.

We used network analysis to characterize the organization of these agri-chains, in particular by calculating network centrality and density. The survivability of companies in the network was also taken into account. We also developed indicators for various sustainability components:

- for the economic impact: net margin of producers, dependence on CAP subsidies, dependence of producers on downstream operators, adaptability of producers;

(continued)

Box 18.1 (continued)

- for the social impact: attachment to farming, jobs created or maintained, prospect of transmission of farms, quality of life.
- for the environmental impact: use of chemicals, extraction of value from by-products.

This 'small spelt' chain is characterized by a highly centralized organization that enables it to perform well in all three domains: environmentally, with very strict specifications for the protected geographical indication; socially, using the BioSolidaire charter; and economically, through the revival of a long forgotten product. Network analysis shows that even though the small spelt chain presents good attributes in terms of conventional sustainability, it has a limited ability to deal with internal and external disruptive events. In contrast, in the case of the diversification of the organic rice chain in Camargue, a central authority counter-balanced by a strong network with diversified connections confers a certain robustness to the agri-chain. This capability is conducive to the agri-chain's growth despite an uncertain outlook (partial dependence of farmers on CAP subsidies for organic agriculture) and a middling performance regarding the environmental and social aspects. However, an evolution of the current situation due to an unfavourable reform of the CAP (reduction in subsidies) could *a priori* weaken the agri-chain's diversification.

Still, it is difficult to establish cause-effect relationships and to conclude that any given type of structuring strengthens sustainability. It is necessary to further deepen the approach to better link the different dimensions. We can also emphasize the need to find a way to go beyond the subjective nature of analyses by defining the critical thresholds of sustainability.

(Based on Bassene et al. 2014)

The production of food can pollute and be expensive in terms of energy and resources (water, land, non-renewable resources), even as the Earth's population is growing inexorably. As for the social effects of the agri-chains, they remain undeniable. Large companies are proactive in this area. Due to several reasons (including careful monitoring by NGOs), companies, especially multinational ones, have realized that they cannot ignore society's opinions and values.

The example in Box 18.2 shows that beyond the economic activity they generate locally, some companies want to intensify their modes of intervention at the environmental and social levels in order to satisfy customers and ensure the sustainability of their business by legitimatizing it. However, goodwill is not sufficient. Irrespective of how 'good' are the good agricultural practices adopted, there always exists an incompressible physical threshold (in terms of energy, materials, or water used). But even though agri-chains generate negative externalities, they also have significant positive social effects. A well-known study by FAO

(2006) clearly identifies the disastrous ecological footprint of the world's livestock, while highlighting its many social functions that remain vital to the existence of many societies. A provocative study by Müller (2007) describes the dilemma between ecology and society with regard to African strawberries imported by Britain in winter, an agri-chain that sustains many poor families. In order to survive over the long term, many agri-chains must prove that they generate social benefits, especially if their environmental footprint is large.

Box 18.2 Assessing the Impacts of Value Chains: A Major Challenge for Multinational Corporations

Denis Loeillet and Catherine Macombe

Multinational corporations do not enjoy a good reputation. The main charges against them arise from their practice of appropriating the majority of the capital available in the least developed countries at the expense of local actors and from the fact that the profit they export from countries where they operate exceeds the amount they invest in them. They also practice discrimination, support dictatorships, follow aggressive commercial practices with local suppliers, grab land, degrade the local environment, capture natural resources, and acculturate traditional societies. There is abundant documented evidence and accounts from whistle-blowers denouncing such abuses. The daily news reminds us often enough that there is substance to these accusations. Examples of misdeeds of criminal value chains surface regularly, such as the collapse of a textile factory building in Bangladesh (2013) or slavery on shrimp fishing boats in Thailand (2014). Beyond the temptation to generalize and stigmatize, highlighting these tragic situations does help change things.

Economic actors and clients are now asking questions about the impacts of value chains in countries in which they operate directly (producer, processor, exporter) or indirectly (downstream customer). They are investing in gathering information on social and environmental impacts of the products they market. The tools companies use, often imprecise and rudimentary, to analyze their social performance are grouped under the banner of corporate social responsibility. Now increasingly codified and labelled, 'Corporate Social Responsibility' (CSR) is sometimes used as a tool for marketing (Laufer 2003), for (mis)communication in the form of 'social washing', and for management. In France, it has been mandatory since 2001 for all companies listed on the stock exchange to file a CSR report in addition to financial reports, and since the end of 2013, also for all companies with more than 500 employees. This requirement faces numerous criticisms, the foremost among them accusing it of generating nothing more than a mere litany of

(continued)

Box 18.2 (continued)

social or environmental compensation (a school here, a dispensary there, some replanting of trees, etc.), leaving the actual impacts of the company's actions, and thus of the value chain itself, obscured or hidden. In a true assessment, it is not so much a matter of assessing the means as it is about gauging opportunities that are ultimately made available through the means that are implemented.

That said, it must be noted that some CSR reports are more comprehensive than others, such as the one by a company that grows and exports fresh fruit from at least three West African countries (Fig. 18.1).

A review of this company's CSR reports reveals a lack of information on contribution to the local countries' budgets, in terms of taxes and fees. It is, however, clear that local people benefit from the company's operations. The amount of training activities alone that are conducted is substantial. Over 700 managers were trained annually at one site alone, and the reports indicate that a total of about 8000 people participated in training programmes. Access to healthcare is also provided through the construction and regular operation of a hospital and health centres, not only for the workers but also for other inhabitants of the area (over 50 % of the consultations). In-kind benefits (for example, water, electricity, educational expenses) are also provided, and represent about 13 % of the monthly salary. While the information provided

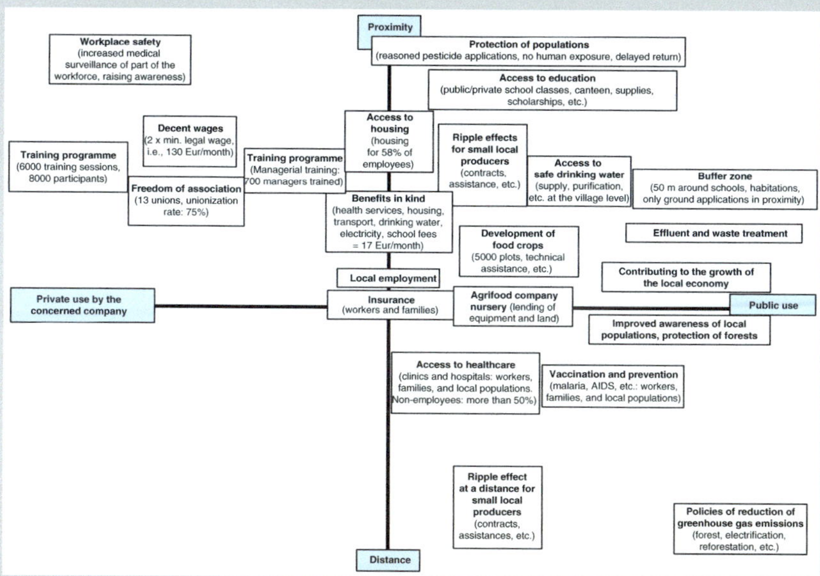

Fig. 18.1 List of actions listed in the CSR reports of 'Company A', operating in the fruit sector in West Africa

(continued)

Box 18.2 (continued)

could be called impact indicators, albeit simplistic ones, we would need more precise ones to outline the extent and intensity of real effects. A case in point is the example of decent wages. Admittedly, wage levels far exceed the official minimum wage, but this does not in any way guarantee that they are decent. In a similar way, we can gauge the contribution to the local economy through, for example, the quantum of salaries distributed, purchase of inputs, and the promotion of small, local entrepreneurships, but direct indicators are as yet lacking.

Thus, agrifood companies consider both ecological and social effects simultaneously. CIRAD, in partnership with its partners from the economic sphere, is engaged in action research projects which combine research and development. CIRAD is trying to meet the expectations of economic actors through these projects, i.e., evaluating the environmental and social impacts of the operation of the agri-chains as they exist, or of changes that are either planned or already implemented.

The economic actors of the agri-chains find life cycle assessments of interest, since they relate particularly well to chain-based reasoning processes. In addition, these are the only methods that reveal and highlight impact transfers between an agri-chain's stages. This is why we will focus on these methods for the rest of this chapter.

18.2 Why Life Cycle Methods?

Do global value chains contribute to sustainable development? 'Given current transnational economic practices (trade, production, etc.), social and environmental issues can no longer be limited to the local context' (Gabriel and Gabriel 2004). It is in this respect that life cycle assessment methods differ from all other methods.

These methods were created as a result of two observations: on the one hand, our planet is threatened by global environmental problems (for example, ozone layer depletion and climate change caused by increased emissions of various gases) that urgently beg for action and, on the other, it is impossible to monitor every pollution source. Engineers have thus proposed indirect assessment methods. A chemical analysis of a combustible material can determine the amount of gas it releases into the atmosphere. Knowing this figure and the quantity of fuel required to produce a given material, we can arrive at the quantity of gas released for a certain production. This kind of calculation for gas is applicable to other components too. The method used for such calculations is known as life cycle assessment. It is applied to the extraction, manufacture, use, and recycling of a product. The estimation of potential effects of concern – environmental damage – is based on the quantity of material (gas and other inputs) used and released into the natural environment. Despite their high cost and complexity, such evaluations have proven to be reliable and helpful.

A method analogous to the one used for environmental life cycle assessment has been developed to assess social effects. Even though similar, these two methods are distinguished by their different historical paths and disciplinary influences. However, methods for assessing environmental life cycles became concerned very early on by issues of human health and incorporate partial information on these aspects. In a similar way, some methods to assess social life cycles encompass the social impacts of environmental damage (for example, O'Brien et al. 1996).

By design, methods for life cycle assessment take into account all the stages that make up the product's life cycle. They thus identify and highlight what are known as transfers of impacts. These can occur when scenario B is selected over scenario A, either because the quantum of damage increases, the nature of the damage is changed, or the target (the people affected) is modified (Box 18.3).

Box 18.3 The Issue of Packaging in the Fruit Sector

Denis Loeillet and Catherine Macombe.

What should be used to package fruit for transportation: wood, cardboard, or plastic? In the latter case, are plastic crates disposable, recyclable, or even reusable after washing? The reuse of plastic crates would require a system of collection, sorting, and washing before sending them back to packing centres, some of which could be located at large distances from the centres of consumption, as in the case of long distance transportation. This is just one example of the complexity of the parameters to be considered and the difficulties faced by operators.

A need has recently emerged for studies that are either centred solely on environmental aspects (predominance of the life cycle assessment approach) or on only social aspects. Although methodologies used in agri-chain approaches and the environmental life cycle approach are very similar, it is only the technical processes that are studied in assessing environmental life cycles. The extension of such an evaluation to the social impacts of agri-chains is much more recent.

18.3 Scope of the Systems Studied

The impossibility of assessing sustainability at the planetary scale forces scientists to not only impose boundaries on the systems to be studied but to also restrict the nature of the impacts or performance to be considered depending on the extent of available knowledge and the specific objectives. The first indispensable step is the

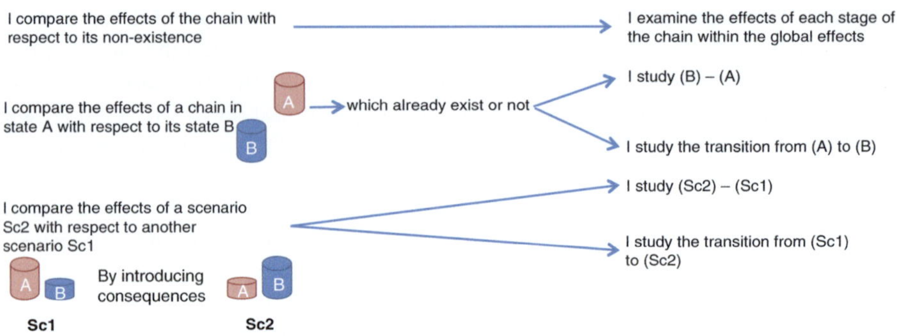

Fig. 18.2 Different types of evaluations of agri-chains using life cycle assessment

'scope and goal definition', which, at the minimum, defines the study's objectives and the system's boundaries.

When an agri-chain already exists, we can evaluate the consequences (or impacts, Fig. 18.2) of its existence in relation to a baseline situation in which the agri-chain does not exist, by considering that it results in a marginal change in society. This approach is known as an 'attributional'[1] approach. The calculations are then undertaken by considering all other things as equal.

But we can also assess the consequences of a change from state A to state B in this agri-chain:

- either by considering that change is marginal in relation to society in its entirety. The situation that prevailed before the change is then compared to the situation after the change. This is known as the comparative life cycle assessment approach;
- or by considering that the change is not marginal for the rest of society. We then consider more elaborate change scenarios that include rebound effects of the change on the baseline situation. This is known as the "consequential" approach.[2]

To understand a change, whether marginal or not, the study undertaken can also choose to include the transitions between states A and B or between scenarios. Figure 18.2 illustrates the different possibilities.

In terms of impacts, the scope of the study is also constrained by the complexity and connectivity of several impact mechanisms. In particular, taking rebound effects into account reiterates the need to integrate our assessments at various scales in order to extrapolate the impacts of a cluster potentially different from the sum of its parts (Box 18.1).

[1] The aim of the attributional approach is thus to attribute impacts to one or the other life cycle phases (and, in this way, identify the phase with the highest impact), starting with the assumption that the impacts will not significantly change the system of the product itself.

[2] The consequential approach thus takes into account the consequences of change on the system of the product itself.

We must choose, *a priori*, two or three boundaries (spatial, temporal, and of actors affected in the case of social life cycle assessment) in order to study the impacts of an action in the agri-chain using life cycle assessment. Determining the spatial boundaries means selecting processes and organizations for which data will have to be collected for evaluation. Care will have to be taken to avoid two pitfalls in this process: to adopt an area that is too large, which would make the study unfeasible, too long, and too expensive; the second is to avoid excluding processes or organizations that cause or receive significant impacts. The spatial extent is determined based on two criteria:

- the magnitude of the change, as judged by those who request for and conduct the study;
- the definition of the goal: assessing the consequences of the existence itself of the agri-chain or only of a change within the agri-chain?

Criterion 1 As far as the magnitude of change is concerned, the boundary of the study varies depending on whether the changes are considered marginal (small scale) or whether they cause impacts well beyond the agri-chain under study (large scale). In the first case, the spatial boundary includes only the agri-chain itself and possibly the usage phase of the agricultural products concerned (Box 18.4).

Box 18.4 Recent Changes in the Technique for Treatment of Leaf Fungus in Banana Plantations in Martinique and Guadeloupe

Denis Loeillet and Catherine Macombe

Until recently, treatment was carried out by aerial spraying that was centralized (one to two logistical points per island) and generalized (compulsory treatment). Its replacement by a ground-based application significantly altered not only the economic impact but also the social and environmental impacts of the treatment of banana plantations. Each farmer, or group of farmers, had to acquire equipment for land-based treatments: spray gun, ATV quad sprayer, backpack sprayer, etc. Logistics for the distribution of treatment products also saw major disruption. As far as farm labour is concerned, the work profile of farmhands changed, with concomitant potential risks of exposure to pesticides. Plot layouts also had to be changed to permit the movement of farm vehicles. Finally, the cessation of a widespread treatment strategy could lead, through the improper use of active substances, to the emergence of disease-resistant strains. This example demonstrates that although changes in an agri-chain, in terms of technical and organizational practices, are significant, they remain more or less confined within a defined boundary.

In the second case, it is appropriate to include, in the scope of the study, the processes and organizations that would be subject to major consequences as a result of changes in the agri-chain. Several models are available to help select processes and organizations for inclusion: general or partial economic equilibrium models and strategic models of organizations affected by change in the strategic arena. Potential transfers of impacts are evident when these processes and organizations are included (Box 18.5). It is indeed possible for efforts adopted in one direction in order to improve a situation to have harmful effects on other populations elsewhere (Missimer et al. 2010).

Box 18.5 Beef Production on Réunion

Denis Loeillet and Catherine Macombe

The study of the development potential of the local meat chain has clearly highlighted the fact that a good understanding of this local production and processing chain requires becoming conversant not only with international trade in livestock feed but also the international trade in breeding animals since they have to be imported to the island. Finally, excluding impacts in terms of image, employment, or other externalities, such as fighting forest fires, promoting a local value chain to counter direct meat imports requires weighing the environmental pros and cons of two types of imports: meat, on the one hand, and livestock feed and breeding animals, on the other. The highly localised scope of the issue finds itself expanded to a larger regional one.

Criterion 2 Any study undertaken to understand the consequences of an agri-chain's existence must focus on all the processes and organizations of the agri-chain. When the consequences of a change need to be understood, it suffices to only study the processes and organizations that will change between the baseline situation and the situation after the change has taken place. This approach is known as a comparative study. The temporal boundary or scope defines the duration of the impacts considered; it differs according to the nature of the impact. As to the scope of the affected actors, it is selected based on various criteria, and varies according to different methods (global boundary in case of environmental life cycle assessment, variable for social life cycle assessment).

18.4 Conclusion

An inventory of field practices for evaluating the impacts of agri-chains highlights the many tools available, which itself is indicative of a strong social demand. A plethora of labels, certifications, and other quality marks rely on many of these evaluation methods and thus reaffirm the importance of taking environmental and social impacts into account.

In spite of their many limitations and the unknowns in their application to agri-chains of the countries of the South, life cycle assessments are promising for two reasons. First, they allow for reliable comparisons, in terms of environmental and social effects, between different scenarios of the same agri-chain. Their main result is probably to change the actors' perception of environmental issues (Heiskanen 2002; Hellweg and Milà i Canal 2014) and social issues by making them consider new criteria. Different life cycle assessments use reasoning based on similar conceptual frameworks. In the simplest cases, the life cycle stages are comparable to various links of the agri-chain, to which we can add a link to reflect the behaviour of users and/or consumers, or even a link to consider the end of life of an asset, for example, recycling of packaging.

Assessment tools can be divided according to four main roles that correspond to entirely different requirements (Macombe and Loeillet 2015): strengthening one's market power, adjusting one's action plan, acquiring knowledge in order to decide, and the need for self-assurance. The way actors employ the tools for evaluating sustainable development, especially life cycle assessment tools, transforms the way they address environmental or social issues. However, the actors, in turn, influence the constitution of the tools. Their various requirements influence the design of the methods in different ways. This is particularly effective in assessing social life cycles, where the variety of methods is evident (Macombe and Loeillet, 2014), but it is no less true in assessing environmental life cycles.

References

Bassene JB, Quiedeville S, Chabrol D, Lançon F, Moustier P 2014 Organisation en réseau et durabilité systémique de deux filières alimentaires (riz biologique et petit épeautre en France). In: Actes des 8e journées de recherches en sciences sociales, Inra/SFER/Cirad, Grenoble, France, 11–12 December 2014, 24 p

FAO (2006) Livestock's long shadow, environmental issues and options. FAO, Rome. 26 p

Gabriel P, Gabriel P (2004) Diffusion du développement durable dans le monde des affaires, Un schéma conventionnel. Rev Fr Gest 152:199–213

Heiskanen E (2002) The institutional logic of life cycle thinking. J Clean Prod 10(5):427–437

Hellweg S, Milà i Canal L (2014) Emerging approaches, challenges and opportunities in life cycle assessment. Science, 344(6188):1109–1113. doi:10.1126/science.1248361. Retrieved 28 April 2016

Laufer WS (2003) Social accountability and corporate greenwashing. J Bus Ethics 43(3):253–261

Macombe C, Loeillet D (eds) (2014). Social LCA in progress. In: 4th SocSem. Pre-proceedings, 4th international seminar in social LCA. FruiTrop Thema, Montpellier, 19–21 November, 2014

Macombe C, Loeillet D (2015) In food supply chain social LCA, what to do? Keynote in session 2.4 Sustainability assessment of food supply chain: socio-economic drivers and impacts. In: Actes de la conférence: LCA for "Feeding the planet and energy for life", Stresa, 6–7 October, Milan, Expo 2015, 8 October

Macombe C, Leskinen P, Feschet P, Antikainen R (2013) Social life cycle assessment of biodiesel production at three levels: a literature review and development needs. J Clean Prod 52:205–216

Missimer M, Robèrt KH, Broman G, Sverdrup H (2010) Exploring the possibility of a systematic and generic approach to social sustainability. J Clean Prod 18(10–11):1107–1112

Müller B (2007) Food miles or poverty eradication? The moral duty to eat African strawberries at Christmas. Oxford Energy and Environment comment, octobre. Oxford Institute for Energy Studies, Oxford. 6 p. Retrieved 28 April 2016, http://www.oxfordenergy.org/wpcms/wp-content/uploads/2011/01/Oct2007-Moralstrawberries-BenitoMuller.pdf

O'Brien M, Doig A, Clift R (1996) Social and environmental life cycle assessment (SELCA). Approach and methodological development. Int J Life Cycle Assess 1(4):231–237

Chapter 19
How to Assess the Environmental Impacts of an Agri-Chain?

Cécile Bessou

The need to assess the environmental impacts of agri-chains arose gradually, in parallel with the recognition of the Green Revolution's limitations, and the emergence of concepts pertaining to the principles of precaution and sustainability. For example, this type of assessment has now become part of the design of agroecological or ecologically intensified systems. Previously considered only as an indirect recipient, the environment today finds itself once again at the heart of agriculture, with environmental assessment issues increasingly preoccupying the research community. These issues can be divided into two complementary parts. The first deals with understanding and quantifying environmental impact mechanisms. In other words, what systems and what agricultural practices result in what environmental impacts and to what extent? The second concerns the aggregation of these impacts. Sustainable development challenges are global in scope, with the economic term 'globalized' echoing the ecological term 'planetary'. Climate change is an apt illustration of the direct links between local actions, quantifiable greenhouse gas emissions at a given location, and unpredictable indirect effects at completely different places and scales. It is therefore necessary to be able to integrate local impacts at much larger scales in order to undertake planetary impact assessments. In other words, what are the interactions and the causal links between activities, agri-chains, and users, which contribute to the building up of one or more global outcomes? What can be deduced in terms of levers of action and responsibilities? Life cycle assessment (LCA) can be used to articulate, at least partially, these complementary issues by integrating impacts all along a supply chain. The challenges, however, remain numerous, particularly in the context of agri-chains, and responses demand a convergence of different methods and tools. The first section of this chapter outlines the challenges of understanding and characterizing

C. Bessou (✉)
CIRAD – UPR Performance of Tree Crop-Based Systems, F-34398 Montpellier, France
e-mail: cecile.bessou@cirad.fr

E. Biénabe et al. (eds.), *Sustainable Development and Tropical Agri-chains*,
DOI 10.1007/978-94-024-1016-7_19

environmental impacts, while the second proposes an overview of available methods and tools. The third section addresses issues of integrating environmental assessment of agri-chains at territorial or larger scales.

19.1 Characterizing the Environmental Impacts of an Agricultural Activity

An environmental impact can be defined as the result of a cause-and-effect link between the utilisation of a resource or an emission and a change of state and consequential effect on the environment. The environment is considered here in its most inclusive sense and encompasses the habitat and its entire gamut of inhabiting species. The impact mechanism involves a source of emission ('pressure'), the transfer and outcome of this emission, which can cause a change in the 'state' of the environment and the exposure of a target that undergoes a final 'impact' (Fig. 19.1). An impact can be assessed at several stages of the impact pathway, with the result that the change of state of the environment may be considered an intermediate impact. The final impact, in the simplest case, reflects an effect felt directly by a target species close to the impact source in space and time. In many other cases, however, the impact may take place at a different time or a different location and involve several species or even the global environment, thus greatly increasing the parameters and dimensions needed to characterize the final impacts. In fact, a source may cause multiple intermediate and final impacts, as illustrated by the nitrogen 'cascade' (Fig. 19.2). A nitrogen source, such as fertilizer used as input in an agricultural field, can lead to various flows towards the environment, each of which can cause a number of impacts such as water eutrophication, climate change, or the destruction of stratospheric ozone. On the other hand, different emissions can contribute to the same impact. Moreover, some impacts, such as effects related to the depletion of resources, will only be felt by future generations, but nevertheless, we must anticipate them now. An impact is termed potential when its characterization does not include all the variables that are expected to occur in the full impact mechanism. This is ultimately the case for all environmental assessment efforts due to the repeated reliance on modelling which, in the broadest sense, represents a simplification of reality (Fig. 19.1). Although life cycle assessments theoretically integrate the full impact mechanism (Fig. 19.1), the results obtained remain, nonetheless, potential impacts because of the simplification of the impact's characterization by a linear model (i.e., no non-linear consideration of the environment's sensitivity).

Thus, the complexity and multiplicity of potential impacts calls for a broad range of analyses, involving various disciplines, primarily those linked intrinsically to emission or transfer phenomena, such as biology, pedology, hydrology, eco-physiology, and agronomy. But other disciplines, more transversal, such as modelling or sociology, are also involved, especially when it is impossible to

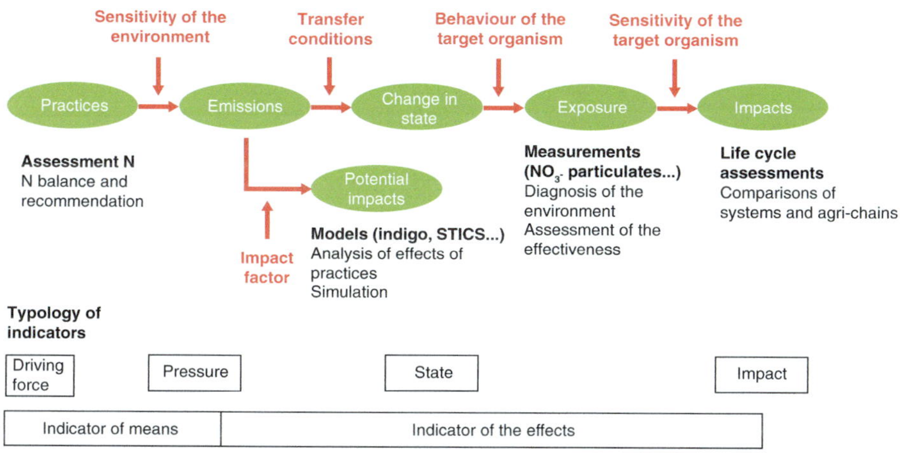

Fig. 19.1 The position of various types of indicators and methods in the causal pathway (According to Bockstaller et al. 2009)
See also Smeets and Weterings (1999) for further details on the assessment framework of the DPSIR impact (Driving Forces-Pressure-State-Impact-Response)

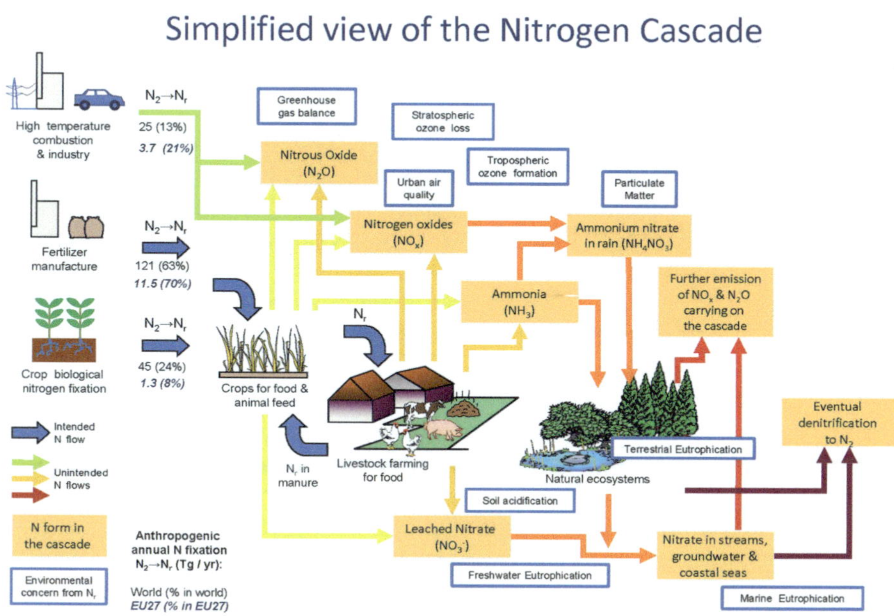

Fig. 19.2 Illustration of the nitrogen 'cascade' (Galloway et al. 2003) in an agricultural landscape (According to Sutton et al. 2011)

discern the mechanisms or when the cause-and-effect links are discontinuous. These shortcomings and lacunae must be overcome, uncertainties modelled and analyzed, and criteria of subjective choices added. We saw in Chap. 18 that an evaluation involves a value judgment when an objective measurement includes too many unknowns. In the course of an examination by INRA of methods to assess the environmental impacts of agricultural practices, the authors made a distinction between the 'estimation' and the 'evaluation' of an environmental impact. The first term refers to the scientific and technical characterization of the impact, while the latter is based on its socio-economic and cultural acceptability (Capillon et al. 2005). Indeed, estimation and evaluation are generally both essential to understand the environmental impacts of agriculture, especially in a multi-criteria approach. Various methods and tools have been proposed that combine estimation and evaluation, using quantitative and qualitative approaches, to varying degrees. Various reviews were published on this subject (Rosnoblet et al. 2006; Singh et al. 2012; Bockstaller et al. 2009). Without attempting to present a comprehensive picture, we recall, in the next section, some relevant and often complementary methods and tools for assessing the environmental impacts of an agri-chain.

19.2 Methods and Tools for Environmental Assessment

Among the many ways to assess the environmental impacts of agricultural systems, we make a distinction here between the 'method', which refers to an approach or more or less operationalized evaluation concepts, and the 'tool', which is an *a priori* established and validated way of applying a method. For example, the assessment method using the agri-environmental indicators (Girardin et al. 1999) of the Indigo® tool is a software application with several validated agri-environmental indicators, especially for field crops in France (Bockstaller and Girardin 2010). Consequently, the indicators, regardless of what they are, can be seen as a method, depending on whether one is interested in the concepts that led to their creation, or as a tool, if they can be applied directly. We retain here the primary meaning of the indicator: an item of more or less integrated information indicative of one or more attributes – that cannot themselves be measured directly – of the object under study. From this general definition, different types of indicators can be defined, categorized, and linked. In particular, an indicator can intervene at every level of the causal pathway (indicators of driving force, pressure, state, impact, or response). It can range from the relatively simple to the very complex[1] and can have a more or less diagnostic capacity, according to a gradient: descriptive, normative, or prescriptive. A field measurement (for example, 30 mg nitrate/litre of water) may be a simple indicator, which becomes normative if this measurement is compared to a reference value (for example, quality threshold <50 mg nitrate/litre of water), or even prescriptive if this value corresponds to a score indicating a more or less favourable practice. The diagnostic power of this indicator depends on the ability of

[1] The complexity of an indicator can refer to scientific as well as practical criteria, for example, implementation cost, utility for the end user, etc. (Girardin et al. 1999; Moebius-Clune et al. 2011).

linking in advance the measured value to an impact pathway and on the robustness of the modelling used. Field measurements lead to an understanding of the mechanisms of emission, transfer, and impact, and improve the robustness of the models, which is an essential requirement for developing and improving environmental assessment methods and tools (Box 19.1).

Box 19.1 The Importance of Field Measurement Devices in the Process of Environmental Assessment

Frédéric Gay

The creation of tools for the agri-environmental assessment of agri-chains requires field equipment to quantify and understand fluxes from production systems to the environment and their associated impacts. The intra- and inter-annual climate variability, the time steps of certain practices (crop rotations), and crop cycle durations (perennial crops) require these fluxes to be measured over the long term. This issue led to the establishment, in the mid-nineteenth century, of the first long-term agricultural trials in the UK and, more recently, of long-term observation and experimentation sites in France, as well as of global networks, for example, the Fluxnet network. However, tropical ecosystems are still under-represented in these networks in comparison with those in temperate zones, despite their significant contribution to global fluxes. In fact, most of the emission factors used in life cycle assessments for determining the carbon footprints of agricultural products are based on data acquired in temperate zones.

There is thus a strong need to establish long-term observation and experimentation sites for tropical agroecosystems. CIRAD is currently involved in several observation sites (one site at Réunion set up by its Recycling and Risk internal research unit; three sites, in Brazil, Thailand, and Costa Rica, set up by the Eco&Sols joint research unit). These sites are designed to monitor the fluxes of water, carbon, and minerals in the soil-plant-atmosphere system on a continuous basis. This form of monitoring relies on extensive and sophisticated field instrumentation, for example, flux towers (Fig. 19.3). The implementation of such research devices raised several issues. The first is to ensure the continued use of the facility over sustained periods, as its principal value lies in obtaining long-term data series. Second, it is essential to have analytical platforms to manage data quality and to make optimal use of it to help increase knowledge on the functioning of tropical agroecosystems. Finally, while these instrumented sites enable a detailed study of the functioning of agroecosystems in which they are installed, they are unable to take the diversity of existing cropping systems into consideration. To this end, CIRAD's research units (internal research units: Agroecology and Sustainable Intensification of Annual

(continued)

Box 19.1 (continued)

Fig. 19.3 Flux tower on a eucalyptus plantation, Eucflux site of the Eco&Sol joint research unit in Brazil

Crops, AIDA, and Performance of perennial cropping systems, Research unit 34) monitor existing networks of farm plots and thus create databases on agri-environmental performances of major tropical agricultural products. It is necessary to link all these field monitoring facilities to sustainability assessment mechanisms of all the agri-chains, as it has been done within most of CIRAD's training and research platforms in partnership.

Due to the complexity of most environmental impact mechanisms, several environmental assessment methods rely on indicators to help explain a reality that is complex and difficult to measure (Boulanger 2004). A system's environmental assessment is thus an organized composite of indicators. The number and complexity of indicators depend on the number and complexity of the environmental impacts being assessed, and more generally, on the selected evaluation approach. By extension, an 'index' is generally obtained by aggregating several indicators, which are sometimes normalized or weighted. In the case of multi-criteria assessments, an index is generally the most aggregated result.

A recent exhaustive review of sustainability assessment in agriculture (Lairez et al. 2015) provides an overview of available methods and tools. An ongoing survey for the Indic database reported the existence of 109 methods and 2639 indicators to assess this sustainability.[2] This plethora can give rise to some confusion (Capillon et al. 2005) as well as to the risk of choosing indicators that are not suited to an appropriate evaluation approach, which could then distort the assessment (Niemeijer and de Groo 2008). Indeed, the utility of a multi-criteria assessment depends on the analysis and interpretation of these indicators. To avoid these pitfalls, the authors recall the guidelines for a proper evaluation approach (Fig. 19.4). It is important to note that the assessment itself takes place only after a comprehensive analysis of the needs and objectives. Furthermore, an assessment does not always require the systematic creation of a new methodology and new indicators, but can use or adapt existing methods and tools. In the case of creating or adapting methods, intermediate steps are required after the preliminary steps, and prior to initiating the assessment. These key steps, especially for the creation of indicators, are: identification of assessment criteria for each specific objective and choice of assumptions for the variables involved; development of indicators (calculation method and definition of thresholds); and validation of indicators and the method (Girardin et al. 1999; Bockstaller and Girardin 2003). An example of implementing indicators using the Indigo® method is presented in Box 19.2.

Box 19.2 Assessing the Environmental Impact of Pesticides in the Caribbean and Modifying Practices

Fabrice Lebellec

Although herbicides are effective against major crop weeds, their side effects are often harmful to human and animal health, and more generally to biodiversity. According to FAO, herbicides make up almost 43 % of all pesticides used globally (2010 data from 58 countries). The predominance of herbicides is even more pronounced in the Caribbean. Of the 200 tonnes of pesticides

(continued)

[2] Pers. com., Thomas-Delille and Feschet 2015.

Box 19.2 (continued)

sold in Réunion and Martinique, more than 75 % are herbicides (*Banque nationale de vente des distributeurs*, 2012–2014 data). A discussion on the use of these herbicides has thus been initiated in order to reduce the contamination that already exists of surface water and groundwater. Of the 21 active substances found in the water in Réunion in 2010, 17 can be linked to herbicides. Such contaminations are not only the result of poor agricultural practices (substances, doses, application, etc.), but also due to local soil, climatic, and topographic conditions, which vary greatly from region to region on the island (for example, rainfall that varies from 600 to 7000 mm/year from east to west, slopes at 0–45 %). Thus, the substitution of one herbicide by another or appropriate application conditions can help reduce the risk of herbicide contamination of the environment.

CIRAD, together with actor groups, has developed two decision-making tools within the framework of the Écophyto National Plan to support and stimulate the process of designing cropping systems with reduced use of pesticides. A design method was formalized for tropical farming systems (Bruchon et al. 2015) to help producers reduce the dependence of their cropping systems on pesticides. This method is presented in the form of a guide which offers producers technical solutions to help substitute pesticides while also including a multi-criteria assessment of the positive and negative effects of these techniques. The second tool (Phyto'aide) is an operational decision-making tool to assist the producer in selecting the right pesticide in case one has to be used. Phyto'aide is based on the results of the I-Phy assessment indicator (Le Bellec et al. 2015), an indicator of the Indigo® assessment method developed by INRA. It assesses the risks of pesticide contamination of the environment and offers customized advice to producers to reduce these risks, depending on the pesticide used and conditions of use.

This tool can now be freely accessed at: http://www.margouilla.net/phytoaide.

Keys to identify the boundaries (or scope) and end-objectives of an assessment were discussed in Chap. 18.

We focus here on the third question in step B: 'How?' (Fig. 19.4 – What are the existing methods or tools to assess the environmental impacts of an agri-chain? What else needs to be developed?)

The main features of the agricultural assessment methods currently available are summarized according to the assessment objectives, and the scales and dimensions of the assessed impacts (Table 19.1).

Only three methods appear to be relevant for the assessment of agri-chains or the life cycle: two methods dedicated to assessing the sustainability of poultry chains (Ovali and Avibio), and the life cycle assessment, designed for the environmental or

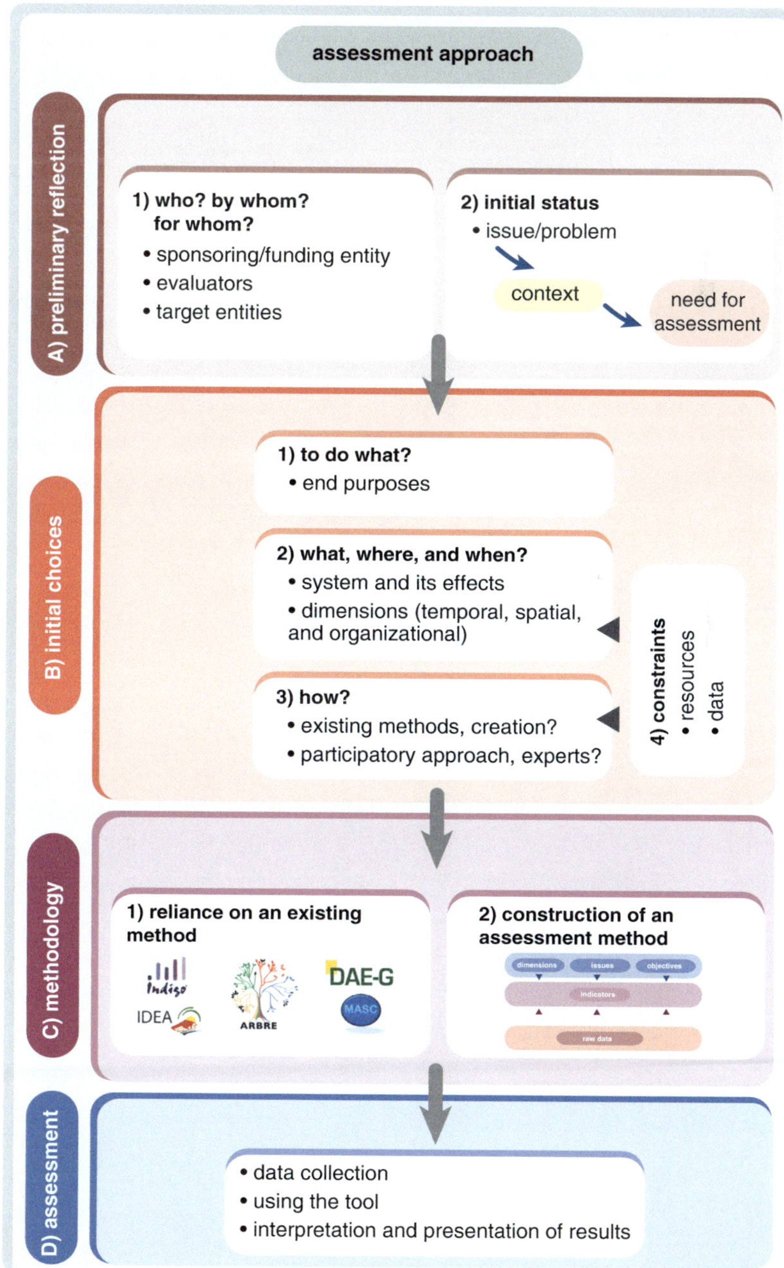

Fig. 19.4 Evaluation approach (Lairez et al. 2015)

Table 19.1 Choosing an assessment method from those that are available

Purposes of assessment	Assessment level: Agricultural system plot	Livestock operations	Farm	Agri-chain	Territory	Life cycle	Several levels
Raising awareness by conducting an academic evaluation		2. Engele^{Porc}	1. IDEA$^{\infty}$ Arbre$^{\infty}$ RAD$^{Rum.}$ Idaqua$^{Pisc.}$	1. Ovali$^{Vol.}$ Avibio$^{Vol.}$			
Providing knowledge by comparing systems	1. MASC^{GC} DEXiPM^{PV}	1. DIAMOND $_{Vol.,Cun.}$	1. SAFE$^{\infty}$ ADAMA$^{\infty}$ Idaqua$^{Pisc.}$ Apoia$^{\infty}$ IndicIADes$^{\infty}$ DAESE$^{\infty}$			3. ACV-E$^{\infty}$	1. MESMIS$^{\infty}$ EVAD^{Pisc}
	2 et 5. Grignon Model^{GC}	2. GEEP^{Porc}	2. MELODIE^{Lait,Porc} DIALECTE$^{\infty}$			4. ACVS∞	
Reporting (regulatory, achievement of objectives)		3. Welfare Quality		1. Ovali$^{Vol.}$, Avibio$^{Vol.}$	2. ClimAgri$^{\infty}$, Territ'eau$^{\infty}$		
		5. GTTT^{Porc} GTE^{Porc}					
Identifying areas for improvement, recommendations	1. MASC^{GC} DEXiPM^{PV}	1. DIAMOND $_{Vol.,Cun.}$	1. ADAMA$^{\infty}$ IDEA$^{\infty}$ Arbre$^{\infty}$ RAD$^{Rum.}$ MOTIFS^{Lait} APOIA$^{\infty}$ IndicIADes$^{\infty}$ Idaqua$^{Pisc.}$	1. Ovali$^{Vol.}$, Avibio$^{Vol.}$	2. ClimAgri$^{\infty}$, EBIOTEP$^{Rum.}$	2. ACV-E$^{\infty}$	1. SSP$^{\infty}$ MESMIS$^{\infty}$ SAFA$^{\infty}$
	INDIGO^{PV} DAEG $^{\infty}$ 2 et 5 Grignon Model^{GC}	2. Cap'2R $_{Lait,BV,Ov.}$, GEEP^{Porc}	2. Dia'Terre$^{\infty}$ DIALECTE$^{\infty}$ DEXEL^{Lait} KUL$^{\infty}$ PISC'N'TOOL$^{Pisc.}$ Composim$^{Porc,BV,Lait,Vol.}$ DECIBEL$^{Porc\ Rum.,\ Vol.}$				
		3. Welfare Quality				4. ACVS$^{\infty}$	
		5. GTTT^{Porc}, GTE^{Porc}					
Designing systems (*ex post* iterative approaches and *ex ante* evaluation)	1. MASC^{GC} DEXiPM^{PV}	2. Engele^{Porc}	2. MOLDAVI$^{Vol.}$	1. Ovali$^{Vol.}$		2. ACV-E$^{\infty}$	1. MESMIS, SSP$^{\infty}$
						4. ACVS$^{\infty}$	
Certification schemes			2. DAEG$^{\infty}$ KUL$^{\infty}$				
			6. Global-Gap ProTerra-Standard				

Lairez et al. (2015)

The sustainable development dimensions assessed: (1) comprehensive approach to sustainable development, (2) environment, (3) animal well-being, (4) social, (5) technical and economic, (6) specifications

Abbreviations: ∞ all production, *Lait* bovine milk, *BV* bovine meat, *Porc* pork, *Rum.* ruminant, *Ov.* sheep, *Vol.* poultry, *Cun.* rabbit, *Pisc.* fish farming, *GC* field crops, *PV* field crops, fruits, vegetables, vines

Details of acronyms and methods can be found in the book by Lairez et al. (2015). See the Plage Internet platform for a guided overview of the methods: http://www.plage-evaluation.fr/webplage/ (in French)

social assessment of supply chains. These methods are fewer in number than those pertaining to operational assessments of crop cultivation or animal husbandry in their totality. The life cycle approach is the most comprehensive in terms of considering flows that constitute the product through its life cycle (Chap. 18). In comparison, agri-chain methods (Table 19.1) mainly incorporate variables related to the organizational footprint of a supply chain in a given territory, i.e., including the key stages in that territory, which are not necessarily exhaustive in terms of including locations, periods, and flows pertaining to the product's life cycle. Environmental assessment thus questions here the very definition of the agri-chain or supply chain, in terms of the integration of the product or territory. The boundaries (or scope) of the agri-chain being assessed thus vary depending on the objectives of the assessment and the methodology (Chap. 18).

Upstream and downstream processes in a life cycle approach can play a significant role in certain agri-chains. In the case of palm biodiesel, for example, depending on the scope of the study, fertilizers may represent 10–30 % of the total greenhouse gas emissions, including from production, transportation, and on-field emissions (Pleanjai et al. 2009; Choo et al. 2011). Conversely, the territorial integration of an agri-chain helps apprehend the significant indirect impacts, especially in the context of a multi-dimensional sustainability assessment. In the case of the Ovali method, poultry chains are examined at a regional scale in France, on the basis of typical cases and sustainable development objectives defined by various actors of the agri-chain. In terms of environmental impacts, for example, Ovali takes into account the integration of production tools within the landscape, which represents a very local impact. This original approach helps, as in the case of the life cycle assessment, prevent some pollution transfers (Lairez et al. 2015).

The assessment of agri-chains, due to an intrinsic need to integrate the scales of time, location, and actors must, *de facto*, provide information about possible shifting of problems and needs for trade-offs. However, Ovali or Avibio do not have the same generic scope as a life cycle assessment, and require a complete adaptation on a case-by-case basis to assess other agri-chains in other territories. The generic framework of the life cycle assessment, on the other hand, is standardized (ISO 14040 series) and allows for an assessment and possible comparisons of agri-chains on a basis that is harmonized, validated, and revised regularly, based on new knowledge and publications. Figure 19.5 shows a recent mapping of methods and tools from the 'life cycle assessment family' (Basset-Mens et al. 2015). More specific methods (choice of emission and characterization models) based on original ISO standards have been proposed for the implementation of the methodological framework, in particular for the agri-chains, for example, AgriBalyse® or EnviFood®. Other partial life cycle assessment methods are focused on an environmental impact, such as climate change. Hybrid methods allow for the combination of life cycle assessment with more territorial approaches, such as input-output methods that produce flows at more integrated organizational scales based on statistics of incoming and outgoing flows. In fact, combinations of methods are increasingly being used. Although the life cycle assessment's generic framework is interesting for the above-mentioned aspects of comprehensive transversal integration and scientific harmonization, it remains restrictive in terms of adaptation for a

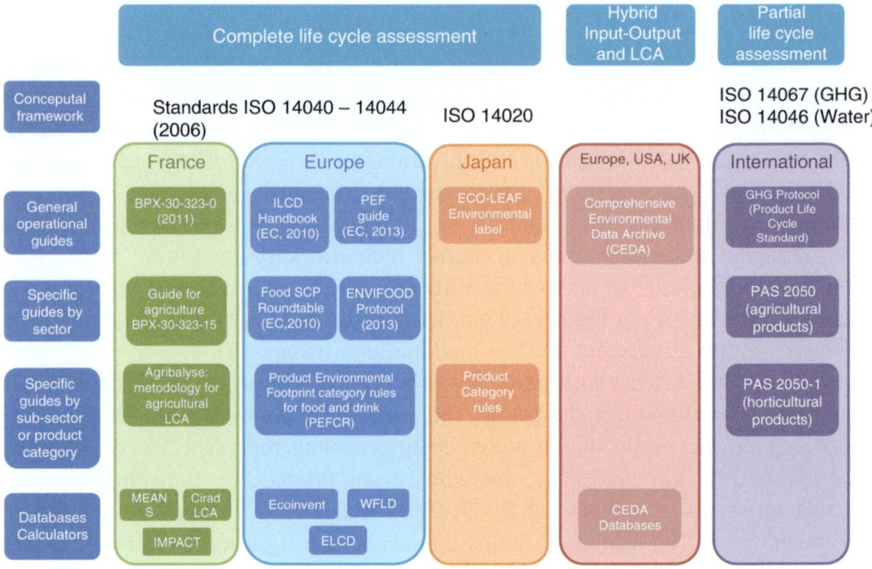

Fig. 19.5 Mapping of life cycle assessment methods (Basset-Mens et al. 2015)

better assessment of local impacts and the integration of socio-economic impacts. Thus, studies that combine life cycle assessment with agri-environmental indicators, for example, can result in an enhanced consideration of local production conditions. Finally, in addition to exploring perspectives of comparing and improving the environmental performance of agricultural systems with the help of an environmental life cycle assessment, it is equally important to maintain, or even improve, their techno-economic performances (yield, product quality and price). Box 19.3 illustrates the necessary combination of environmental and techno-economic assessments in the case of irrigated rice in Thailand (Thanawong et al. 2014). However, many challenges need to be overcome in order to make various combinations functional in terms of impacts, for example, for combining environmental and social life cycle assessments, or in terms of a larger scale, such as of a territory or country.

Box 19.3 Combining Environmental and Techno-Economic Assessments to Quantify the Efficiency of Agricultural Systems: The Case of Irrigated Rice in Thailand

Sylvain Perret and Kwansirinapa Thanawong

Northeast Thailand is the main production area of the famous fragrant rice (Jasmine or Hom Mali rice) which is both exported as well as consumed locally. This rice is mainly grown in rainfed paddy fields without water

(continued)

Box 19.3 (continued)

supply control, during the rainy season. An increase in area through controlled irrigation could make its production more secure during the dry season, and also boost overall production. While such projects exist, their implementation is hampered by water scarcity and a lack of adequate intensification skills amongst producers. They are, besides, multi-active farmers and are rarely available in the dry season to grow rice. A study, based on a diversity of 43 systems in the Lam Sieo Yai basin, compared rice cultivation systems in environmental and techno-economic terms, based on indicators of life cycle assessment and of water and energy consumption. The study was conducted in 2010 on three types of systems: wet-season uncontrolled rainfed (Rw), wet-season controlled irrigated (Iw), and dry-season controlled irrigated (Id) systems.

The results highlight the wide heterogeneity of the techno-economic and environmental performances, despite the fact that agricultural practices were relatively homogeneous and used little agrochemical inputs. Such significant differences are a result of very different yields, ranging from 2625 kg/ha (Iw) to 2375 kg/ha (Rw) and 2188 kg/ha (Id), essentially dependent on ways of managing irrigation water. The dry-season systems (Id) depend almost exclusively on irrigation water, resulting in additional costs and impacts associated with pumping water and the labour needed for water management. The other two systems (Iw, Rw) are mainly supplied by rainwater. The productivity of most production factors is also higher in wet-season systems.

Farmers using dry-season irrigated systems (Id) must produce twice as much rice (0.41 kg) to receive a net income of 1 baht (€ 0.025), compared to 0.23 kg and 0.25 kg required for wet-season systems (Iw and Rw respectively). Emissions of pollutants are quite similar across these systems with the exception of methane, which was much lower in wet-season rainfed (Rw) systems due to specific management of water and organic residues, more concentrated during the pre-planting cropping calendar. These systems produce the lowest environmental impacts, per hectare and per kg of rice produced. Dry-season irrigated systems (Id) consistently emit more nitrates, phosphates, and pesticides. Global warming potential (GWP_{100}) is the highest in Id systems (5.55 kg CO_2-eq/kg of rice) compared to those of the wet-season (Iw) (4.87) and wet-season rainfed (Rw) (2.97). Finally, these wet-season rainfed systems (Rw) are proving to be the most efficient ones across all the impact categories. Gross income per kg CO_2-eq emitted works out to € 0.1 (Rw), € 0.0625 (Iw), and € 0.055 (Id). The results of this research highlight the weak techno-economic and environmental performances of innovative dry-season irrigation systems, which are promoted by the authorities to strengthen the fragrant rice chain, and which appear to be ill-adapted to the needs of producers (labour problems, including of availability, in the dry season), to regional constraints (scarcity of and competition for water), and to regional and global environmental issues.

19.3 Challenges of Environmental Assessment at Integrated Scales

The environmental assessment of an agri-chain involves questions of scale in terms of its spatio-temporal representativeness and the consideration of endogenous variables at higher organizational levels.

As shown above, an agri-chain can either be seen as being anchored to a territory (Ovali) or not (life cycle assessment). A life cycle assessment allows for the integration of impacts of activities occurring at different locations; in the case of an agri-chain, the considered plot is virtually isolated from its surroundings. It is necessary to have this spatio-temporal disconnection when integrating various processes occurring at different times and locations all along an agri-chain. In order to describe an agri-chain, the spatio-temporal representation of the life cycle assessment must always be considered in conjunction with the assessment objectives and the data collected. Comparing a local agri-chain with an import-oriented agri-chain or comparing two local agri-chains with different agricultural practices represents fundamentally different challenges. In order to compare agri-chains that are 100 % local, specific developments are required for the life cycle assessment and combinations with other methods may be needed. In reality, the connectivity of plots, activities, and agri-chains may influence the outcome of the impact. The connectivity of rivers or aquifers across a watershed has been the subject of specific studies in the environmental assessment of agricultural production (Gascuel-Odoux et al. 2009). However, these studies have not yet been crossed-referenced with agri-chain approaches. They could, for example, be correlated with life cycle assessments using regionalization factors by taking into account the variability in the availability of water in the different watersheds around the world (for example, 'water scarcity index', Pfister et al. 2009). This would allow for the weighting of flows towards the environment by characterizing them in connected spatial dynamics. The variable time scales between different and more or less long-term impacts also pose problems of integration that must be addressed (Capillon et al. 2005), again depending on the objectives of the study and the resulting choice of the assessment method.

The other major issue of an assessment at an integrated organizational level is the consideration of endogenous variables at each higher level of organization (Capillon et al. 2005). In other words, superposition, linking, or extrapolation of indicators cannot be done without a preliminary assessment of possible effects due to synergies, antagonisms, thresholds etc., whether at the scale of a plot, farm, or territory, or between transfer processes, impact mechanisms, or sustainability dimensions. Aggregation is more common at the scale of the agri-chain and there exist different aggregation criteria and methods (Lairez et al. 2015). Thus, the co-construction of the Gamede model (Box 19.4) with actors of a livestock chain in Réunion has helped apprehend the challenges and expectations of different actors, and has resulted in a combined environmental and techno-economic assessment at the scale of farms that are representative of the agri-chain.

Box 19.4 Simulation of Changes in Practices at the Farm Level and Consequences on the Impact of a Product at the Agri-Chain Level: The Case of Dairy Farms in Réunion

Jonathan Vayssières, Mathieu Vigne, François Guerrin, and Philippe Lecomte

A participatory modelling approach was undertaken in Réunion from 2003 to 2007, resulting in the co-construction with a group of six dairy farmers of the Gamede computer model that simulates the management of a dairy farm integrating pastures and sugarcane plots (Vayssières et al. 2009). These six farms are located in different agroecological zones of the island and are representative of the diversity of agricultural practices observed there (Vayssières et al. 2011b). Their greenhouse gas emission levels show impacts that vary from farm to farm (Fig. 19.6).

The Gamede model was used with the cattle farmers to explore different ways of reducing their environmental impacts: better utilization of the manure produced on the farm to replace mineral fertilizers; better utilization of forage available on the farm to substitute feed concentrates; and improved reproductive performance of milk cows. This work has demonstrated that these three ways potentially represent a reduction in greenhouse gas emissions on Réunion of 0–35 % depending on the dairy production system considered, which corresponds to a reduction in the contribution of dairy farming to climate change by an average of 12.3 % for all farms on the island.

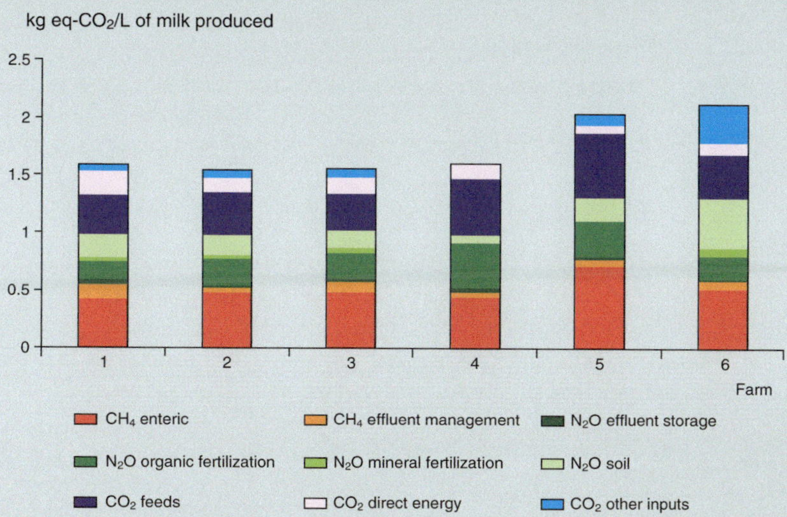

Fig. 19.6 Greenhouse gas ballances of six dairy farms in Réunion that are representative of the diversity of production systems on the island

(continued)

Box 19.4 (continued)

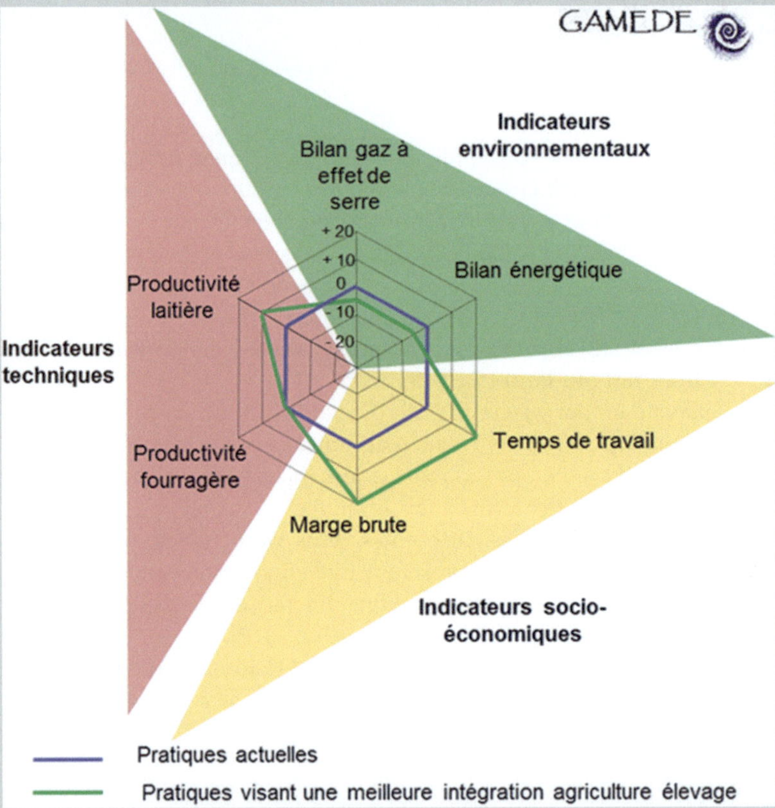

Fig. 19.7 Effect of improved crop-livestock integration on the environmental, technical, and socio-economic sustainability of a dairy farm in Plain of Palmistes, Réunion (% change with reference to practices observed in 2004)

Cattle farmers expressed a wish to complement the environmental assessment with a technical and socio-economic assessment. Simulations have shown that improved crop-livestock integration would reduce the environmental impact, while increasing the farms' gross economic margins (Fig. 19.7). The increase in labour time remains the main barrier to adopting these more environmentally friendly practices.

The method of calculating the environmental indicators selected (greenhouse gas and energy ballances) is based on a life cycle assessment (Vayssières et al. 2011a). These two indicators take into account indirect emissions and energy consumptions, which occur upstream of the farm, due

(continued)

Box 19.4 (continued)

to the production of various inputs (Fig. 19.7). Thus, the Gamede model simulates a fine modification of agricultural and livestock practices at the operational level and its consequences on the environmental impact of a product at the level of the supply chain, which is milk in this case. The Gamede model targets only a few environmental impacts and does not take other indirect environmental impacts into account (especially those pertaining to packaging and marketing of milk, and waste recycling). This is a promising first step towards combining a dynamic simulation model with environmental life cycle assessment.

The aggregations of impacts of farms or agri-chains at larger scales in order to assess overall impacts are less frequent. Life cycle assessment essentially allows for the assessment of global impacts. Linear characterization, linked more or less virtually to regional conditions through regionalized characterization factors, calculates the impacts that are added at the global scale, all things being equal in the case of ‘attributional’ life cycle assessment (Chap. 18). In this integration, interference with other systems or other agri-chains at higher organizational levels is not taken into account. The analysis of the ‘consequential’ life cycle, on the other hand, allows for the accounting of rebound effects and thus includes, at least partially, this interference with other systems or other agri-chains. Hybridizations with input-output approaches can also provide answers at higher organizational scales. However, it is not always possible to decipher and characterize endogenous variables. In parallel, the more top-down approaches, for example, those pertaining to ecological footprints or planet boundaries, introduce a different perspective, justifying a comprehensive approach to global sustainability issues. The concept of planet boundaries (Rockström et al. 2009) enhances our understanding of the state of the earth in terms of global ecosystem functions (state of resources, climate regulation, etc.) and the extent of human impact in order to avoid exceeding these tipping points and precipitating potentially irreversible impacts. The scientific community working on life cycle assessments is currently exploring the use of this comprehensive approach as a benchmark.

Given its multidimensional nature, an agri-chain directly encourages researchers to address the dual methodological challenge of assessing local impacts as accurately and precisely as possible, while analyzing these impacts at the global level in terms of connection, interaction, and relative contribution. A line of reasoning based on non-isolated systems is an unavoidable prerequisite towards a possible sustainable global development.

References

Basset-Mens C, Perret S, Biard Y, Bessou C (2015) Methodological support for analysis and development of inclusive and sustainable value chains – environmental dimension. ISS-FANSSA-BX11. Project for the European Commission – Devco, 43 p

Bockstaller C, Girardin P (2003) How to validate environmental indicators. Agric Syst 76 (2):639–653

Bockstaller C, Girardin P (2010) Mode de calcul des indicateurs agri-environnementaux de la méthode Indigo® V2. ENSAIA/Inra.

Bockstaller C, Guichard L, Makowski D, Aveline A, Girardin P, Plantureux S (2009) Agri-environmental indicators to assess cropping and farming systems. A review. Agron Sustain Dev 28(1):130–149

Boulanger PM (2004) Les indicateurs de développement durable : un défi scientifique, un enjeu démocratique. *Séminaire* « Développement durable et économie de l'environnement », 27 April, Paris, Iddri, Les séminaires de l'Iddri, 12, 24 p

Bruchon L, Le Bellec F, Vannière H, Ehret P, Vincenot D, De Bon H, Marion D, Deguine JP (2015) Guide tropical, Guide pratique de conception de systèmes de culture tropicaux économes en produits phytosanitaires. In: Le Bellec F (ed) Cirad, Paris, 210 p

Capillon A, Gabrielle B, Girardin P, Guichard L, Guillaume B, Hubert A, Leiser H, Soulas G, van der Werf H (2005) *Méthodes d'évaluation des impacts environnementaux des pratiques agricoles*. Département environnement agronomie, Inra, Paris, 51 p

Choo YM, Muhamad H, Hashim Z, Subramaniam V, Puah CW, Tan YA (2011) Determination of GHG contributions by subsystems in the oil palm supply chain using the LCA approach. Int J Life Cycle Assess 16(7):669–681. doi:10.1007/s11367-011-0303-9. Retrieved 19 June 2016

Galloway JN, Aber JD, Erisman JW, Seitzinger SP, Howarth RW, Cowling EB, Cosby BJ (2003) The nitrogen cascade. Bioscience 53(4):341–356. Retrieved 19 June 2016, http://bioscience.oxfordjournals.org/content/53/4/341.abstract

Gascuel-Odoux C, Massa F, Durand P, Merot P, Troccaz O, Baudry J, Thenail C (2009) Framework and tools for agricultural landscape assessment relating to water quality protection. Environ Manag 43(5):921–935

Girardin P, Bockstaller C, van der Werf H (1999) Indicators: tools to evaluate the environmental impacts of farming systems. J Sustain Agric 13(4):5–21

Lairez J, Feschet P, Aubin J, Bockstaller C, Bouvarel I (eds) (2015) Agriculture et développement durable. Guide pour l'évaluation multicritère. Éducagri éditions/Éditions Quæ, Dijon/Versailles, 228 p

Le Bellec F, Vélu A, Fournier P, Le Squin S, Michels T, Tendero A, Bockstaller C (2015) Helping farmers to reduce herbicide environmental impacts. Ecol Indic 54:207–216

Moebius-Clune BN, Idowu OJ, Schindelbeck RR, van Es HM, Wolfe DW, Abawi GS, Gugino BK (2011) Developing standard protocols for soil quality monitoring and assessment. In: Bationo A, Waswa B, Okeyo JM, Maina F, Kihara JM (eds) Innovations as key to the green revolution in Africa. Springer, Dordrecht, pp 833–842

Niemeijer D, de Groo RS (2008) A conceptual framework for selecting environmental indicator sets. Ecol Indic 8(1):14–25

Pfister S, Koehler A, Hellweg S (2009) Assessing the environmental impacts of freshwater consumption in LCA. Environ Sci Technol 43(11):4098–4104. doi:10.1021/es802423e. Retrieved 19 June 2016

Pleanjai S, Gheewala SH, Garivait S (2009) Greenhouse gas emissions from the production and use of palm methyl ester in Thailand. Int J Glob Warming 1(4):418–431

Rockström J, Steffen W, Noone K, Persson Å, Chapin FS III, Lambin E, Lenton TM, Scheffer M, Folke C, Schellnhuber H, Nykvist B, De Wit CA, Hughes T, van der Leeuw S, Rodhe H, Sörlin S, Snyder PK, Costanza R, Svedin U, Falkenmark M, Karlberg L, Corell RW, Fabry VJ,

Hansen J, Walker B, Liverman D, Richardson K, Crutzen P, Foley J (2009) Planetary boundaries: exploring the safe operating space for humanity. Ecol Soc 14(2):32

Rosnoblet J, Girardin P, Weinzaepflen E, Bockstaller C (2006) Analysis of 15 years of agriculture sustainability evaluation methods. In: IX ESA congress. Bibliotheca Fragmenta Agronomica 2006. Warsaw, Poland, pp 707–708

Singh RK, Murty HR, Gupta SK, Dikshit AK (2012) An overview of sustainability assessment methodologies. Ecol Indic 15(1):281–299. doi:10.1016/j.ecolind.2011.01.007. Retrieved 19 June 2016

Smeets E, Weterings R (1999) Environmental indicators: typology and overview. EEA, Copenhagen, 19 p

Sutton MA, Howard CM, Jan W, Bealey WJ, Billen G, Bleeker A, Bouwman AF, Grennfelt P, van Grinsven H, Grizzetti B (2011) The challenge to integrate nitrogen science and policies: the European Nitrogen Assessement approach. In: Sutton MA, Howard CM, Erisman JW, Billen G, Bleeker A, Grennfelt P, van Grinsven H, Grizzetti B (eds) The European nitrogen assessment. Sources, effects and policy perspectives. Cambridge University Press, Cambridge, pp 82–96

Thanawong K, Perret SR, Basset-Mens C (2014) Eco-efficiency of paddy rice production in Northeastern Thailand: a comparison of rainfed and irrigated cropping systems. J Clean Prod 73:204–217

Thomas-Delille E, Feschet P (2015) INDIC: méthodes d'évaluation et indicateurs du développement durable, Working paper Mars-Août 2015 du GIS Grande Culture GCHP2E et INRA UMR LAE, Nancy-Colmar, France, 4p

Vayssières J, Guerrin F, Paillat JM, Lecomte P (2009) GAMEDE: a global activity model for evaluating the sustainability of dairy enterprises. Part I – whole-farm dynamic model. Agric Syst 101(3):128–138

Vayssières J, Thévenot A, Vigne M, Cano M, Broc A, Bellino R, Diacono E, De Laburthe B, Bochu JL, Tillard E, Lecomte P (2011a) Évaluation des inefficiences zootechnique et environnementale pour intensifier écologiquement les systèmes d'élevage tropicaux. Rev Elev Med Vet Pays Trop 64(1–4):73–79

Vayssières J, Vigne M, Alary V, Lecomte P (2011b) Integrated participatory modelling of actual farms to support policy making on sustainable intensification. Agric Syst 104(2):146–161

Chapter 20
Instruments to Assess the Social Impacts of Value Chains

Catherine Macombe and Denis Loeillet

The objective of the methods described in this chapter is to answer the question: What are the main social impacts of the implementation of a project in a value chain? Or even: What are the consequences of a project that affect individual and/or collective wellbeing? Some methods rely on performance criteria (the practice of child labour, benefits in kind, etc.) while those based on impacts seek to calculate, estimate, and directly assess social consequences, and consequently the actors' experiences (e.g., progress in health, in solidarity, etc.). These methods are very diverse due to the nature of the social impacts they take into account: from calculating the number of jobs created to assessing a whole range of social processes that affect a value chain's actors.

Methods are presented here based on a temporal perspective, in three possible situations: pre-project, during the project, or post-project, as discussed in different parts of this chapter. In practical terms, it is rare for a single action to be subjected to all three types of assessment: *ex-ante*, *in itinere*, and *ex-post*. The *in itinere* and *ex-post* phases are often ignored due to time constraints, since they happen too late to influence the project design. *Ex-ante* methods, on the other hand, are useful in influencing projects (for example, the design of a new cropping system), and even in calling the viability itself of the project into question.

C. Macombe (✉)
UMR ITAP, Irstea – Centre de Montpellier, 361, rue Jean-François Breton, BP-5095 34196 Montpellier, Cedex 5
e-mail: catherine.macombe@irstea.fr

D. Loeillet
CIRAD – UPR Geco, F-34398 Montpellier, France
e-mail: denis.loeillet@cirad.fr

E. Biénabe et al. (eds.), *Sustainable Development and Tropical Agri-chains*,
DOI 10.1007/978-94-024-1016-7_20

20.1 *Ex-ante* Assessment Methods

Although several checklists exist to help determine whether a project deserves to go ahead or not, few deal with real social consequences. Project formulation methods can draw up theories of change[1] based on interviews, but they do not apply directly to value chains whose main objective is to make a profit.

The PROSUITE project (FP7[2] of the EU) has tried to implement a new *ex-ante* assessment method to check the sustainability of new technologies. This project proposes the use of impact methods (Weidema 2006) to assess social wellbeing, and acknowledges that 'the complete cause-effect pathways [...] are still in an early phase of development as far as impacts on social wellbeing are concerned' (Prosuite Report 2013). CIRAD and Irstea have recently proposed such causal pathways (Feschet et al. 2012) and have suggested innovations that link variables characteristic of the assessed action to significant health impacts.

We have categorized *ex-ante* assessment methods into four groups (Fig. 20.1).

The first group of methods is based on the assumption that there are people who know what will happen if such a change occurs. The second group assumes the existence of a previously proven relationship between two variables, with a strong likelihood that this relationship will endure in the future. We can thus use it to predict a future state. The third group also uses the relationships between risk and human health that, for example, was formalized through (partial) consensus among researchers as a kind of accounting process (similar to environmental life cycle assessment). The fourth group predicts the characteristics of a future situation by combining existing economic and technical documents as well as interviews with actors.

20.1.1 Methods of the First Group: Ask Those Who Know

Either the people being asked are already living in a scenario that is sought to be created, they hold a view on the consequences of the impending change, or, if they are experts, they are requested to quantify impact indicators. People are assumed to be knowledgeable and willing to respond. The choice of the type of indicators can be left to researchers, a panel of experts, participants, or to a group comprising a mix of these various entities. The protocol often includes steps to prioritize impacts and assign an overall rating.

In most cases, participatory protocols (Mathé 2014) help organize data collection. Participatory methods and tools for dialogue used in agriculture were

[1] A theory that explains how a cause, or a set of related causes, results in consequences in case of a change. This is thus the description of a pathway.

[2] Framework Programme (full name: Framework Programme for Research and Technological Development).

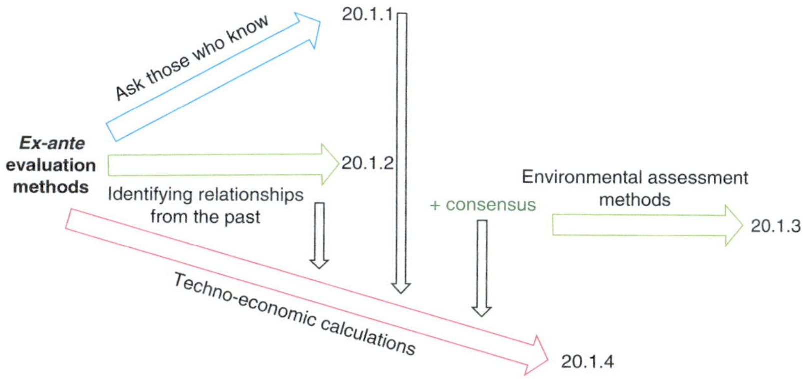

Fig. 20.1 The four groups of *ex-ante* assessment methods and their linkages

described in detail in a report by Bousset et al. (2005) of the EU Seamless project. More than 27 basic tools are available. In practice, a protocol for interacting with participants usually combines two tools, but can always use more. This is the reason the number of possible protocols described exceeds 500.

These methods incorporate the actors' representations with a commendable accuracy. All types of consequences can be described by an indicator and can thus be assessed. However, studies of this kind are lengthy and consume time (at least 6 months) and energy. They can lead to unpredictable reactions if the topic under consideration is too sensitive. Furthermore, the results obtained cannot be generalized. That is why such methods are used mainly when the effort required can be justified by the need to facilitate appropriation or build a consensus.

20.1.2 Methods of the Second Group: Identifying Relationships from the Past and Assuming They Will Hold in the Future

If a single criterion has to be used to gauge wellbeing, the consensus will be to choose health (Weidema 2006). This is the reason we will use it to illustrate this section. For a long time, epidemiologists (such as McCartney et al. 2013) have agreed on the idea that the average health status of a population depends primarily on socio-economic factors. It is also accepted that some major agri-chains in countries of the South are changing the socio-economic situation in those countries. Researchers from CIRAD and Irstea have demonstrated two such relationships (Feschet et al. 2012; Bocoum et al. 2015). The first links average wealth per person to the average life expectancy in a poor country (Preston pathway). The second links an increase in income inequality to worse health (Wilkinson pathway). Each pathway has a corresponding formalized relationship whose applicability is subject to its conditions of use. The following example illustrates the conditions of use of the Preston pathway.

Feschet et al. (2012) confirmed a correlation between the level of wealth and average life expectancy (Preston effect), but more importantly, they demonstrated that a variation in the level of wealth induced a change in life expectancy. If the change in the agri-chain under study generates a new added value, it will lead to increased life expectancy in the country, assuming that:

- the country is poor, with per capita GDP being less than 10,000 US $, i.e., approximately Mexico's per capita GDP;
- the added value is generated by a value chain that represents at least about 1 % of the country's GDP, or which enjoys a strong institutional legitimacy in its sector (for example, ability to impose new labour standards);
- the minimum duration of the value chain's activity is 4 years. There is, indeed, a time lag between the distribution of new income and an improvement in health;
- at least 60 % of the added value generated is distributed directly in salaries or through social benefits (health coverage, housing etc.).

CIRAD applied these research findings to a banana chain in Cameroon (Feschet et al. 2012). These methods have the advantage of being quick and generalizable provided, of course, that the case concerned complies with the conditions of use. They require little primary data. In contrast, the number of simple relationships that we can finally identify is quite limited, and each method relates only to a single scale.

20.1.3 Methods of the Third Group: Assessment of Health Using Environmental Assessment Methods

These methods group together a set of relationships between exposure to a risk (carcinogenic substance, climate change, etc.) and human health. They are built from a large number of assumptions and unknowns (even within the modelling itself) for which researchers need to make choices. These methods are thus, at least partially, based on consensus. They are generically referred to as 'human health impact calculations' in environmental life cycle assessments (Spriensma and Goedkoop 2001). As soon as an environmental life cycle assessment is carried out, it is recommended that its impact on human health be calculated. Indeed, this operation does not require any data other than that used to calculate other impacts. However, results obtained estimate potential damage to the health of an 'average human being', which is not relevant to specific populations. It must be noted that results could vary according to the type of environmental life cycle assessment method used (Dreyer et al. 2003).

20.1.4 *Methods of the Fourth Group: Techno-economic Calculations*

Sometimes the best way to anticipate a future state is to rely on economic or technical documents. For example, the workers/hectare ratio of two competing value chains is the starting point to assess the number of people who will be affected by the replacement of one value chain by the other. When there is a social matrix showing that each new farm generates six jobs in the value chain, we can use it to predict how many jobs will be created by the setting up of *n* new farms, etc. These methods allow the reconstruction of the value chains' technical systems so as to ensure the plausibility of the data gathered elsewhere (multi-angulation). They are, however, demanding in terms of time and technical and accounting data.

This group is not linked to any single method. However, its principles rely on development theory as a combination of capital (Guellec and Ralle 1996) and have been formalized under the designation 'capacities social life cycle assessment' (Garrabé and Feschet 2013; details in Gillet and Loeillet 2013). They are essential to generate intermediate data for the final calculation of social impacts. Since accounting reports are progressively becoming more and more available in the countries of the South, this group of methods is set to become increasingly more useful.

20.2 *In itinere* Assessment Methods (Monitoring)

Monitoring involves observing, in the course of implementing the action, if the change takes place as foreseen. All methods that are based on monitoring are created from criteria that illustrate the situation at the time of the assessment. They help diagnose possible deviations. However, the causes of variation are not always explicit. By definition, monitoring systems help minimize plan deviations, even as they help identify and report on the project status, compare with the plan, analyse deviations, and implement appropriate corrective actions (Hazir 2014). Methods based on performance are suitable for monitoring.

Monitoring requires the development of a dashboard of indicators based on criteria that will help decision makers take action. Ideally, the indicators must be calculated or obtained before the change occurs, and then at regular intervals until its conclusion.

Although several guides exist to monitor the assessment of a value chain's sustainability, many of them overlap. They include a range of other standards and self-assessment tools. Figure 20.2 illustrates the three groups that we distinguish.

The first group, *ad hoc monitoring*, includes criteria specifically created for the case being studied in order to monitor only the ongoing change. Although the World Bank suggests that the monitoring criteria must be created by the value chain's stakeholders, this is often not the case in practice.

The second group, a complete dashboard of indicators for the agrifood sector, applies to all farms and agricultural value chains. It assumes that social effects can

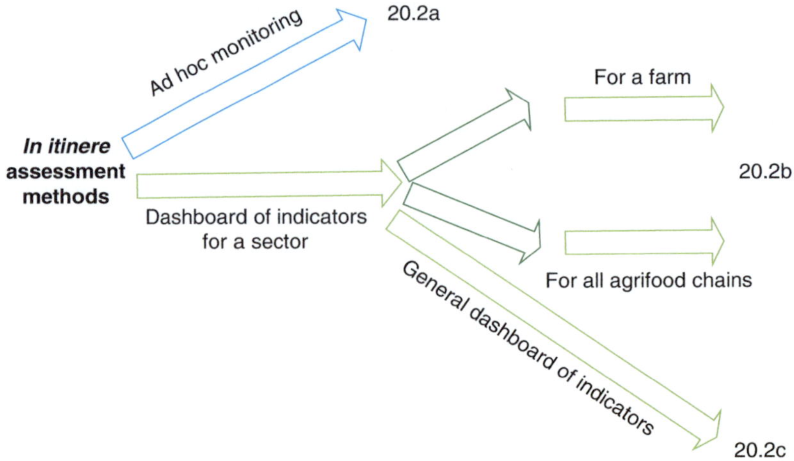

Fig. 20.2 The three groups of *in itinere* assessment methods

be apprehended by studying the value chain alone; other value chains that could potentially be competing with it can be ignored.

The same assumptions form the basis for the creation of the general dashboard of indicators of the third group. The aim, however, is to create indicators that are understandable and relevant across sectors. The indicators are normally obtained from conceptual frameworks developed by the United Nations. It is often observed that the criteria used are the same as found in reports resulting from the corporate social responsibility approach. In addition, several stages of the value chain are monitored and queried, as in the life cycle method. That is why these methods are called 'corporate social responsibility life cycle' methods.

20.3 *Ex-post* Assessment

The *ex-post* assessment of the value chain aims to create an end state of all the social impacts of the change. It should therefore be undertaken only after these effects have begun to be discerned, which is not always the case. Another difficulty is to correctly attribute a particular impact to a change. Two conditions must be met for this to happen. The first is to determine the difference between the end state and the baseline state (which was established in the past, prior to the change). The second is to confirm that the effects cannot be attributed to a cause other than the change being studied. The utility of these methods therefore depends on the quality of the baseline state. The quality of the baseline state, in turn, depends on the existence of proper monitoring, i.e., the monitoring of criteria that are suitable for describing the action, which also serve as levers for decision makers. The groups of

Fig. 20.3 The two groups of *ex-post* assessment methods

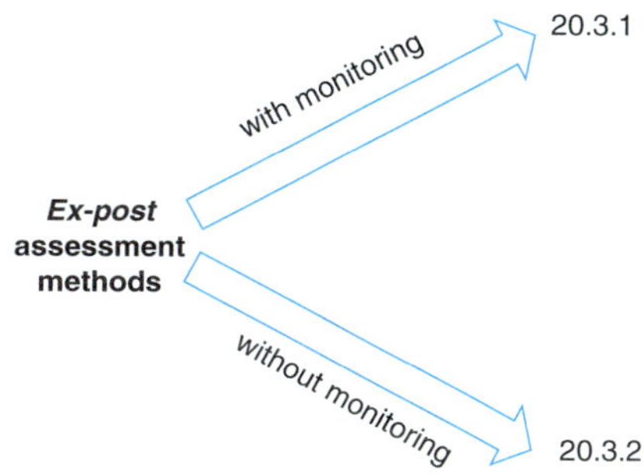

methods are classified in Fig. 20.3 based on this criterion, depending on whether any monitoring takes place or not.

20.3.1 Ex-post *Methods When Monitoring Takes Place*

This is the most desirable situation. In fact, monitoring provides data at different time intervals. Most can be used to describe short- and medium-term effects (for example, we can assess the final state of the desired social effects that are described in the theory of change). On the other hand, effects that only manifest over the long term – and especially the unforeseen ones – cannot be captured through monitoring.

Figure 20.4 shows an example of the monitoring and *ex-post* assessment process. During monitoring, criteria q and N are documented during three surveys, at times t_0, t_1, and t_2, with the survey at t_0 providing the baseline state. In order to carry out the *ex-post* assessment at time t_2, the evaluator compares the value of each criterion between t_0 and t_2. The evolution of the state between t_0 and t_2 is not only caused by the deliberately implemented change but also by other unforeseen events that may have occurred in the period between t_0 and t_2. The contribution of these unforeseen events to the change must also be investigated. The stakeholders' views are an important resource for attributing effects to unforeseen events, and to interpret the significance of differences between values.

20.3.2 Ex-post *Methods for Reconstructing the Baseline State*

There are instances where no monitoring was done, or the data was lost. In such a case the reconstruction of the baseline state is a long and difficult process that actors can find very complicated. It is essential to ensure that these efforts are justified. Information is cross-referenced to reconstruct the past. All sources are cross-referenced and combined (accounting records, technical reports, interviews of key people, photographs, newspapers, articles, historical reconstitutions, notes etc.) to

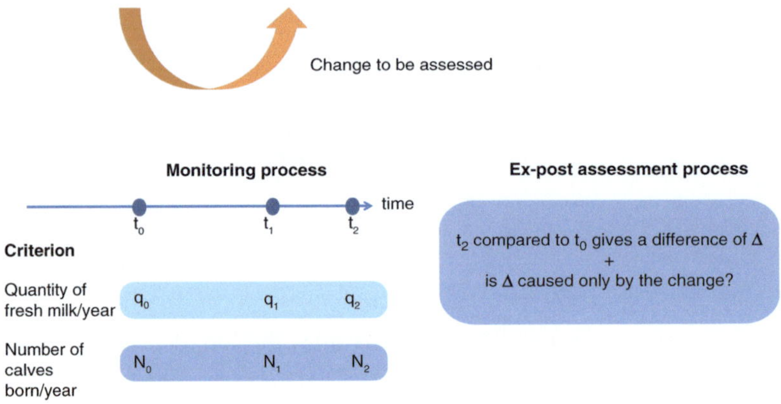

Fig. 20.4 Example of a successive process of monitoring and *ex-post* assessment

Fig. 20.5 General mapping of social impact assessment methods

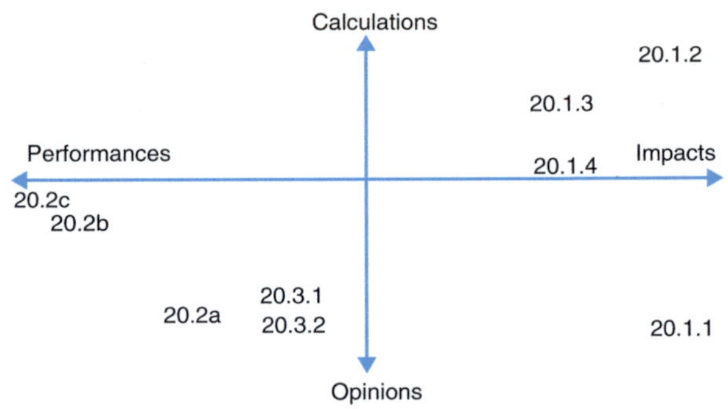

describe the original situation, while accounting for the fact that people tend to rationalize *a posteriori*.

20.4 Conclusion on Social Assessment Methods

The diversity of social impact assessment methods can be represented (Fig. 20.5) along two axes:

- methods based on impacts (consequences) versus those for assessing performance, because they have different objectives;
- methods based on calculations or on the views of stakeholders.

Methods based on participation hold a promising future. If such a method is conducted properly, it encourages the participation of populations in a local societal project, beyond the method's ability to explain significant social impacts to that society. Indeed, participatory methods develop a sense of ownership of the social project concerned. In contrast, the advantages of methods based on computation, undertaken mainly within distant offices, are genericity and speed, while the disadvantage is that such methods can only be applied to a small number of social impacts. Although methods based on performance are useful in tracking major criteria (number of employees, etc.) in a value chain, they reveal little about the social consequences of the value chain's existence, or of the changes occurring therein. Impact assessment is not meant for choosing a scenario from a predefined set. Its role is to provide keys to modify and adapt a given scenario by working up the links of causality. The goal is to enhance the positive social impacts while, more importantly, mitigating adverse ones. In this sense, social impact assessments of value chains play a significant role in fostering innovation by helping design value chains that actively contribute to sustainable development.

References

Bocoum I, Macombe C, Revéret JP (2015) Anticipating impacts on health based on changes in income inequality caused by life cycles. Int J Life Cycle Assess 20(3):405–417

Bousset JP, Macombe C, Taverne M (2005) D 7.3.1. Participatory methods, guidelines and good practice guidance to be applied throughout the project to enhance problem definition, co-learning, synthesis and dissemination. Deliverable of the seamless European project (project no. 010036) of the 6th framework programme

Dreyer LC, Niemann AL, Hauschild MZ (2003) Comparison of three different LCIA methods: EDIP97, CML2001 and eco-indicator 99. Does it matter which one you choose? Int J Life Cycle Assess 8(4):191–200

Feschet P, Macombe C, Garrabé M, Loeillet D, Benhmad F, Rolo SA (2012) Social impact assessment in LCA using the Preston pathway. The case of banana industry in Cameroon. Int J Life Cycle Assess 18(2):490–503. doi:10.1007/s11367-012-0490-z Accessed 5 May 2016

Garrabé M, Feschet P (2013) Un cas particulier: l'ACV sociale des Capacités. In: Macombe C (ed) ACV sociales Effets socio-économiques dans les chaînes de valeurs. Cirad, Paris, pp 97–117

Gillet C, Loeillet D (2013) Pratique de l'évaluation sociale du cycle de vie d'une filière. In: Macombe C (ed) ACV sociales Effets socio-économiques dans les chaînes de valeurs. Cirad, Paris, pp 119–139

Guellec D, Ralle P (1996) Les nouvelles théories de la croissance. La Découverte, coll. Repères, Paris, p. 128

Hazir Ö (2014) A review of analytical models, approaches and decision support tools in project monitoring and control. Int J Project Manag 33(4):808–815

Mathé S (2014) Integrating participatory approaches into social life cycle assessment: the SLCA participatory approach. Int J Life Cycle Assess 19(8):1506–1514

McCartney G, Collins C, MacKenzie M (2013) What (or who) causes health inequalities: theories, evidence and implications? Health Pol 113(3):221–227. doi:10.1016/j.healthpol.2013.05.021 Accessed 5 May 2016

Prosuite Report (2013) 7th framework programme of the European Union. http://www.prosuite.org/web/guest/home;jsessionid=2338C218E98000A099407613F12DE976. Accessed 5 May 2016

Spriensma R, Goedkoop M (2001) The eco-indicator 99: a damage oriented method for life cycle impact assessment, Methodoloy report, report no. 1999/36A. Ministerie van Volkshuisvesting, Amersfoort, p. 87

Weidema B (2006) The integration of economic and social aspects in life cycle impact assessment. Int J Life Cycle Assess 11(Suppl 1):89–96

Chapter 21
Agri-Chains and Evaluation of Sustainability

Catherine Macombe, Denis Loeillet, and Cécile Bessou

Part 4 of this book focuses on the tools that are available to researchers and practitioners to assess the contribution of agri-chains to sustainable development. Depending on the case, an agri-chain can be an excellent vector of sustainable development or a catalyst for human and ecological disasters. This is why it is essential to assess the real impacts (consequences) of the agri-chains' activities, without limiting oneself to their performances whose relationships to real consequences are unknown. Assessment helps in designing or redesigning an agri-chain, and in order to undertake it, it is necessary to define boundaries correctly and to carefully choose the appropriate impacts to be assessed. An approach based on these considerations applies to all methods. However, we have focused here on life cycle assessment methods as they are gaining recognition and benefiting from a growing interest in them, and they enjoy special attention at CIRAD.

Indeed, life cycle assessment methods are very relevant to the study of agri-chains because they focus on an object (the life cycle) pertaining to the analysis of the 'chain' entity. And despite their flaws and limitations, these methods help broaden the actors' understanding on what should be considered (from an environmental and social point of view) in designing or redesigning an agri-chain. It must, however, be understood that different uses call for different methods, and only some of them have been covered in Part 4 of this book.

C. Macombe (✉)
UMR ITAP, Irstea – Centre de Montpellier, 361, rue Jean-François Breton, BP-5095 34196 Montpellier, Cedex 5
e-mail: catherine.macombe@irstea.fr

D. Loeillet
CIRAD – UPR Geco, F-34398 Montpellier, France
e-mail: denis.loeillet@cirad.fr

C. Bessou
CIRAD – UPR Performance of Tree Crop-Based Systems, F-34398 Montpellier, France
e-mail: cecile.bessou@cirad.fr

E. Biénabe et al. (eds.), *Sustainable Development and Tropical Agri-chains*,
DOI 10.1007/978-94-024-1016-7_21

There are many key messages that environmental life cycle assessment methods hold for us. To begin with, the methods presented deal with systems that are not isolated, and attempt to encompass the most general level possible. They rely on understanding the mechanisms of environmental impacts – many of which are still to be identified – in order to quantify them. Furthermore, several questions regarding the method remain unresolved: How to correctly integrate local impacts on a broader scale, or even on the planetary scale? With which levers of action and what responsibilities? These methods are yet to find proper anchoring in scientific disciplines, given the subjective choices and the large number of assumptions that must be made in order to undertake calculations. Much work still needs to be done to achieve this.

We have highlighted a host of social assessment methods, and the subset of them mapped in Chap. 20 illustrates but a small part of it. Social assessment methods are sure to see further progress and development in the coming years. New and original combinations of the basic tools, as well as of their variants, will probably emerge, especially as concerns their implementation within a participatory framework. At the same time, social science research on agri-chains (including that being conducted at CIRAD) will help specify the conditions and relationships between the various factors involved (for example, between shared goods, inequality of access to such goods and the health of the people; on a different scale, between cropping systems and the health of agricultural workers).

In practical terms, two key methods are available for designing or redesigning agri-chains. If the appropriation by actors of an agri-chain or territory is a major goal, then the assessment will have to be carried out within a participatory framework. However, this process is very long and demanding. If, on the other hand, the assessment is intended only to identify key impacts, then social life cycle assessment methods, based on the evaluation of the future state using calculations and surveys, seem more appropriate. In both cases, whether a challenging participatory process or a simple and rapid assessment, the design phase is important because it has to take into account the real causes of the impacts to be avoided (for negative impacts) or those to be enhanced (for positive impacts).

We make a distinction between methods for environmental assessments from those for social assessments. However, this distinction does not presuppose the adoption of a sustainable development design based on the three 'formalized' pillars: economic, environmental, and social. On the contrary, research work in progress is calling into question this model based on three distinct pillars.

Large companies now view the assessment of their positive and negative impacts as a strategic issue, both at the local level and for their customers and shareholders. This is evident from the fact that key actors, mainly global distributors, have decided to make their supply chains sustainable within a few years, especially in the agricultural sector. A challenge for the research community is, therefore, to assist these actors in assessing the impacts of their global or local supply chains in order to improve their social and environmental quality. Helping them create documented scenarios in terms of social and environmental implications can change the lives of these agri-chains' most vulnerable populations.

Part V
Can Agri-Chains Act as an Arena of Regulation of Sustainable Development?

Chapter 22
Agro-industrial Strategies and Voluntary Mechanisms for the Sustainability of Tropical Global Value Chains: The Place of Territories

Jean-Philippe Tonneau, Stéphane Guéneau, Marie-Gabrielle Piketty, Isabel Drigo, and René Poccard-Chapuis

It is only since the turn of this century that modernized large-scale industrial agriculture began growing rapidly in tropical countries in order to meet an increasingly sustained global demand for food, fibres, fuel, etc. Despite gains of productivity (Byerlee et al. 2014), the growth of agro-industrial livestock farming and crop cultivation has mainly taken place through the clearing of land at the expense of primary and secondary forests (Phalan et al. 2013). This substitution results in the loss of biodiversity and contributes to climate change (DeFries et al. 2010). Mechanized farming uses large quantities of chemicals which contaminate water and soils. The setting up of agro-industrial projects deprives local populations of the right to use the land (Alden 2012), and generates conflicts between these populations and agro-industrial companies (Gerber 2011).

NGOs, lobby groups, and consumer and media groups denounce the negative social and environmental consequences of tropical agro-industrialization. To ensure that the purchase of agricultural goods does not tarnish their reputations,

J.-P. Tonneau (✉)
CIRAD – UMR Tetis, Montpellier, France
e-mail: jean-philippe.tonneau@cirad.fr

S. Guéneau
CIRAD – UMR Moisa – SHIS, Brasília, Brazil
e-mail: stephane.gueneau@cirad.fr

M.-G. Piketty
CIRAD – UPR Green, Montpellier, France
e-mail: marie-gabrielle.piketty@cirad.fr

I. Drigo
Nexus Socioambiental, São Paulo, Brazil
e-mail: isabel.drigo@gmail.com

R. Poccard-Chapuis
CIRAD – UMR Selmet – EMBRAPA, Belém, Brazil
e-mail: rene.poccard-chapuis@cirad.fr

E. Biénabe et al. (eds.), *Sustainable Development and Tropical Agri-chains*,
DOI 10.1007/978-94-024-1016-7_22

large multinational processing and distribution companies seek to control the origin of their raw materials through responsible procurement policies, which require their suppliers to conform to sustainability standards. Investors have to demonstrate, even provide guarantees, that they are managing environmental and social impacts of agro-industrial activities. To be able to give these guarantees, these companies have to implement and adopt new forms of private governance. These are the strategies of CSR and responsible investment, voluntary standards and sustainability certifications, moratoriums, and 'zero deforestation' agreements (Gibbs et al. 2016).

Analyses of these voluntary mechanisms arrive at very contrasting conclusions. While some authors see the development of tropical agro-industry through these voluntary mechanisms as an opportunity to strengthen environmental and social protection (Rudel et al. 2009), others are more critical and point out that these mechanisms respond only very partially to the upheavals caused by agro-industries in the organization of territories (Linder 2013). These critics also note that the measures announced are not always implemented. These differences of perception show that the issue still requires further exploration. What is the real effectiveness of environmental and social commitments made by global value chains? Can these strategies enable a transition from an economy based on agricultural frontiers to one that is more sustainable and integrated territorially? What is their real capacity to regulate the social and environmental dimensions of development? How do these private-sector strategies fit into governmental regulatory frameworks?

To answer these questions, we analyze these phenomena in two case studies, one in Gabon and one in the Brazilian Amazon. In Gabon, we study the implementation of the Olam company's CSR policy on a rubber plantation. In Brazil, we analyze the impacts of environmental commitments made by the soya bean and beef global value chains. For both case studies, we present the context, the features of the standardization mechanisms, and the limitations of these mechanisms, and propose a few alternative paths to follow to overcome the problems identified.

22.1 CSR Approach on a Rubber Plantation in Gabon

The sustained global demand for natural rubber has led to a rapid expansion of rubber plantations. Multinational companies are turning increasingly to sub-Saharan Africa in search of available, relatively unprotected, and sparsely populated land on which to set up agro-industrial projects (Deininger 2011).

Most African countries, and in particular Gabon, encourage these investments. Foreign acquisitions of land since 2005 for agro-industrial investment projects on Gabonese soil totalled 400,000 hectares (Schoneveld 2014). In late 2010, the Gabonese government signed a US $ 1.5 billion contract with Olam, a Singaporean multinational company, which became the first private land owner in Gabon, with exploitation rights over nearly 3 million hectares (Oyono et al. 2014) for various agricultural productions, free-trade forest zones (subsequently given up), and a fertilizer factory. Olam has obtained a forest license in Gabon's Woleu-Ntem Province, in the form of a renewable 50-year lease, in order to establish an industrial rubber plantation with an initial area of 36,000 ha, to be set up in 5 years.

22.1.1 Sustainability Standards and Olam Rubber's CSR Strategy

In the case of this rubber plantation, the identification of environmental and social issues, a mandatory requirement in Gabon, was carried out as part of the environmental and social impact assessment (ESIA) before the start of the project by EnviroPass, a Gabonese consultancy firm. These issues are presented in Box 22.1.

Box 22.1 Environmental and Social Issues Identified as Part of the Environmental and Social Impact Assessment (ESIA)

Jean-Philippe Tonneau and Stéphane Guéneau

According to the ESIA, the concession extends over sparsely populated areas and is largely composed of secondary forests whose maturity is more or less advanced. The ESIA observes that biodiversity loss, even though not altogether negligible, can be relativized. It does not mention that 'flush' cutting has prevented regeneration and has had a negative effect on the potential biodiversity. The impact of population growth on hunting, including in nearby reserves (Minkébé National Park), is not mentioned. Nevertheless, the study highlights the impact risks of the plantation on the disappearance of the megafauna. The effect of machinery on soils, risks of runoff and contamination of surface water and groundwater, and the generation of industrial waste are among the environmental issues highlighted by the ESIA.

The main social issues identified concern the working conditions of the plantation's workers and the shrinking of spaces used by the local people for their agricultural and forest activities (hunting, gathering, etc.).

Studies requested by Olam to complement the ESIA on the presence of large mammals focused only on the concession area.

To address these issues, Olam Rubber has implemented CSR measures (Box 22.2), based on the Olam group's plantations code (Olam 2014), sustainability standards being developed within the framework of the Sustainable Natural Rubber Initiative (Warren-Thomas et al. 2015); and standards already existing in other global value chains.

Box 22.2 Some Aspects of the Implementation of Olam's CSR Strategy at Its Rubber Plantation in Gabon

Jean-Philippe Tonneau and Stéphane Guéneau

At the Environmental Level

Olam Rubber decided to reduce the usable area of the concession from 36,000 to 28,000 hectares in order to avoid encroaching on village areas. A buffer zone of 5 km between the plantation and the villages was created so that the local people can retain most of their rights to land or to other resources.

Measures to protect the most important ecosystems for biodiversity, including riparian forests, have been implemented. Forests on very steep slopes have been preserved. Specific collection and storage facilities have been established for waste processing and recycling of used oils and fuels.

At the Social Level

A clinic was set up and protective equipment is provided to the employees most at risk. Road safety measures have also been implemented. Land has been made available inside the plantation to allow employees to practice their own agricultural activities.

Under the principle of free and prior informed consent, an individual and collective compensation programme was agreed upon for the loss of forest use rights. Three main types of actions were foreseen: the provision of basic infrastructure (pumps, roads, etc.); priority in the hiring of local labour given equal skills; support for income-generating activities (subsistence farming projects, for example).

22.1.2 The Limits of the Normative Approach to CSR: The Problem of Indirect Effects

CSR measures, when implemented,[1] apply only within the plantation's boundaries and in its immediate vicinity. And yet, the effects of the plantation far exceed this limited area. An industrial plantation requires a large workforce (Balac 2001). Wages paid by companies and the social conditions they create by providing relatively high salaries and favourable conditions of work, including in terms of infrastructure, draw a large migrant population seeking employment. The relative wealth of Gabon compared to neighbouring countries is an added attraction. The region will thus experience a population boom.

[1]For example, for legal reasons and due to the local lumber chain's capacity constraints, the sale of cubic metres of timber felled were very limited. Proceeds from this activity were destined for a social actions fund.

In very similar African situations, the total number of migrants has been observed to be four to ten times the number of plantation workers[2] (Assembe-Mvondo et al. 2015). In the case of Olam Rubber's plantation in Gabon, migration can be estimated at twice or more the current population of the capital of the department, the city of Bitam, which has about 15,000 residents. These people will need to be housed and fed. No doubt, Olam Rubber has planned housing construction and land grants programmes for its employees. But urban planning issues (type and location of housing, access to services such as roads, schools, hospitals, police) and food security issues concern the territory as a whole. Access to food can, of course, be guaranteed through wages. However, the cost of food will increase and part of the non-native population, especially those who will not find salaried jobs, will have to undertake and depend on subsistence farming (Oosterveer et al. 2014). Access to land will thus represent a major challenge. Gabonese villagers are very attached to their customary usage rights and are unwilling to cede them without very high consideration. Access to land is therefore likely to be a major source of tensions, which in other circumstances and other places has led to serious conflicts.

For the local people and authorities, these issues are the responsibility of Olam, in a natural continuation of the compensation programme put in place to defuse initial conflicts with local residents neighbouring the plantation. For the company, these compensation programmes have now already ended or are nearing completion. Faced with a weak and ineffective State and very limited local entrepreneurship, Olam is being asked to manage complex situations, for which it certainly bears part of the responsibility but, in its perception, only an indirect responsibility. The weakness of the State is often perceived by companies as an enabling factor allowing low-cost investments through tax exemptions, which are the cornerstone of many investment agreements. Olam too has benefited from these exemptions. But the State is depriving itself of the means necessary for it to be able to exercise its regulatory functions. The company is then 'called upon' to respond to unfulfilled expectations.

22.1.3 The Search for Alternatives

Faced with these crucial challenges, the decisions of Olam's management and of the State and local administration and authorities seem insufficient. The CSR measures implemented can guarantee an 'acceptable' agricultural production, by ensuring that there are no abuses – which some refer to as social or environmental crimes – but in no way can they ensure sustainable development.

Ideally, the company contributes to the local economy through its purchases, its subcontracting, the distribution of wages, and, above all, the taxes it pays (Bloch

[2]In Cameroon, the population living on the site of the Hevecam plantation, which extends over 18,000 hectares, consists of over 20,000 people, including 550 employees, spread out over 17 camps and 3 villages. It can reach up to 35,000 people at certain times.

and Owusu 2012). Ideally, the State invests in infrastructure by drawing on the taxes paid by agro-industrial companies and the commercial and industrial activities that these companies generate. But in the absence of taxation, can the State play its development role and ensure the necessary infrastructure? Can the State fulfil its responsibilities of regulation given the risk of social and environmental crises that may paralyze economic activities?

The public-private association must be recast on a more collaborative and more balanced basis. To this end, territorial development approaches could be mobilized. These approaches are aimed at fostering synergies between individuals, civil society, institutions, businesses, and the State (central and local authorities) for collaborating on joint projects (Wallerstein 1992). The territories appear as entities promoting the integration of various existing projects into a coherent whole: a territorial project, which always originates from a compromise. This compromise requires a dialogue organized around more or less participatory approaches, reconciling the emergence of an endogenous project with externally imposed policies, guidelines, and standards.

Admittedly, the conditions to initiate this dialogue are far from ideal: civil society is poorly organized and subject to many pulls and pushes; local authorities are not fully independent of the central government; and decentralization has yet to take place. This is the ambiguous context in which strategies of CSR, responsible investment, and voluntary standards are implemented. These strategies are internal to companies. They thus depend ultimately on the companies' own interests. These strategies can only be successful if local balancing powers are also created, which can transmit and benefit from campaigns in consumer countries, keeping in mind that not all markets have the same requirements.

This work is not easy and will take time. In the case of Olam, we proposed the creation of a regional development agency, funded by the State, the company, and donors. Such an agency could promote the region's sustainable development by providing support for strategic long-term thinking; support for the drafting and updating of planning documents (development master-plan); research funding; and support for the development of projects. As of now, this proposal has not found a favourable response, despite the urgency to take the social and environmental issues at the territorial level into consideration.

22.2 Deforestation and Commitments of Global Value Chains in the Brazilian Amazon: Limits and Alternatives for a Sustainable Territorial Development

During the last decade, the annual rate of deforestation in the Brazilian Amazon has slowed down significantly, from 2.7 million hectares in 2004 to 0.5 million hectares in 2014. The changes in the strategy of major global value chains, particularly soya bean and beef, have played a role in this development, in association with, on the one hand, the strengthening of command-and-control public policies and, on the other, pressure from civil society, NGOs, and lobby groups (Nepstad et al. 2006).

22.2.1 *Commitments of Soya Bean and Beef Global Value Chains in the Brazilian Amazon*

22.2.1.1 The Soya Bean Moratorium

In 2006, a moratorium was introduced and signed by the major buyers of soya bean produced in the Brazilian Amazon. This agreement prohibits the marketing of soya bean grown on land cleared by deforestation after 24 July 2006. The moratorium establishes a principle of market exclusion: producers who do not respect the moratorium are added to a blacklist which the buyers consult before every purchase. The moratorium, initially signed for 2 years, has since been successfully renewed every year. The monitoring is undertaken annually through analysis of satellite imagery in all municipalities with more than 5000 hectares of soya beans planted.[3] When soya bean cultivation is detected on areas deforested after 2006, aerial photos are taken for confirmation, and the concerned property's data are recorded.

In 2014, the last monitoring detected 47,028 hectares of soya bean planted on land deforested after 2006, which represents 1.11 % of total deforestation recorded in the same period in the three states concerned. The success of the moratorium depends on several factors, among them, a limited number of buyers who exert strong control over the producers. From 2014, in some states such as Para, the global value chain's main traders and producer associations have adopted additional measures, i.e., buying only from producers who have registered their lands with the Rural Environmental Cadastre (CAR)[4] and who are not on any embargoed IBAMA[5] list (green soya bean protocol).

22.2.1.2 The Commitments of the Beef Global Value Chain

In the case of the beef global value chain, after years of strong disagreement between its stakeholders and environmental movements, an agreement (Cattle Agreement) was reached by the major industries of the sector (Mafrig, Bertin, JBS) and the association of beef exporters (ABEG). Slaughterhouses and export companies agreed not to buy animals from lands cleared from deforestation carried

[3]These muncipalities are all located in three states: Mato Grosso, Para, and Rondônia.

[4]The Rural Environmental Cadastre (CAR) is the record of the environmental status of rural property (legal reserve, permanent preservation areas, etc.). It is mandatory since 2012.

[5]Brazilian institute responsible for monitoring compliance with environmental legislation.

out after July 2008 and from properties not registered in the Rural Environmental Cadastre (CAR) system. Buyers now also have to check that their suppliers are not on any list of embargoed properties or use slave labour.

At the same time, a multi-stakeholder working group on sustainable cattle farming in Brazil (GTPS) was established in 2007 with the support of the International Finance Corporation (IFC). It consists of representatives of livestock farmers, industries, distributors, public institutions, banks, NGOs, and social movements. This group is a member of the Global Roundtable on Sustainable Beef (GRSB), a transnational mechanism of multi-stakeholder governance on sustainability of the beef industry.

22.2.2 The Limits of Voluntary Commitments: A Partial Compliance with Legal Obligations

The primary objective of the soya bean moratorium is to prevent the expansion of soya bean cultivation into the Amazon rainforest. This objective has been achieved. But this first step does not ensure sustainability. The soya bean moratorium takes into account only some of the legal obligations. It does not guarantee that the properties that have stopped clearing the forest areas after 2006 are in compliance with the Brazilian Forest Code. This code stipulates in particular that the owners must retain between 50 and 80 % of the surface area of their land as forest reserves. The moratorium also does not guarantee that owners with properties that are deficient in legal forest reserves or permanent preservation areas will restore deforested lands ('environmental liability'). The Forest Code has also not yet clearly established the possible modes and speeds of recomposition of legal reserves and permanent preservation areas. Indeed, the moratorium is deeply unfair as it ensures equal access to the market for owners who have fully cleared their property as well as for those who have respected the Forest Code or are engaged in a process of restoration. Furthermore, the moratorium does not prevent deforestation for purposes other than for cultivating soya beans. Finally, some authors highlight the induced effects (or leakage effects) on other regions, such as the states of Maranhão, Tocantins, Bahia, and Piauí, which are not affected by the moratorium and where large surface areas have been planted with soya beans at the expense of the native vegetation of the Cerrado and the Caatinga (Gibbs et al. 2016).

Similarly, the commitments of stakeholders in the beef global value chain do not differentiate between properties that respect the forest code from those that do not, and do not take into account the livestock farmers' 'environmental liabilities'. In addition, they do not guarantee that the animals sold by a 'legal' property did not in fact originate from properties where deforestation occurred after 2009 or which are unregistered in the CAR. Indeed, animals slaughtered on a property are usually born on other properties, specializing in the cow-calf business (Poccard-Chapuis et al. 2005). The full traceability of animals does not seem easy to organize in the short term for the whole of the Amazon.

The discussions concerning Brazil's beef sector within the framework of the Global Roundtable on Sustainable Beef are making some headway, but the actors of the production sector show no willingness to commit to anything beyond the agreements already signed and resist the idea of a monitoring mechanism controlled from outside (Drigo 2013; Piketty et al. 2015.). Finally, as is the case for soya beans and, indeed, for all agro-industrial chains, there is no immediate prospects of markets rewarding additional efforts for sustainability. The global value chain's actors emphasize the responsibility of the State in the operational implementation of the Forest Code and monitoring mechanisms.

22.2.3 The Search for Alternatives: The Case of Paragominas

The story of the municipality of Paragominas is emblematic because in 2 years it went from being on the blacklist to a status of 'exemplary', the first environmentally friendly municipality ('Green Municipality') of the Brazilian Amazon (Piketty et al. 2015). The soya bean moratorium was enforced after it was signed in 2005 and deforestation started slowing and has continued to do so ever since. However, it remained high enough that in 2008 this municipality was one of the 36 municipalities on the federal government's blacklist. As a consequence, the municipality had to contend with several types of restrictions: embargo, 'vigorous' monitoring by IBAMA, which led to the closure of most of the sawmills and charcoal-powered furnaces, restricted access to credit, etc. In this context of crisis, under the leadership of the municipal team, representatives of rural producer associations and civil society organizations signed an agreement with the objectives of 'zero deforestation', 'zero fire' and of recording over 80 % of the territory in the CAR. With the support of private funding – including from Vale, the Brazilian mining giant, as part of its CSR policy – and technical support from NGOs, these objectives were met within 2 years. In doing so, the municipality regained access to credit and the pressure from IBAMA eased off.

These few successes aside, analyses of changes in land use that have occurred since 2009 highlight several limitations of the mechanisms implemented (Piketty et al. 2015). On the one hand, the residual deforestation occurs mostly due to family farming, which is not covered by the agreements of the global value chains and is excluded from the ongoing agrarian transition in the region. On the other hand, the intensification and diversification processes of land use apply mainly to the most favoured areas (close to roads and with clayey soils). Thus, the intensification project funded by Vale as part of its CSR (Pecuaria Verde project) involves a very small number of producers (6 out of 400), all of whom are in favourable situations. The prospects for a wider dissemination of technical proposals tested in this project are therefore quite limited. Finally, there is no action to restore degraded lands. This restoration would require additional investments.

22.2.4 Lessons Learnt and Possible Avenues for the Future

Multi-level governance mechanisms implemented in recent years in the Brazilian Amazon, especially in Paragominas, have therefore more than met the expectations of reducing deforestation. Along with the global value chains' voluntary commitments, these results are a witness to an improved synergy between the various levels of public governance. But these initial efforts are not sufficient in themselves to promote sustainable territorial development (Piketty et al. 2015). A municipality cannot be called 'green' just because deforestation stops; other elements have to be considered too: restoration of pastures, maintenance of soil fertility, reforestation, etc. With deforestation banned, space is now constrained, and Amazonian territories have to rapidly establish production and land use systems that contribute to sustainable development in its economic, social, and environmental dimensions. All the producers, including family farmers, have to be involved in this endeavour. Such an objective raises technical issues (Which technical models to adopt?), institutional issues (How to promote particular models? What technical support to provide?), and political issues (How can the selected models be made acceptable?). The main difficulty is to impart a certain coherence between the different dimensions of sustainable development, between projects involving actors with diverging interests, and between different sectoral policies. The aim is to define other governance mechanisms at the level of the global value chains and, above all, at that of the territories. By proposing a coherent framework for action, structured and accepted by the people, these mechanisms should be able to make the territory more attractive to public and private investment, thus allowing the shortcomings of the current trajectories to be rectified.

22.3 Conclusion

The analysis of how global value chains manage the environmental and social fallouts of agro-industrial development reveals the fragility of the mechanisms at work, whether they be CSR strategies or arrangements negotiated between the productive sector and NGOs. The case study of Olam Rubber in Gabon illustrates the difficulties agro-industrial companies have in managing all the environmental and social impacts of their projects when the State is weak. The company cannot be expected to replace the State to manage a complex set of indirect effects as a result of population growth in the region. The case of the Brazilian Amazon shows that voluntary measures put in place in the soya bean and beef global value chains do not guarantee the rule of law and sometimes just transfer the problems to other agri-chains in other places. This set of effects tarnishes what is otherwise a very positive assessment of the fight against deforestation in the Amazon.

Voluntary mechanisms are put in place around a given product and its value chain and are unable to manage territories undergoing constant change, even if these changes are caused by agro-industrialization. One has to look beyond the governance of the global value chains to explore new models of sustainable territorial development, along with new arrangements between the State, global value chains, and other actors.

In line with the work of Hirschman (1958), the challenge is to improve economic and spatial links between the company and the territory, and to invent new ways of management. 'Invent' is the correct term because, as of now, no solutions or even convincing experiments exist to overcome all the obstacles identified: the weakness of the State, little involvement of elected officials, lack of territorial planning capacities, lack of trust between the different actors, lack of knowledge, predominance of sectoral rationales, increased emphasis on private interests, lack of an organized civil society, lack of defined and accepted regulations, etc.

In this regard, the contrast between the Gabonese and Brazilian situations is revealing. The Brazilian government has clearly demonstrated its political will. In addition, its capacity to monitor, regulate, mobilize, and sanction is much higher. Brazilian civil society is better organized and more visible, both at the local and international levels. Decentralization is well advanced. These characteristics explain why the experiment of territorial governance in Paragominas generated synergies between different levels of public intervention (federal, state, municipal) and closer links between public and private actions. This is absolutely not the case in Gabon. And yet, the mechanisms of governance in Brazil are failing to ensure a transition to sustainable land uses, and succeeding only in containing large-scale deforestation.

The difficulty of funding innovative or alternative operations, which are inherently high risk, is obvious. Sustainable territorial development demands profound changes in current models of economic growth, hence the substantial need for the development of skills. To drive these changes, information and monitoring systems have to be based on indicators that highlight the specific needs of the territory. Only then will it be possible to transcend simplistic visions such as of reducing deforestation or the immediate effects of an agro-industrial investment, however important they may be.

In our case studies – but the observation can easily be extended to many other situations –, the economic power of agro-industries dictates the rules. The territorial development approaches will have to face this reality by taking the agro-industrial strategies into account and, most importantly, by seeking possible synergies. What interests are common to territorial management and agro-industrial management? The answer to this key question will probably require significant changes in territorial development concepts and approaches, mainly implemented so far in remote or underdeveloped areas without much industrial activity.

References

Alden WL (2012) Looking back to see forward: the legal niceties of land theft in land rushes. J Peasant Stud 39(3-4):751–775

Assembe-Mvondo S, Cerutti PO, Putzel L, Eba'a AR (2015) What happens when corporate ownership shifts to China? A case study on rubber production in Cameroon. Eur J Dev Res 27(2):1–14

Balac R (2001) Dynamiques migratoires et économie de plantations. In: Tapinos GP, Hugon P, Vimard P (eds) La Côte d'Ivoire à l'aube du XXIe siècle, Défis démographiques et développement durabl. Karthala, Paris, pp 195–231

Bloch R, Owusu G (2012) Linkages in Ghana's gold mining industry: challenging the enclave thesis. Res Policy 37(4):434–442

Byerlee D, Stevenson J, Villoria N (2014) Does intensification slow crop land expansion or encourage deforestation? Global Food Security 3(2):92–98

DeFries RS, Rudel T, Uriarte M, Hansen M (2010) Deforestation driven by urban population growth and agricultural trade in the twenty-first century. Nat Geosci 3:178–181

Deininger K (2011) Challenges posed by the new wave of farmland investment. J Peasant Stud 38 (2):217–247

Drigo IG (2013) Towards sustainable and certified beef? The reasons why and mechanisms through beef producers are beginning to change their production practices, Final Post-Doc Report. NESA/FEA/US, São Paulo, 88 p

Gerber JF (2011) Conflicts over industrial tree plantations in the South: who, how and why? Glob Environ Chang 21(1):165–176

Gibbs HK, Munger J, L'Roe J, Barreto P, Pereira R, Christie M, Amaral T, Walker NF (2016) Did ranchers and slaughterhouses respond to zero-deforestation agreements in the Brazilian Amazon? Conserv Lett 9:32–42. doi:10.1111/conl.12175 Accessed 25 May 2016

Hirschman AO (1958) The strategy of economic development. Yale University Press, New Haven, 218 p

Linder JM (2013) African primate diversity threatened by "new wave" of industrial oil palm expansion. Afr Primates 8:25–38

Nepstad DC, Stickler CM, Almeida OT (2006) Globalization of the Amazon soy and beef industries: opportunities for conservation. Conserv Biol 20(6):1595–1603

Olam (2014) Building Sustainable Supply Chains. In: Corporate responsibility & sustainability report 2013. Olam, London, 38 p

Oosterveer P, Adjei BE, Vellema S, Slingerland M (2014) Global sustainability standards and food security: exploring unintended effects of voluntary certification in palm oil. Glob Food Sec 3 (3–4):220–226

Oyono PR, Morelli TL, Sayer J, Makon S, Djeukam R, Hatcher J, Assembe S, Steil M, Douard P, Bigombé P (2014) Affectation et utilisation des terres forestières: évolutions actuelles, problèmes et perspectives. In: Les forêts du bassin du Congo. État des forêts 2013 (OFAC), Weyrich éditions, Neufchâteau, pp 215–240

Phalan B, Bertzky M, Butchart SH, Donald PF, Scharlemann JP, Stattersfield AJ, Balmford A (2013) Crop expansion and conservation priorities in tropical countries. PLoS One 8:e51759

Piketty M-G, Poccard-Chapuis R, Drigo I, Coudel E, Plassin S, Laurent F, Thâles M (2015) Multilevel governance of land use changes in the Brazilian Amazon: lessons from paragominas, state of Pará. Forests 6(5):1516–1536

Poccard-Chapuis R, Thalês M, Venturieri A, Piketty M-G, Mertens B, da Veiga JB, Tourrand J-F (2005) A cadeia produtiva da carne: uma ferramenta para monitorar as dinâmicas nas frentes pioneiras na Amazônia brasileira? Cadernos de Ciência & Tecnologia 22(1):125–138

Rudel TK, Defries R, Asner GP, Laurance WF (2009) Changing drivers of deforestation and new opportunities for conservation. Conserv Biol 23(6):1396–1405

Schoneveld GC (2014) The geographic and sectoral patterns of large-scale farmland investments in sub-Saharan Africa. Food Policy 48:34–50

Warren-Thomas E, Dolman PM, Edwards DP (2015) Increasing demand for natural rubber necessitates a robust sustainability initiative to mitigate impacts on tropical biodiversity. Conserv Lett 8(4):230–241

Wallerstein N (1992) Powerlessness, empowerment and health: implications for health promotion programs. Am J Health Promot 6(3):197–205

Chapter 23
The Standardization of Sustainable Development Through the Insertion of Agricultural Global Value Chains into International Markets

Emmanuelle Cheyns, Benoit Daviron, Marcel Djama, Ève Fouilleux, and Stéphane Guéneau

Projects to develop industrial agriculture for exporting biomass are expanding in the South: doubling of oil palm cultivation areas between 1990 and 2005 in Malaysia (1.7–3.4 million ha); an increase of almost 12 times, from 0.6 to 7 million ha, between 1975 and 2006 in Indonesia of the same crop; a five-fold increase in sugarcane plantations in Brazil from 1.4 to 7 million ha between 1960 and 2013; and expansion of soya bean cultivation at a compounded annual rate of 16.8 % between 2000 and 2005 in the Brazilian Amazon (Costa et al. 2007; FAOSTAT). In sub-Saharan Africa, since 2005, over 22 million ha of arable land has been allocated to the expansion of agro-industrial projects funded by foreign investments and intended primarily to supply the international market. Of this total, the palm oil chain in West and Central Africa accounts for 22 % and the sugar chain in southern Africa for 13 % (Schoneveld 2014).

This research has received financial support from the ANR through the Normes project (ANR-06-PADD-013) and the Prigoue project (ANR-11-ECLA-0009).

E. Cheyns (✉) • B. Daviron • È. Fouilleux
CIRAD - UMR Moisa, Montpellier, France
e-mail: emmanuelle.cheyns@cirad.fr; benoit.daviron@cirad.fr; eve.fouilleux@cirad.fr

M. Djama
CIRAD – UMR Moisa – Universiti Putra Malaysia, Kuala Lumpur, Malaysia
e-mail: marcel.djama@cirad.fr

S. Guéneau
CIRAD – UMR Moisa – SHIS, Brasília, Brazil
e-mail: stephane.gueneau@cirad.fr

E. Biénabe et al. (eds.), *Sustainable Development and Tropical Agri-chains*,
DOI 10.1007/978-94-024-1016-7_23

Since the late 1990s, an increasing number of civil-society actors and documented studies have reported the negative effects ('externalities') that this biomass production for distant markets entails. They note, in particular, the degradation of resources and social violence on the concerned agricultural frontiers: deforestation, biodiversity loss, pesticide pollution, displacements of populations, isolation of persons, expropriation of traditional lands, violation of human rights, criminalization of the political actions of local communities, etc.

While some authors draw attention to the difficulty of internalizing the effects that international trade makes invisible through a distance effect (Princen 1997; Clapp 2014), others think that this agriculture is part of a 'tropical agricultural revolution' (Nepstad and Stickler 2008) and see it as more of an opportunity than an environmental or social threat. This optimism is based on the new opportunities that a global governance could offer through voluntary sustainability standards.[1] Multi-stakeholder mechanisms, designed to create standards and certify sustainable production practices, were in fact implemented in early years of the century to establish responsible rules of corporate behaviour (Utting 2002; Fransen and Kolk 2007), and in this way regulate the environmental and social impacts of corporate agriculture.

Can these new mechanisms of private or transnational standardization, structured around global value chains, solve the problems of degradation of resources and social violence caused by this form of agriculture? Are voluntary sustainability standards capable of making long-distance trade in biomass compatible with issues of sustainability? To answer these questions, we will discuss, in the first part of this chapter, in a long-term perspective, the social and environmental effects induced by industrial agriculture inserted into international markets. We will then discuss, in the second part, the emergence of standardization mechanisms as new forms of regulation, promoted primarily by NGOs and economic actors and not by the State. In the third part, we will show that these standards form part of a liberal and market-centric approach that does not permit the inclusion of the diversity of perspectives and ignores some of the environmental and social criticism. We will explain why they are proving incapable of dealing with at least some of the negative externalities of biomass-exporting industrial agriculture.

[1] The standards can include labeling processes but do not necessarily have to (for example, the so-called 'B to B' or 'Business to Business' standards). Our purpose goes beyond the issue of the label of quality and its visibility to the consumer, and focuses instead on the standardization of production practices, open to a variety of processes to indicate quality to the market.

23.1 Massive Social and Environmental Effects of the Long-Distance Trade in Biomass

23.1.1 A Historical Perspective of the Long-Distance Trade in Biomass and Its Social and Environmental Consequences

The long-distance trade[2] in biomass has a long history. Already occupying a key role in supplying the Netherlands in the seventeenth century, it acquired a new dimension starting in the nineteenth century with the transportation revolution. Industrial England thus became a huge importer not only of food (half of food calories consumed in England in 1913) but also textile fibres, leather, wood, and rubber. Continental Europe soon followed suit.

This demand gave rise to agricultural frontiers, i.e., farming in the forests and on the steppes, in ever more remote locations on all five continents. At the environmental level, these agricultural frontiers resulted in the destruction of biodiversity-rich ecosystems and depletion of soil fertility, often followed by massive erosion. The rationale of the agricultural frontier is identical to that of mining. The belief that the supply of resources is endless leads to the exploitation of the soil, without thought to the maintenance of its fertility, and to the colonization of new lands when it is fully depleted. Thus, the heart of Brazil's coffee production shifted from the Paraiba Valley, in the state of Rio de Janeiro, to west of Sao Paulo in the early to the late nineteenth century (Monbeig 1952). During this shift, the Atlantic Forest (*Mata Atlantica*) was replaced by coffee trees, which, after half a century, in turn vacated their place to a meagre pasture. At the social level, the agriculture frontiers were based, with their own specific forms of violence, on a large scale expropriation of land and other assets from populations of hunter-gatherers or pastoralists who previously lived in these regions. This violence has often taken the form of genocide, whether in the United States against the native Indians, in Argentina with the War of the Patagonia, in German South-West Africa with the extermination of the Herero, or in Australia against the Aborigines of Tasmania (Lindqvist 1998). These agricultural frontiers were accompanied by very large migrations,[3] mainly from Europe and Asia. In European colonies in Asia, forms of slavery thus reappeared insidiously. Indeed, the plantation system was dependent on the huge labour pool that was India and China, with the figure of the coolie replacing that of the slave. Even though they were not officially treated as a commodity and migrated of their own free will, the coolies were subjected to a quasi-military discipline and an extremely paternalistic relationship was imposed on them (Stoler 1985; Barral

---

[2] Long-distance trade precedes the formation of national markets. It predates the trade in 'export crops', which marks a boundary between national markets.

[3] Except in Africa, which experienced only marginally the dynamics of agricultural frontiers before 1914.

2015). In Africa and Latin America, extreme forms of violence accompanied the extraction of rubber, with the infamous 'adventures' of King Leopold in Congo (Hochschild 1998) and the Anglo Peruvian Amazon Co. Ltd. in Peru (del Pilar Otero 2009).

In the twentieth century, the proliferation of synthetic products from coal, and later from oil, partially changed the situation. Agricultural production experienced a revival in Europe and the United States through motorization and the massive use of mineral fertilizers and pesticides. The industrial countries' requirements of biomass from distant countries were thereby partially curbed. The demand did not disappear altogether but reoriented itself primarily on products originating in the tropics that non-tropical countries could not easily produce themselves: coffee, cocoa, cotton, natural rubber, banana, wood, and oilseed/protein crops. The rapid industrialization of Japan, and later of China, helped prop up this demand. In response, the old agricultural frontiers continued their progress in South-East Asia and West Africa. New frontiers were opened up, such as in the Amazon or Borneo, and continued destroying ecosystems. While the violence induced by agricultural frontiers no longer involved genocide, it still occurred as frequently because the agricultural system was based on the expropriation of lands and other assets from local populations, as happened not so long ago in Brazil with the gunning down of Chico Mendes, leader of Amazonian rubber tappers, or is happening today in the Meta region of Colombia (Maher 2015).

Nowadays, tropical biomass production for distant markets relies also on chemical products (fertilizers, pesticides), following the European and North American agricultural model. Just like in Europe, the use of these inputs is accompanied by environmental and social effects whose true magnitude is revealed only over time: pollution of the air, water courses, groundwater, or even coastal waters (example of chlordecone use on banana in Guadeloupe and Africa); poisoning of workers applying pesticides; etc.

23.1.2 The Difficulty of Internalizing the Negative Externalities of the Long-Distance Trade in Biomass

How can the existence and magnitude of the negative consequences of biomass production for distant markets be explained? There are several possible answers.

A first answer is that it is indeed in the interest of long-distance trade to externalize its costs, i.e., to let other societies or economies bear the social and environmental costs of the production of certain products. This is the interpretation of analyses in terms of 'unequal' ecological exchanges (Cabeza-Gutés and Martinez-Alier 2001). But this reading hardly reflects the processes that enable such a relationship to exist. Coercion and force have certainly been, and continue to remain, key elements of long-distance trade, especially on agricultural frontiers, but they cannot be the only causal factors. Today, part of this trade takes place voluntarily without any coercion, through the sole medium of market transactions between actors who can be assumed to be honest, even if, as economic theory tell us, they seek above all to maximize their own profit. Why then are these costs not internalized as the dominant economic thinking would seem to suggest?

The analyses of Thomas Princen prove very useful in this context (Princen 1997, 2001). He notes that a growing market implies a constant search for new 'frontiers', if we broaden the notion of frontier to all situations, all resources, or waste sinks over which no legal authority exist. This allows users to claim rights over these frontiers without assuming responsibility for them. In an agricultural or other frontier, the goal is to extract resources from a given location in the most profitable way possible, then to be able to abandon the location to move on to another. According to this perspective, the forest and the savannah, as indeed also the atmosphere and groundwater, can be considered to be frontiers.

According to Princen, what makes these frontiers possible is the ability to 'shade' external costs. While shading is enabled by problems of information, knowledge, and ability to interpret, it must also be seen as the result of the strategic action of States and of companies engaged in oligopolistic competition. These companies have an incentive to maximize the private and collective gains of their commercial activities and to ignore their costs. This becomes easy since the gains are usually immediate and relatively predictable, while external costs are often difficult to estimate, especially if they only become apparent much later on. What also makes the existence of these costs possible is the fact that the end consumers do not have to bear the consequences of their choices and, indeed, do not even receive any information about the costs of their choices. This situation can be analyzed in terms of the distance between the decisions of production and those of consumption that Princen (1997) calls 'distancing'. The distance, which allows external costs to be shaded, however, goes far beyond the notion of geographical distance, and must also be understood, according to this author, in three additional dimensions:

- a 'cultural' dimension when consumers are foreign to the places of production and have a very limited knowledge and ability of understanding and interpretation;
- a 'strategic' dimension linked to asymmetries in powers of negotiation that result from monopolistic and oligopolistic production and trade structures;
- and, finally, a dimension related to the high number of intermediaries ('multiple agencies') between production and consumption, which decreases each actor's responsibility for the state of resources.

These different elements allow the distance to shade (or even completely hide) externalities. They make any intention to internalize external costs very problematic because, in a strategic action of firms and States, the exploitation of remote resources is, by its very nature, conducive to the concealment of these costs and dilution of responsibilities.

Such a reading tends to view these negative effects as inherent to long-distance trade (of which international trade is one of the more common – but not the only –

form[4]) and therefore is quite pessimistic about the possibilities of solving the problem without a more radical approach, one which calls into question this mode of trade itself. However, since the early 2000s, there have been several initiatives that claim to address these issues specifically and make long-distance commodity chains more sustainable at the social and environmental levels. Can these standardization mechanisms restore the visibility of effects 'shaded' by distance and strategic actions of firms and States? Can they help reduce the distance, as defined by Princen (1997), in all its four dimensions (geographical, cultural, strategic-asymmetric, and multiple-agency)?

23.2 The Rise of Private Regulation in Global Value Chains: Sustainability Certification as an Alternative to Public Policy?

23.2.1 The Proliferation of Transnational Voluntary Mechanisms: Recourse to the Market and Stakeholders

Sustainability standardization mechanisms emerged in an attempt to address the problems generated by the long-distance trade in agricultural commodities. Inspired by the model of the Forest Stewardship Council (FSC, created in 1993) for certifying sustainable forests, several roundtables and other multi-stakeholder initiatives to certify sustainable agricultural commodities emerged during the 2000s: Roundtable on Sustainable Palm Oil (RSPO) in 2003, Roundtable on Responsible Soy (RTRS) in 2005, Better Sugar Cane Initiative (BSCI, later renamed Bonsucro) in 2006, Better Cotton Initiative (BCI) in 2007, Roundtable for a Sustainable Cocoa Economy (RSCE) in 2007, Roundtable on Sustainable Biomaterials (RSB) in 2007, and Global Roundtable for Sustainable Beef (GRSB) in 2012. These initiatives are part of an increasingly powerful and structured movement at the global level, represented by the ISEAL alliance (International Social and Environmental Accreditation and Labelling Alliance), which promotes voluntary sustainability standards as forms of regulation capable of controlling globalization and curbing its negative effects (Loconto and Fouilleux 2014).

These sustainability standardization mechanisms are based on voluntary commitments by firms; conformance to them is not imposed by States. They are usually structured around a particular product and the voluntary commitment by various stakeholders of the concerned global value chain (producers, buyers, processors,

[4] 'Long-distance' trade, in Princen's reading, can also take the form of intra-national trade, involving, for example, many intermediaries and/or oligopolistic structures that generate strong power asymmetries.

retailers), as well as by NGOs, banks and sometimes governments. The principle is to agree, through a multi-stakeholder process, on the contents of the sustainability standards that producers will have to meet and on their implementation modalities. The participants draft specifications that list production practices considered as guarantors of sustainability. These practices are codified through principles, criteria, and measurable indicators of sustainability. Producers who decide to adopt the standard voluntarily (or who are forced to do so in order to be able to sell their products) then have to be audited by a third-party certification body that verifies compliance of their practices with the specifications. The certification authority issues a certificate of compliance to the producer, which allows the latter to demonstrate the sustainable quality of his products to his customers. The certificate is also transmitted throughout the global value chain to its various operators. The standards become effective because of market incentives (buyers seeking out sustainable products) and market sanctions (the 'highest' sanction for a producer is his exclusion from the market for certified products). Agricultural commodities are thus each placed at the core of a sectoral standardization mechanism, and social and environmental issues are treated as externalities to be internalized via the standard and the market.

These initiatives are driven by thinking at the international level that does not believe in the structural and functional capacities of governments to address problems caused by globalization. States and international organizations are seen as incapable of managing complex problems, developing interdisciplinary approaches, taking the opinion of the people into account, or even enforcing national laws for the protection of nature and people. In contrast, the growing willingness and action of private actors, civil society organizations, and companies to address these problems are being hailed as the seeds of governance without government (Roseneau and Czempiel 1992). The mechanisms for multi-stakeholder sustainability standardization are apt examples of such initiatives. Gereffi et al. (2001) describe them as NGO-industrial complexes. The launch of the RSPO (Roundtable on Sustainable Palm Oil), through a joint initiative of the multinational firm Unilever, the World Wild Fund for Nature (WWF), the Swiss retailer Migros, the firm Aarhus United UK, and the association of Malaysian producers (MPOA), joined in 2004–2005 by HSBC bank and the International Finance Corporation, is a good illustration (Box 23.1).

23.2.2 New Public-Private Frontiers of Interaction

Nevertheless, the private nature of these mechanisms does not preclude governmental involvement in certain aspects. For example, the governments of the Netherlands, Germany, and Switzerland, all consumer countries of these agricultural commodities, play key roles in supporting these private initiatives through their policies and cooperation actions. The governments of producer countries too are associated in various ways. Although States are not considered full stakeholders in these mechanisms (they often have no voting rights in a general meeting), the governments of producer countries are involved through their active participation in working groups during annual conferences, or even through their financial participation in (and control over) industrial plantation companies, which are full members in these mechanisms, as is the case for the RSPO in Asia (Box 23.1).

Box 23.1 Emergence and Functioning of a Roundtable: The Roundtable on Sustainable Palm Oil (RSPO)

Emmanuelle Cheyns, Benoit Daviron, Marcel Djama, Ève Fouilleux, and Stéphane Guéneau.

The sharp increase in the production of palm oil in Indonesia and Malaysia in the 1990s and the serious forest fires in Indonesia in 1997–1998 began to attract the attention of NGOs. Their subsequent studies have documented the effects of the expansion of industrial oil-palm plantations on deforestation, biodiversity loss, emission of greenhouse gases, as well as on the living conditions of the resident populations: loss of traditional lands, pollution of drinking water, forced evictions, damage to nature, conversion to an export-dependent economy that fragilizes households, dependence of smallholders on monopsony buyers, excessive indebtedness, etc. These NGOs have also denounced the employees' working conditions on some plantations, similar to contemporary practices of slavery: recruitment debts, torture, intimidation, confiscation of identity documents, and child labour for the children of workers (for some recent studies, see Colchester 2011; Jiwan 2011).

The campaigns waged by the NGOs in the early 2000s targeted importers, retailers, and European banks, in a new strategy of engagement, not of States but of private actors downstream of the agri-chains, to convince them to put pressure on their suppliers. In 2002, some of these companies responded favourably to a proposal from WWF to create a multi-stakeholder roundtable. The following year, WWF, Unilever and other companies, and representatives of palm oil producers in Malaysia launched the RSPO initiative, which led to the creation of a voluntary 'sustainable palm oil' standard to be certified by a third party. In 2004, the RSPO was registered as an association under Swiss law with its secretariat in Kuala Lumpur. The first companies were certified in 2008, most of them belonging to Asian groups and consortiums. In 2015, just under 10 % of palm oil traded on the world markets was RSPO certified.

To become a full member of the RSPO, one has to belong to any of the seven categories of stakeholders specified in its bylaws, i.e., have or represent a specific issue or interest. These categories are: oil palm growers (134 members in 2015), palm oil processors and traders (391 members), consumer goods manufacturers (513 members), retailers (59 members), banks and investors (13 members), environmental or nature conservation NGOs

(continued)

Box 23.1 (continued)

(29 members), social or developmental NGOs (12 members). Governments and research institutions are not considered members at the same level as they do not, in theory, represent a specific interest. They can acquire affiliate member status, giving them the right to attend general assemblies, but, unlike other members, enjoy no voting rights. That said, the Indonesian and Malaysian governments have held key positions in groups of national interpretation of the standard, in the working groups, and through the membership of plantation companies (representing the interests of growers) of which they are part or majority owners.

The 'RSPO sustainable palm oil' standard is codified in a set of specifications, which began to be drafted by the stakeholders in 2003. In 2005, the specifications consisted of a list of 8 principles, 39 criteria, and 111 indicators, approved by the members in general assembly. It was subsequently revised in 2013.

The RSPO also institutionalizes activities and mechanisms such as:

- a system for the audit of plantations by certification bodies accredited by the RSPO. These certifiers verify the application of criteria and indicators, and grant – or withhold in the case of non-compliance – a certificate that is tradable on the international market;
- procedures for the payment of compensations and reparations by members who have not complied with the requirements to identify and protect high conservation value areas, i.e., areas that cannot be used to produce oil palm;
- a paralegal mechanism to handle complaints filed against RSPO members, for example for non-compliance with the specifications. The majority of complaints are lodged against plantation companies and pertain to their non-compliance with the principle of free consent of the local people, the violation of land rights, and for not respecting the procedure to identify 'high conservation value areas';
- a mechanism for the resolution of land disputes between local communities and companies, not through an arbitration mechanism but through procedures of amicable mediation and negotiation between the parties concerned.

The RSPO has a budget of seven million Euros (2014 annual data), 31 % of which is funded by the subscription fee its members pay (annual membership fee is 2000 Euros) and 68 % by a levy on the trade in RSPO oil certificates (1 US $/T for a total of 4.7 million Euros in 2014). This budget funds the staff expenses of the Executive Secretariat in Kuala Lumpur and the liaison office in Jakarta, the meetings of the executive committee, and the activities of multi-stakeholder working groups on compensation procedures, revision of specifications, certification procedures, etc.

Despite the ambivalence of States towards them, these standards have institutionalized a form of hybrid governance of agri-chains, linking public and private regulations in a variable manner depending on the economic sector concerned and the dynamics of national and global policies. These new interactions between public and private sectors affect global value chains in several ways. First and foremost, the sustainability standards simultaneously reflect a kind of privatization and a new form of regulation of agri-chains, now increasingly being subjected to the upheavals of deregulation and withdrawal by the State. In the wake of privatizations, sustainability standards sometimes substitute for public policy. An example is the adoption by Bolivia of the Forest Stewardship Council (FSC) as the preferred instrument to manage its forests (Carey and Guttenstein 2008). In other cases, these standards promote the privatization of public services, such as of agricultural extension, when public structures that used to provide these services are dismantled.

These privatization phenomena do not reflect a decline of the State's authority or represent the only competition between modes of public and private governance. It is mainly the countries of the North that have promoted the dissemination of private sustainability standards by providing expertise, intellectual reasoning within international organizations, sometimes funding (Bartley 2007; Gulbrandsen 2014), or by accrediting private standards for access to their markets (Directive of the European Union on biofuels). And countries in the South have adopted them proactively to obtain competitive advantages or to ensure access to markets (Carey and Guttenstein 2008). For example, the Brazilian government created favourable conditions for those actors already inserted into international markets to access these certification mechanisms (Guéneau and Drigo 2013; Nepstad et al. 2014). Other countries, both in the South as well as in the North, often promoting the agendas of various economic interest groups, oppose them or take advantage of the potential competition between different standards. For example, in a comparative study on the implementation of two certification schemes (for forests with the FSC and for fisheries with the Marine Stewardship Council (MSC)), Gulbrandsen (2014) shows that States (mainly from the North) take varying positions depending on their strategic and industrial interests in these agri-chains. Furthermore, the creation of competing national standards (public and mandatory) is one of the main strategies adopted by countries to reduce the scope of private sustainability standards or even to mandate more flexible specifications for companies. For example, the Indonesian government adopted regulation governing the production of palm oil (Indonesian Sustainable Palm Oil) in 2011 to make this activity conform to sustainable development requirements and, at the same time, reduce the scope of the voluntary RSPO sustainability standard (Roundtable on Sustainable Palm Oil). While this national regulation has, in the perception of some actors, the advantage of being mandatory in nature (more widely binding) and leading to a certain rationalization of disparate national laws, it also excludes major criteria of the private voluntary RSPO standard, such as the protection of forest areas and the recognition of land rights of local and indigenous communities.

In the next section, we will examine how these private voluntary standards were built and how they tackle issues of sustainability. Have these standards adequately addressed social and environmental criticisms expressed on the public stage as well as the concerns of the populations affected by the expansion of industrial-scale farming for exports?

23.3 Creating Standards and Qualifying Sustainability: Do the Standards Live Up to Their Promise?

These mechanisms for the creation of voluntary and private standards – which aim to take on a common good such as sustainability – base their legitimacy on multi-stakeholder composition. This aspect is presented as evidence of the integration of a wide variety of stakeholder interests and thus of legitimacy. It is based on the assumption, at the core of a wider political model of liberal pluralism, that a public good will emerge from this process of negotiation between specific interest groups and of finding a balance between them. It also relies on a participatory rhetoric in decision making, through the implementation of a process of dialogue claimed to be horizontal, on openness to voluntary participation, and on decision making based on the consensus of members (Cheyns and Riisgaard 2014). These characteristics of governance by stakeholders have important implications for addressing social and environmental issues. On the one hand, they lead to a depoliticization of the debate and therefore to the exclusion of certain qualifications of the common good, thus calling into question the professed goals of openness and inclusiveness. On the other hand, they exclude from consideration some relationships that people have with their environment, thus concealing part of the damage and the externalities.

23.3.1 Creating Standards: Inclusion of the Political Perspectices in Question

An ever growing body of work shows that the search for a quick agreement between stakeholders with disparate and divergent interests leads to a reduction in the diversity of the perspectives within debates and the exclusion of many participants. Furthermore, those directly affected by the expansion of large-scale cropping systems, whether designated as vulnerable, disadvantaged, marginalized, indigenous peoples, workers or smallholders, feel that their concerns are not heard, even if they are expressed during the negotiations of these standards (Fransen and Kolk 2007; Utting 2002; Cheyns 2011, 2014; Schouten et al. 2012). How have these perspectives and these participants been excluded?

A first set of studies note an imbalance of power between stakeholders, and therefore between the interests being represented. These studies either highlight the poor representation of vulnerable groups or point out the difficulty these initiatives have in regulating the balance of power between stakeholders, especially between organizations from the North, which generally assume key positions, and those from the South (Utting 2002; Fransen and Kolk 2007; Guéneau 2012; Fouilleux 2013), or even between different organizations from the South with access to varying levels of resources. The representation of family farmers (or smallholders) and workers in industrial plantations has frequently been usurped by other stakeholders, such as plantation companies, international NGOs, parastatal organizations, or even banks (Nelson and Tallontire 2014; Cheyns 2014), even though their interests are far from being identical. The substitution of their voices by plantation companies tends to perpetuate existing local hierarchies, especially the 'domestic' subordination (Boltanski and Thévenot 1991) of family farmers and workers to plantation companies (Cheyns 2014), even though this is denounced by these categories of actors as forms of paternalism (Barral 2015).

Other studies highlight another criticism, more fundamental, that challenges the claim that this liberal model of balance of stakeholder interests is a model of inclusiveness. These studies have expressed doubt about the ability of these roundtables to accommodate categories of participants other than of those who express and defend an interest. They show that these initiatives have reduced the expression of pluralism to a pluralism of interests. These roundtables have thus ruled out the possibility for participants to engage in qualifying a common good (What does 'sustainable' mean? What meaning to assign to this term?), open to a pluralism in the sense of what is just and unjust in practice, as well as to the actors' critical capacities (Thévenot 2006). In a continuing effort to minimize dissension, debates are framed in a way that restricts participants to the expression of individual preferences or the need for and the development of final results, measurable and accepted by the market, rather than to a possibility of conducting a substantial discussion on the very content of sustainability (Cheyns 2011; Schouten et al. 2012; Djama and Verwilghen 2012). These studies reveal a process of depoliticization of the debate and the action, based on a two-pronged approach: the exclusion of a political debate open to the actors' critical capacities, and the promotion, at the same time, of goal-oriented decision-making processes based on technical elements. Negotiating on the basis of a list of technical criteria of sustainability that are backed by specific interests is emblematic of this restrictive framework. Rather than debate the principles of sustainability and uncover divergences of perspectives and visions, which raise fears of excessive discord and of the mechanism's collapse, participants are directed to the elaboration of technical criteria backed by the various categories of interests.

Citation 'What is sustainability? If you start discussing it, you will never agree. It will take forever. Nobody has the same vision!' (nature conservation NGO, 2009, the Netherlands).

Lead NGOs of these mechanisms justify this exclusion of a process to qualify the common good (or of topics they deem 'too political') on the basis of an imperative of pragmatism, one that originates from a desire to arrive at an agreement with the largest possible number of industry representatives. This concern has led to the hasty endorsement of the principles of 'industrial' efficiency and 'market' competition (Boltanski and Thévenot 1991), and, in particular, to the naturalization of the growth of market demand, presented in advance as a condition to discuss sustainabiity options (Cheyns 2011).

Citation 'The activity of the industry is to produce. The industrialists say, "We produce what the market demands." So we have to assume that there is a demand for soya beans and palm oil and it will have to be fulfilled. (...). Ideally, we should be able to say, "Stop the production of soya bean!" But, pragmatically, there is no way this will happen. There are too many interests involved. If companies can make money with soya bean, they are not going to stop. NGOs like us must be pragmatic and establish a set of criteria to continue to allow production of what the market demands and, at the same time, to preserve a part of nature or social values. That is the balance that we have to find' (nature conservation NGO, 2009, the Netherlands).

This restrictive framing of the debate very quickly leads to the sidelining of participants who want to question the existing models of production, trade, and consumption. Such was the case, for example, of the Federation of Workers of Family Farming in Southern Brazil (Fetraf-Sul), which, concerned by the expansion of soya bean monocultures, intensification, and the use of GMO seeds, took advantage of its membership of the Roundtable on Responsible Soy (RTRS) to question soya bean production models and criticize the domestic and market subordination of farmers by firms. The federation finally chose to resign from the roundtable when the organizing committee informed it that these questions on 'monoculture versus agricultural diversification' and 'GM versus non-GM' models could not be part of the RTRS debate because of the very fundamental and wide differences between these production models (Cheyns 2011; Fouilleux 2013). This refusal to discuss production models (and a diversity of practices) has also often been justified by the participants by a single-product sectoral framing ('We are here to discuss only soy', RTRS workshop in 2006).

In a similar manner, the formatting of the debate in the RSPO in terms of interests being represented has prevented the issues of injustice that local communities and farmers want to raise to be addressed. These mainly concern access to land, production models, and contracts between firms and smallholders (Cheyns 2014). This became particularly evident when farmers, representing their union, attempted to criticize their dependence on a 'paternalistic' and 'neo-colonial' model, one that was highly asymmetric with regard to contracts that bind them to specific plantation or processing companies in monopsony situations. At the same time, their request for more equity in the agri-chain was ignored. They had hoped that this would have led to a reduction in inequalities in access to and use of resources. These interventions by farmers and rural communities were deemed inadmissible and it was suggested to them that they engage in case-by-case negotiations on the margins of the collective debate in a process of bilateral dialogue with the relevant production companies (Cheyns 2011, 2014; Köhne 2014). But the refusal to address the issue of inequalities in these mechanisms has led to their exacerbation in some highly asymmetric national contexts. Selfa et al. (2014) show that the framing of the Bonsucro standard in the form of a list of technical and legal criteria – such as 'compliance with national regulations' regarding access to land – in an approach claimed to be apolitical or neutral, only strengthened and exacerbated the unequal access to land in Colombia, precisely because the standard, by providing a new legitimacy, excluded the addressing of this issue in its definition of sustainability.

23.3.2 A Restricted Vision of Sustainability

What meaning has been assigned to sustainability in these standards' specifications? Organized each around a particular product, these mechanisms have, through their specifications, perpetuated the principles of industrial and technical efficiency prevailing in earlier decades (Cheyns 2011; Nelson and Tallontire 2014). They reiterate the principles of division of labour, through agricultural specialization (by type of biomass) and specialized companies (by techno-economic operations), in an effort to maximize production.

The specifications thus define sustainability through a rationalization and separation of agricultural production and natural (or forest) spaces. Sustainability is therefore based on the ability of producers to intensify agricultural production in order to reduce pressure on ecologically sensitive areas and to maximize the biological productivity of these areas (forests, high conservation value areas). A form of revival of productivism (Fouilleux and Goulet 2012), this approach puts the maximization of yields per hectare and productivity at the heart of sustainable practices, in an implicit land-sparing decision. In the case of the RSPO, for example, its principle 3, 'Commitment to long-term economic and financial viability', reflects a goal of high yields of palm bunches per hectare. The latest initiative, the Sustainable Natural Rubber Initiative, adopts the same orientation by including a minimum per-hectare yield criterion in order to obtain the 'sustainable' label, through seed optimization, stand density, and fertilizers.[5]

In addition to this complete separation of productive and 'natural' spaces, the desire to maximize land productivity has resulted in the use of pesticides to be continued to be allowed, including of pesticides banned in Europe (such as paraquat) or banned by the Stockholm and Rotterdam Conventions.[6] The decision to

[5] http://snr-i.org/file/file/SNR-i_KPI_document_June_EN_for_Ref.pdf, retrieved 30 May 2016.

[6] In the RTRS standard, the use of pesticides listed by the Stockholm and Rotterdam Conventions is prohibited, but the use of paraquat has been allowed 'at least until 2017' (RTRS was created in 2006), and 'which deadline could be extended if enough evidence is put forward before June 2016 demonstrating that there are no alternatives in the market', http://www.responsiblesoy.org/documentos/estandar-rtrs-para-la-produccion-de-soja-responsable/, retrieved 30 May 2016.

In the case of the RSPO standard, since 2013, these pesticides (paraquat, lists of the Stockholm and Rotterdam Conventions) are subject to 'plans of minimization' but are permitted in certain situations defined by the groups responsible for implementing the standard at the national level. Until 2013, they were allowed without any restrictions pertaining to specific situations (RSPO was established in 2003 and the specifications adopted and implemented in 2005), http://www.rspo.org/file/revisedPandC2013.pdf, retrieved 30 May 2016.

allow GMO seeds in the RTRS standard (responsible soy) was similarly supported, including by nature conservation NGOs, on the grounds that their use would help agricultural intensification. Finally, in the RSPO, current debates tend to promote models of partnership between agro-industries and smallholders, encouraged by the Indonesian government, in which agro-industries relieve smallholders from the responsibility of maintaining and managing their plots in order to optimize yields. Thus, forms of large-scale and monoculture production, promoted in the last few decades, are recast as sustainable agriculture. The adoption of these models is expected to continue to increase the production of agricultural commodities and their trade on international markets, on the condition of preserving untouched areas that maximize, on their part, the conservation of biodiversity and carbon storage. But this is being done at the expense of other forms of non-specialized and/or more localized farming, such as mixed livestock/crop farming, agroecology, agroforestry, and diversified family farming, practiced by family farmers and supported by family farming organizations, smallholders, and civil society[7] (Cheyns 2011; Selfa et al. 2014; Nelson and Tallontire 2014).

At the social level (distinct from the environmental level in the list of criteria), sustainability is generally handled in two related ways in these initiatives. On the one hand, agro-industrial production companies highlight their contribution to the social pillar through value creation and the employment, foreign exchange, and infrastructure they generate, and thereby emphasize their contribution to poverty reduction. This macro-economic vision of well-being, which reflects the interest in agro-industrial forms of production and in increased supply on international markets, however, is challenged by civil society and those NGOs that are not members of these roundtables (Escobar and Cheyns 2012; Schouten et al. 2012). On the other hand, lists of social criteria, largely endorsed by the 'social' NGOs, are now starting to find place in the specifications of these sustainability standards. An analysis of the criteria of seven standards (including those of roundtables for sustainable soya bean, oil palm, biomaterials, and sugarcane) shows that these social criteria are based to a large extent on a human rights approach (Sirdey and Cheyns 2015). This approach is a necessary response to the numerous violations reported on the ground (dispossession of land, forced labour, etc.) that can hardly be overlooked. But the question of the ability of workers and local communities to effectively assert their rights within the ambit of such mechanisms, even though of crucial importance, is not given sufficient consideration (Silva-Castañeda 2015). Furthermore, such an approach would draw attention to other political constructions of social sustainability that have been excluded. For example, the standards accord little importance to reducing the vulnerability of populations through a sustainable improvement in their abilities and their livelihoods. In a similar way, they have excluded the political constructions that aim to reduce inequalities in the distribution and access

[7] See, for example, the campaign launched in 2008 by the Friends of the Earth International federation.

to resources, even though they are at the core of the political project of peasant movements and civil society (Sirdey and Cheyns 2015; Guéneau et al. 2015).

Thus, by naturalizing the principles of industrial and market efficiency and by preventing their critical scrutiny, these initiatives exclude many civic demands of peasant movements (and of other NGOs). They exclude, in particular, demands that pertain to the redistribution of value, access to resources, and independence of farmers vis-à-vis firms to reduce structural asymmetries of power, arising from situations of concentration (monopsony of the buyer of palm bunches, oligopolies of seed suppliers, etc.). This is the case, for example, of the demand of family farmers, who seek an independent access to credit markets (outside the purview of the concerned agri-chain), which would allow them to withdraw from contracts with monopsony buyers (Cheyns 2014; Selfa et al. 2014; Nelson and Tallontire 2014). Furthermore, the decision to undertake specialized agriculture for distant markets is made at the expense of diversified-agriculture production models close to those of family farmers. These decisions have led individuals and organizations to refuse to participate in these initiatives in the first place or to resign from them, to challenge them publicly, or assert alternative forms of production and trade in other arenas more open to their participation (Escobar and Cheyns 2012; Nelson and Tallontire 2014; Cheyns and Riisgaard 2014). This was the case, for example, of the campaign of international and South American NGOs, which began denouncing the Roundtable on Responsible Soy in 2008, or of South American forums bringing together family farming organizations to promote another model of agriculture, re-embedded in local territories and independent of firms, thus reducing the various dimensions of distance, not only geographic, but also those pertaining to the value chain's structure.

23.3.3 Qualifying, Measuring, and Making Reparations: The Incommensurability of Damages

Finally, one wonders whether sustainability is fully measurable by these standards. If the standards are supposed to expose the deleterious effects of agricultural production destined for long-distance trade and internalize them, one has to ask whether the instrument itself has the ability to assess everything, measure everything. How to expose and assess the damage? How to equate environmental and social damage to a form of reparation? In the case of the RSPO, Cheyns (2011) and Silva-Castañeda (2012) have shown that the form taken by the standards has led to the exclusion of forms of knowledge and information that cannot be measured by this instrument. These authors thus highlight the difficulty to take into account damage to the environment and to collective or subsistence way of life, expressed by local populations, either at the time of the creation of the standard or in the way the auditors (and members of the internal mechanisms for complaints resolution) consider the forms of evidence that make sense and exclude others.

Studies have emphasized the exclusionary effects of the limited forms of knowledge and information that these mechanisms recognize. Technical, legal, and statistical forms of information, and technical and scientific skills, accepted as principles of neutrality or objectivity, are often used to disqualify the voices of local communities and their knowledge and narratives. This is the case of the narratives that portray monographic experiences, that are based on contextualized real-life stories, or that refer to personal and ancient footprints on nature (lived through personal connections with old trees or rivers, for example) which only make sense for those who dwell in the environment and share a link with it (Bain and Hatanaka 2010; Cheyns 2011, 2014; Silva-Castañeda 2012; Nelson and Tallontire 2014). In the case of the RSPO, local communities find themselves excluded from the debate when trying to report violations of their daily lives and in their physical and personal environment caused by the expansion of oil palm plantations, and any consequent loss of traditional land or destruction of their environment, which they do with great emotion and using familiar markers to which non-locals cannot relate. In this way, the local people communicate a strong attachment that embarrasses and is foreign to other participants. Since they do not communicate in terms of an interest, their input is unusable in any trade-off between interests (Thévenot 2015; Cheyns 2014; Silva-Castañeda 2015).

Similarly, by favouring systems of monetary compensation for the damage done to people and nature, these initiatives are based on the possibility of relocating populated areas (and uprooting of populations) which, though understandable from a macro-environmental standpoint, rarely compensates for the violations against people living in and attached to specific places. The internalization of costs implies the use of a common metric to compare and equate the gains and costs, and to calculate their distribution between different actors. Various authors have however pointed out this approach's inherent limitation: a 'radical incommensurability' of some forms of damage to the 'dwelled-in environement', where memories are deposited (Centemeri 2014). In her study of the complaints of local communities to the RSPO, Silva-Castañeda (2015) shows how the transformation of community concerns (pertaining to personal and intimate attachments to the environment built over time) into monetary compensations hardly meets their expectations of undoing or rectifying the damage caused to the places they live in, where the environment is almost a constitutive part of their person (Centemeri 2014). The negotiating processes that the RSPO proposes require a transformation of attachments to the environment and of actions that these communities see as a violation of their rights (for example, land rights) into mutually beneficial options, both for the communities and for the companies with which they are in conflict. This is a case of 'win-win' negotiations in the perception of the RSPO, with, for example, the creation of productive projects with the aim of sharing the potential benefits between the two parties (instead of engaging in land restitution). These 'mutually beneficial' options, however, can only be based on a metric of economic interests (Silva-Castañeda 2015).

The inability of transnational mechanisms to accommodate these categories of participants, by taking into consideration their attachments to the environment and familiar relationships with it, is also a reason why many local voices are excluded and why certain damage is rendered invisible.

23.4 Conclusion

These voluntary standardization mechanisms aim to make biomass production for distant markets compatible with the challenges of sustainable development. Regardless of the particular definition of sustainable development one accepts, it is expected that, at a minimum, the degradation of resources (through removal or pollution) would stop as also the social violence resulting from production.

Through the greater role they assign to NGOs, private sustainability standards certainly help raise the visibility on the international stage of local disputes, environmental damage, and human rights violations (Colchester 2011). This is even truer in countries with authoritarian regimes, where relatively weak social movements struggle to uncover certain local effects. These movements find, sometimes unexpectedly, a new space for raising awareness and support. But, despite their ambition and claims of inclusion, these standards exclude some participants and perspectives from the wider public debate.

These standards also encourage the emergence of unexpected coalitions between heterogeneous participants, between NGOs and firms in the downstream sectors, for example. They may, at times, even strengthen reformist elements working within the State apparatus (Cashore and Stone 2014). That said, they do not seem at present able to disrupt local and national interest regimes of the political and economic elites (McCarthy et al. 2012) that structure the global value chains in question.

Private standardization mechanisms exclude from consideration social and environmental concerns raised in the broader public debate and by the very people who are affected. While these mechanisms do allow a form of supervision and monitoring from a distance, they are unable to internalize all the local concerns or accord them visibility. Indeed, the agreements made around standards favour and assign value to production models based on the specialization and division of labour, associated with an enlargement of the commercial space and thus of distance. They also perpetuate the prevalence of oligopolistic and monopsonic structures, maintaining a strategic distance of power asymmetries. Finally, the cultural distance – which we prefer to rephrase as the 'incommensurability of damage' to the environment and to people's lives and livelihoods (Centemeri 2014) – has also not been seriously addressed by these mechanisms. By maintaining structural forms of distance (Princen 1997), these mechanisms have, at the end, completely excluded alternative constructions of sustainability in which the goal was precisely to reduce the various dimensions of distance by promoting less specialized agricultural production, a reduction of asymmetries by greater independence of the producers from firms, or by relocated supply channels, closer to the consumers, not only from the point of view of the territory, but also of the number of intermediaries or of shared knowledge.

For now, we have to ask ourselves whether meeting the challenges of sustainable development does not require us, even before we choose one or more standardization instruments, to call into question the logic itself of producing biomass 'at a distance', in its different dimensions?

References

Bain C, Hatanaka M (2010) The practice of third-party certification: Enhancing environmental sustainability and social justice in the global South. In: Higgins V, Larner W (eds) Calculating the Social: Standards and the Re-configuration of Governing. Basingstoke, Palgrave MacMillan, pp 56–74

Barral S (2015) Capitalismes agraires : économie politique de la grande plantation en Indonésie et en Malaisie. Presses de Sciences Po, Paris, 240 p

Bartley T (2007) Institutional emergence in an era of globalization: the rise of transnational private regulation of labor and environmental conditions. Am J Sociol 113(2):297–351

Boltanski L, Thévenot L (1991) De la justification. Les économies de la grandeur. Gallimard, Paris, 496 p

Cabeza-Gutés M, Martinez-Alier J (2001) L'échange écologiquement inégal. In: Damian M, Graz JC (eds) Commerce international et développement soutenable. Economica, Paris, pp 159–185

Carey C, Guttenstein E (2008) Governmental use of voluntary standards: innovation in sustainability governance. In: Network report 079. Iseal Alliance & Trade Standards Practitioners, London, 32 p

Cashore B, Stone MW (2014) Does California need Delaware? Explaining Indonesian, Chinese, and United States support for legality compliance of internationally traded products. Regul Gov 8(1):49–73

Centemeri L (2014) Reframing problems of incommensurability in environmental conflicts through pragmatic sociology. Environ Value 24(3):299–320 http://www.whpress.co.uk/EV/EV2412.html

Cheyns E., 2011. Multi-stakeholder initiatives for sustainable agriculture: limits of the "inclusiveness" paradigm. In: Ponte S, Gibbon P, Vestergaard J (eds) Governing through Standards: Origins, Drivers and Limitations. Palgrave Macmillan, London, pp 318–354

Cheyns E (2014) Making "minority voices" heard in transnational roundtables: The role of local NGOs in reintroducing justice and attachments. Agric Hum Values 31(3):409–423

Cheyns E, Riisgaard L (2014) Introduction to the symposium. The exercise of power through multi-stakeholder initiatives for sustainable agriculture and its inclusion and exclusion outcomes. Agric Hum Values 31(3):439–453

Clapp J (2014) Financialization, distance and global food politics. J Peasant Stud 41(5):797–814

Colchester M (2011) Palm oil and indigenous peoples of South East Asia: land acquisition, human rights violations and indigenous peoples on The Palm Oil Frontier. Rome/Moreton-in-Marsh (United Kingdom), International Land Coalition/Forest Peoples Programme, 36 p

Costa MH, Yanagi SNM, Souza PJOP, Ribeiro A, Rocha EJP (2007) Climate change in Amazonia caused by soybean cropland expansion, as compared to caused by pastureland expansion. Geophys Res Lett 34(7):L07706

del Pilar Otero D (2009) Fronteras, etnocidio y comercio mundial. La Amazonía peruana-colombiana durante el boom cauchero. Revista de Estudios Avanzados 11:79–98

Djama M, Verwilghen A (2012) Figures de l'expertise dans un dispositif de normalisation en agriculture durable. L'exemple de la certification « huile de palme durable » (RSPO). In: Alphandéry P, Djama M, Fortier A, Fouilleux E (eds) Normaliser au nom du développement durable. Éditions Quæ, Versailles, pp 175–192

Escobar M, Cheyns E (2012) NGOs acting along global supply chains: Between market and politics. An assessment of campaigns and regulatory devices in the palm oil and soy sectors. In: Workshop: responsible supply chains and networks: challenges for governance and sustainability, Stockholm, 22–24 November 2012, 31 p
Fouilleux E (2013) Normes transnationales de développement durable. Formes et contours d'une privatisation de la délibération. Gouvernement et action publique 2(1):93–119
Fouilleux E, Goulet F (2012) Firmes et développement durable : le nouvel esprit du productivisme. Études rurales 190:131–146
Fransen LW, Kolk A (2007) Global rule setting for business: a critical analysis of multistakeholder standards. Organization 14(5):667–684
Gereffi G, Garcia-Johnson R, Sasser E (2001) The NGO-industrial complex. Foreign Pol 125:56–65
Guéneau S (2012) La participation comme processus de construction d'une gouvernance environnemental globale : le cas du FSC. In: Alphandéry P, Djama M, Fortier A, Fouilleux E (eds) Normaliser au nom du développement durable. Éditions Quæ, Versailles, pp 121–138
Guéneau S, Drigo IG (2013) Interações entre regulação pública e privada: a influência da certificação florestal privada sobre a renovação da ação pública na Amazônia brasileira. In: Carneiro MS, Amaral Neto M, Ramos Castro EM (eds) Sociedade, floresta e sustentabilidade, Instituto Internacional de Educação do Brasil/NAEA, Belém, 17–42
Guéneau S, Carneiro MS, Toni F (2015) Le Brésil veut-il d'un modèle de croissance agricole durable ? In: Pachauri RK, Paugam A, Ribera T, Tubiana L (eds) Regards sur la Terre 2015. Construire un monde durable. Armand Colin, Paris, pp 221–232
Gulbrandsen LH (2014) Dynamic governance interactions: evolutionary effects of state responses to non-state certification programs. Regul Gov 8(1):74–92
Hochschild A (1998) King Leopold's ghost: a story of greed, terror, and heroism in colonial Africa. Houghton Mifflin, Boston
Jiwan N (2011) Oil palm expansion and the need for human right standards in Indonesia. In: Chao S, Colchester M (eds) Human rights and agribusiness. Plural legal approaches to conflict resolution. Moreton-in-Marsh (United Kingdom)/Bogor (Indonesia), Forest Peoples Programme/Perkumpulan SawitWatch, pp 93–106
Köhne M (2014) Multi-stakeholder initiative governance as assemblage: roundtable on sustainable palm oil as a political resource in land conflicts related to oil palm plantations. Agric Hum Values 31(3):409–423
Lindqvist S (1998) Exterminez toutes ces brutes. In: L'odyssée d'un homme au cœur de la nuit et les origines du génocide européen. Le Serpent à Plumes, Paris, 234 p
Loconto A, Fouilleux E (2014) Politics of private regulation: Iseal and the shaping of transnational sustainability governance. Regul Gov 8(2):166–185
Maher D (2015) Rooted in violence: civil war, international trade and the expansion of palm oil in Colombia. New Pol Eco 20(2):299–330
McCarthy J, Gillespie P, Zen Z (2012) Swimming upstream: local Indonesian production networks in globalized palm oil production. World Dev 40(3):555–569
Monbeig P (1952) Pionniers et planteurs de Sao Paulo. Armand Colin, Paris, 376 p
Nelson V, Tallontire A (2014) Battlefields of ideas: changing narratives and power dynamics in private standards in global agricultural value chains. Agric Hum Values 31(3):409–423
Nepstad DC, Stickler CM (2008) Managing the tropical agriculture revolution. J Sustain For 27 (1–2):43–56
Nepstad DC, McGrath D, Stickler C, Alencar A, Azevedo A, Swette B, Bezerra T, DiGiano M, Shimada J, da Motta RS, Armijo E, Castello L, Brando P, Hansen MC, McGrath-Horn M, Carvalho O, Hess L (2014) Slowing Amazon deforestation through public policy and interventions in beef and soy supply chains. Science 344(6188):1118–1123
Princen T (1997) The shading and distancing of commerce: when internalization is not enough. Ecol Econ 20(3):235–253

Princen T (2001) Consumption and its externalities: where economy meets ecology. Glob Environ Pol 1(3):11–30
Rosenau JN, Czempiel EO (dir.) (1992) Governance without government: order and change in world politics, vol 4. Cambridge University Press, Cambridge, 324 p
Schoneveld GC (2014) The geographic and sectoral patterns of large-scale farmland investments in sub-Saharan Africa. Food Policy 48:34–50
Schouten G, Leroy P, Glasbergen P (2012) On the deliberative capacity of private multi-stakeholder governance: the roundtables on responsible soy and sustainable palm oil. Ecol Econ 83:42–50
Selfa T, Bain C, Moreno R (2014) Depoliticizing land and water "grabs" in Colombia: the limits of Bonsucro certification for enhancing sustainable biofuel practices. Agric Hum Values 31 (3):409–423
Silva-Castañeda L (2012) A forest of evidence: third-party certification and multiple forms of proof – a case study of oil palm plantations in Indonesia. Agric Hum Values 29(3):361–370
Silva-Castañeda L (2015) What kind of space? Multi-stakeholder initiatives and the protection of land rights. Int J Soc Agricul Food 22(2):67–83
Sirdey N, Cheyns E (2015) La durabilité sociale dans les normes transnationales volontaires de développement durable : tensions entre droits, inégalités et capacités. In: Colloque Recherche et régulation. La théorie de la régulation à l'épreuve des crises, Paris, 10–12 June 2015
Stoler AL (1985) Capitalism and Confrontation in Sumatra's Plantation Belt, 1870–1979. Yale University Press, New Haven, 304 p
Thévenot L (2006) L'action au pluriel : sociologie des régimes d'engagement. La Découverte, Paris, 312 p
Thévenot L (2015) Making commonality in the plural, on the basis of binding engagements. In: Dumouchel P, Gotoh R Social bonds as freedom: revising the dichotomy of the universal and the particular. Berghahn, New York, pp 82–108.
Utting P (2002) Regulating business via multistakeholder initiatives: a preliminary assessment. In: Voluntary approaches to corporate responsibility: readings and a resource guide (NGLS). NGLS, Geneva, 61–130.s

Chapter 24
Ecosystem Services, Payments for Environmental Services, and Agri-Chains: What Kind of Regulation to Enhance Sustainability?

Estelle Biénabe, Céline Dutilly, Alain Karsenty, and Jean-François Le Coq

No agricultural production system anywhere in the world can today afford to ignore environmental issues, a reality that is widely reflected in the increasingly significant agroecological movement, which is pushing for the promotion of agri-environmental practices, control over pollution and emissions of greenhouse gases, a halt to deforestation, and, in general, the prevention of the artificialization of natural areas, the adoption of ecological compensation mechanisms, etc. The political and social recognition of what we shall hereafter refer to as ecosystem services, which sometimes have a global scope (global public goods such as biodiversity and climate), complements that of services usually expected from agriculture at local and national scales: income, employment, food security, and local development. This chapter contributes to this part of the book, which examines the role of agri-chains as arenas of regulation of sustainable development, by addressing this issue more specifically through the use of ecosystem services and payments for environmental services, as a complement to approaches built within the agri-chains. We first clarify these concepts and show how payments for

E. Biénabe (✉)
CIRAD – UMR Innovation, F-34398 Montpellier, France
e-mail: estelle.bienabe@cirad.fr

C. Dutilly
CIRAD – UMR Selmet, Montpellier, France
e-mail: celine.dutilly@cirad.fr

A. Karsenty
CIRAD – UPR BSEF, Montpellier, France
e-mail: alain.karsenty@cirad.fr

J.-F. Le Coq
CIRAD – UMR Art-dev – UNA, Heredia, Costa Rica
e-mail: jean-francois.le_coq@cirad.fr

E. Biénabe et al. (eds.), *Sustainable Development and Tropical Agri-chains*,
DOI 10.1007/978-94-024-1016-7_24

environmental services have evolved from an initial, very 'Coasian'[1] concept of 'polluted pays' to a public action instrument that aims to promote agroecological transitions at large territorial scales. After noting the benefits and limitations of environmental and sustainable labels in relation to payments for environmental services, we propose an original and integrative approach to such payments, which combines agri-chain approaches (labels, zero deforestation in particular) and territorial approaches to reconcile the conservation and development of territories by ensuring the sustainability of agricultural activities.

24.1 Ecosystem Services, Environmental Services, and Payments for Environmental Services: Concepts to Be Defined

24.1.1 Origins of the Concepts

The concept of ecosystem services was introduced in the 1970s by the ecological and economic disciplines, with the aim of encouraging a global perspective of environmental problems. It was first popularized in the late 1990s by Costanza et al. (1997), who calculated an overall economic value of all ecosystem services provided by the biosphere (even though this term was not used), and, subsequently, by the Millennium Ecosystem Assessment (2005). The latter proposed a typology of ecosystem services[2] based on a comprehensive schematic framework of the links between these services and human well-being. Metaphors to illustrate the dependence of human societies on ecosystems (Norgaard 2010), ecosystem services are now increasingly being used to highlight the benefits people derive from nature, resulting in the development of approaches to recognize, measure, and integrate these services in private and public decision making.[3]

Payments for environmental services are, in fact, a set of empirical practices and mechanisms to manage water and soil at the watershed scale or to protect forests.

[1] Applying the 'Coase theorem' to the identification of payments for environmental services suggests that 'in a world where transaction costs are zero, and where property rights are clearly defined, a free play of negotiations results in an optimum independent of the initial allocation of rights' (Bertrand and Destais 2002).

[2] Provisioning services, i.e., products (food, water, energy, etc.) obtained from ecosystems; cultural services and amenities (spiritual, recreational, aesthetic benefits, etc.); regulation services linked to ecosystem processes (climate regulation, water and air purification, etc.); services that help maintain conditions suitable for life on earth, i.e., those necessary for the provision of all other services (nutrient cycling, formation and retention of soil and humus, creation and maintenance of natural habitats, etc.).

[3] The Economics of Ecosystems and Biodiversity (TEEB) initiative, which proposes, in continuation of the Millennium Ecosystem Assessment, an economic framework and methodological tools for the conservation of ecosystem and biodiversity services, is a prominent illustration.

They are all based on the direct incentives for conservation offered to local actors whose practices are potentially harmful to ecosystems and who have 'property rights'[4] that permit such practices. The use of the term 'payments for environmental services' gained currency with Costa Rica's national PSA[5] programme in 1996. Other programmes were implemented at the same time or later, including China's Grain for Green programme in 1999, Mexico's payment for hydrological environmental services programme (PSA-H) in 2003, Ecuador's *Sociobosque* programme in 2008 as well as many local initiatives.[6]

Since payments for environmental services were based on direct incentives, contracts, and conditional remunerations, institutions and development agencies quickly showed interest in them. They are *a priori* considered to be more efficient and better able to mobilize private sector resources than 'command and control'[7] measures and integrated conservation and development[8] programmes (Gómez-Baggethun and Muradian 2015). S. Wunder, a CIFOR researcher, proposed a definition of payments for environmental services in 2005 that was simple and seemingly clear. An initial ideal type of payment for environmental services emerged from his definition that was subsequently widely used: 'a voluntary transaction where a well-defined environmental service (or a land-use likely to secure that service) is being 'bought' by a (minimum one) environmental service buyer from a (minimum one) environmental service provider if and only if the environmental service provider secures environmental service provision (conditionality).' This definition, based on a commercial exchange (buyer, supplier), has led to a persistent misunderstanding on the nature of the instrument.

The fact that this definition was proposed in 2005 led to a confusion between the environmental services of PES[9] and the 'ecosystem services' of the Millennium Ecosystem Assessment, which were popularized in the same year, and contributed to the misunderstanding regarding the instrument. Few researchers clearly differentiate between ecosystem services ('benefits that men derive from nature') and

[4] The concept of property rights is used here in its Anglo-Saxon sense ('bundle of rights' proposed by Schlager and Ostrom 1992) and is not limited to strict ownership as defined in the Civil Code with the *usus*, *fructus* and *abusus* attributes. 'Property rights' also include the rights of access, extraction, management, exclusion, etc. With regard to land issues, E. Le Roy (1996) proposed to use the concept of 'land control' for property ('maîtrises foncières' in French).

[5] The 'A' corresponds to *ambiental*, which means 'environmental' in Spanish.

[6] One oft-quoted pioneering local experiment is that of Vittel, but many more experiments have been recorded (Ezzine de Blas et al. 2015).

[7] 'Command and control' measures are those in which the State fixes rules, usually coercive, and enforces them. Examples are bans on deforestation, prohibitions in protected areas, etc.

[8] Beginning in the 1980s, integrated conservation and development projects were rolled out with the aim of reconciling the management of protected areas and the interests of local populations.

[9] Depending on the authors, PES stands for payments for environmental services or payments for ecosystem services.

environmental services, which are human practices that affect nature.[10] Ecosystem services obviously constitute a reductive and anthropocentric notion, but their function is primarily to convince society to become aware of the value of protecting nature. The concept of ecosystem services lends itself poorly to an economic analysis insofar as it encompasses marketable or potentially marketable goods (wood, fibre, water, food) as well as positive externalities (by definition without a market) such as climate and water cycle regulation, biodiversity, etc. Within the context of payments for environmental services, remuneration obviously concerns the actions of actors (or the practices they give up), and the amount of remuneration is more likely to be determined based on the producers' loss of earnings (or opportunity costs) than on an overall economic value of ecosystem services (which, in any case, is rarely possible to calculate). In addition, there exist some uncertainties regarding the level of ecosystem services that payments for environmental services would like to promote and the practices actually paid for: the relationship between the maintenance of forest cover and the quantity and/or quality of the water downstream of a watershed is sometimes complex. Remuneration thus pertains to 'proxies', practices that are presumed to be favourable to different ecosystem services (water quality, biodiversity, carbon sequestration, etc.), which are often associated together in the objectives of payment for environmental services, since they cannot be precisely measured. This led Farley and Costanza (2010) to suggest that even though ecosystem services are often poorly defined in the context of payments for environmental services, this does not constitute a serious problem.

24.1.2 'Commodification' of Nature?

Contrary to the suggestion of the 2005 definition by Wunder, there is no 'market for ecosystem services'.[11] Payments for environmental services are designed to promote the production of only those ecosystem services that are 'positive externalities', i.e., those services which, by definition, have no market. These particular ecosystem services are public goods (biodiversity, climate regulation, scenic beauty, etc.) or collective goods (water quality in a watershed), and do not lend themselves to privatization.[12] Yet 'an essential pre-condition for price-making

[10] One can note the definition of environmental service by Aznar and Perrier-Cornet (2003): 'an intentional contribution to the management of a given space with an environmental and collective goal' (which entails a remuneration as part of 'payments for environmental services') or by Karsenty and Ezzine de Blas (2014): 'the services that men render amongst themselves to maintain or increase certain ecosystem services'.

[11] There do exist markets for certified reductions of greenhouse gas emissions (commonly known as carbon credits), but they are quasi-monetary instruments and not benefits derived from nature.

[12] The fact that a water company can directly benefit from the water quality of a watershed does not imply that it prevents other users of the watershed to also benefit from this quality (non-rivalry and non-exclusivity).

markets is the existence of well-defined and enforced property rights over the good or service to be exchanged' (North 1977, p. 710). Payments for environmental services are therefore not 'market instruments', even if the level of remuneration results from 'bargaining', to quote the apt term used by Boisvert (2015). As noted by Wunder and Vargas (2005): 'If [an urban water utility] thinks the price for watershed protection charged by upstream farmers is too high, usually it cannot just go to the next three watersheds for better offers.'

This does not prevent either the selection of the 'providers' of environmental services through competitive mechanisms (reverse auctions, to select the lowest bidders to maintain certain habitats on their lands) as is done in Australia and the United States. Nor even some payments for environmental services being financed by the sale of carbon credits, marketable goods created by a specific mechanism (including measurements and certification) and which are contingent on the payment for environmental services instrument (Karsenty et al. 2014). But this does not mean that payments for environmental services are instruments to 'commodify nature' (Karsenty and Ezzine de Blas 2014). Non-market definitions do exist; for instance, that of Muradian et al. (2010, p. 1205), which incorporates not only the incentive nature of the instrument, but also the political and institutional challenges of its implementation: 'a transfer of resources between social actors, which aims to create incentives to align individual and/or collective land use decisions with the social interest in the management of natural resources.' In 2015, Wunder proposed a new ideal-typical definition of payments for environmental services that avoided the use of business language this time, and even steered clear of the initial idea of a 'well-defined environmental service'. This proposal, however, clearly did not bring an end to the debate.[13]

Like any instrument, payments for environmental services can generate adverse effects. In principle, they are meant for actors who have (ownership) rights and who are willing to suspend those that are harmful to the environment. However, it is common for incentives to be offered as payments for environmental services even though regulations already exist to prohibit the concerned practices that are harmful to the natural environment. Such a situation risks eroding the civic spirit in this domain. Will actors choose, in the future, to conform to restricting regulations only if they are paid to do so?[14] More generally, payments for environmental services may lead actors to adopt the following reasoning: if there is no other motive than a (monetary) interest in conserving nature, then we can legitimately be 'irresponsible' the moment this starts costing us something. In other words, will a generalization of payments for environmental services reduce the likelihood of selfless conservation action (Karsenty 2013)?

[13] Wunder now only retains the following characteristics: (1) voluntary transactions, (2) between service users, (3) and service providers, (4) who comply with agreed rules of natural resource management, (5) to generate off-site services (Wunder 2015, p. 9).

[14] However, we must differentiate between the cases of poor populations (who do not have the means to comply) and those of enterprises (which tend to chase windfall profits).

24.1.3 A Co-evolution of the Concept and Implementation Practices

It is common to portray payments for environmental services as being based on the 'polluted pays' principle, as against the better known 'polluter pays' principle on which environmental taxes are founded. With regard to water, several qualified initiatives on payments for environmental services (often retrospectively) have contributed to this understanding. The case of Vittel corresponds well to the 'polluted pays' scenario: it has, from the early 1990s, paid farmers whose farms border the water source to stop using pesticides and to permanently modify their farming and livestock rearing practices. Practitioners have thus often sought out beneficiaries of the service, in order to get them to fund payments for environmental services. Many public or private companies that generate hydro-electric power or distribute municipal water levy specific charges that are included in users' bills to fund a compensation scheme for farmers located upstream of the watershed. If the contractual (and thus voluntary) nature of payments for environmental services remains an essential feature of the instrument, funding by end users is often enforced through the billing system.

Successes in implementing private governance mechanisms to supply quality water to catchment areas are more the exception than the rule. In the case of these payments for water-related environmental services, the beneficiaries are limited in number and can be identified. This allows bilateral agreements between a group of farmers and a company or local authority (one can think of this as a short value chain). However, payments for environmental services pertaining to biodiversity or carbon, which generate global services, do not have specific beneficiaries and require different institutional mediations at national and international levels, i.e., on the international market of emission permits, with the involvement of international funding entities, national or local organizations that offer contracts for the provision of services and remuneration, etc. The vast majority of current payments for environmental services are publicly governed, with the State setting payment levels, identifying the beneficiaries, and, as part of government regulations, defining the environmental responsibilities included in the instruments.[15] This broadly relativizes the 'polluted pays' principle to which payments for environmental services are often equated.

Furthermore, although crop cultivation and livestock rearing have always been based on the provision of agricultural products and derivation of value from them, and although agricultural production has always benefited from various ecological functions (for example, pollination), until recently agriculture was mainly considered a threat to nature and the environment. However, the idea that agriculture can also contribute to the preservation of the environment, which was already at the heart of discussions on multi-functionality in the 1990s, is seeing a growing and

[15] According to Vatn (2014), the contribution of public funding towards payments for environmental services amounts to at least 90 %. This form of funding is partly based on taxes.

general interest subsequent to the Millennium Ecosystem Assessment. The agroecosystem is not only considered as a provider of provisioning services (agricultural products), but also of other services (carbon sequestration, maintenance of open landscapes for biodiversity, etc.).[16] And, in line with the 'greening' of agricultural policies of the countries of the North, the rationale of compensation for environmental services rendered is increasingly being accepted by the agricultural sector.

24.2 Payments for Environmental Services and Environment Labels in the Countries of the South: Some Key Lessons Learned

24.2.1 Payments for Environmental Services in Countries of the South: Experiences from Costa Rica and Mexico

The impact assessments of payments for environmental services in Costa Rica (PSA) and Mexico (PSA-H), which are among the most iconic and the oldest of their kind, validate and improve the analysis of payments for environmental services in the countries of the South. They show that the environmental additionality of payments for environmental services,[17] contrary to expectations, is often limited, (Alix-Garcia et al. 2012; Le Velly et al. 2015b; Legrand et al. 2013). Moreover, analyses of PSA-H's implementation in Mexico reveal leakage effects, i.e., the shifting of activities that affect the environment outside the area of implementation of the mechanism. Thus, in Chiapas, the programme of payment for environmental services (PSA-H) led farmers to give up their practices of fallowing and slash-and-burn and to compensate for lower soil fertility and the development of diseases by using the payments received to buy fertilizers and pesticides, with the negative ecological effects of these forms of intensification of agricultural practices in areas adjacent to forests under contract. These developments had not been anticipated by the mechanism. Similarly, in Yucatan State, payments help overcome credit constraints, leading some farmers to invest in cattle with significant risks of rebound effects at the end of the conservation contracts. These indirect effects seriously call into question the sustainability of actions taken as a result of the payments for environmental services if they are not designed in an integrated manner with cultivation and livestock activities. On the other hand, Legrand et al.

[16] For more information, see the reference fact sheets for environmental services and agriculture (http://www.gred.ird.fr/programmes-de-recherche/programmes-acheves/serena, in French).

[17] Additionality seeks to measure the extent to which land uses promoted by payments for environmental services would not have been adopted without them.

(2013) highlight the many beneficial spill-over effects of payments for environmental services in Costa Rica. By considering contracts protecting agricultural land on which a forest is regenerated as admissible, the funding has helped change the use of land in ways that are favourable to conservation. Payments for environmental services have also contributed significantly to an increase in environmental awareness in both these countries, and to the acceptance of the ban on deforestation enshrined in forest laws (Legrand et al. 2013; Shapiro-Garza 2013), even though questions may remain about the future attitudes of the beneficiaries vis-à-vis the regulations if these payments were to stop in the future.

Furthermore, although payments for environmental services have been promoted with an explicit reference to environmental additionality, the negotiation processes of their implementation has led to the inclusion of social objectives. In the case of payment for environmental services in Costa Rica, beneficiaries are not identified on the basis of criteria of deforestation risk or maximizing environmental services, and payments are not differentiated according to opportunity costs or the ability to provide environmental services. The underlying logic was to limit the risk of environmental blackmail by those excluded from the programme and to provide equal opportunities to receive payments for environmental services as a form of social justice (Karsenty and Ezzine de Blas 2014). However, even though payments for environmental services include modalities to favour small and medium producers, several studies have shown that the main beneficiaries are wealthy landowners, most of whom do not even live on the properties for which they receive payments for environmental services, and whose primary sources of income are non-agricultural activities (Zbinden and Lee 2005). In the case of Mexico, on the other hand, McAfee and Shapiro (2010) show how the payment for environmental services have been developed as a hybrid instrument in the course of multi-actor negotiations in the early years of its implementation, by including a form of subsidy to fight rural poverty and target the most marginal communities. Payments for environmental services in Mexico have mainly benefited poor communities, helped contain migration flows, and maintain some rural populations (Le Velly et al. 2015a). Moreover, a real environmental labour market (green labour) has developed in the State of Mexico. Since it may be socially unacceptable to be seen to be receiving a direct payment for conservation, collective payment is redistributed on condition that there is participation in collective work for forest maintenance, such as the construction of firebreaks.

24.2.2 Complementarity Between Payments for Environmental Services and Labels

Payments for environmental services, mainly implemented in countries of the South, are now acquiring a certain measure of political legitimacy. They are evolving under the impetus of international funding entities and large agrifood companies. New mechanisms are being established, enabling cultivation and livestock rearing activities to provide environmental services. The issues of sustainability of agricultural activities began to be managed through quality standards (eco-labels) well before the advent of payments for environmental services. And zero deforestation commitments by companies, an example of other agri-chain approaches promoted for agricultural activities, are emerging as a new way of introducing payments for environmental services (Boucher 2015). A few countries, such as Indonesia and Côte d'Ivoire, have started supporting such approaches.

While the funding and governance of payments for environmental services are mainly under State purview, monetary premiums for labels are provided privately by agri-chains and managed by them. In the case of labels, the remuneration of producers depends on the quantity of products sold, the level of premiums, and the nature of beneficial practices mandated (included in the specifications), while in the case of payments for environmental services, usually on a specific land use, established on the basis of a multi-year contract (Le Coq et al. 2011). The performance of labels in providing ES is very variable as has been observed for PES. For example, Quispe (2007) shows that despite significant changes being observed in practices adopted by organic coffee producers, the changes are more limited in the case of farmers certified by UTZ Certified, Rainforest Alliance, and Starbucks.

Moreover, although labels can have a significant effect on the individual practices that farmers adopt, it is much more difficult to incorporate territorial planning in the definition of specifications (Biénabe 2013). Thus, for coffee, while the factor of shade management is included – with high variability – in the contract specification, the distance to wooded areas, which is also significant in the provision of environmental services, is not (Soto and Le Coq 2011). More generally, there are many and substitutable environmental degradation factors in a territory. This also applies to zero deforestation approaches. Even a successful effort to manage the production conditions of farmers in an agri-chain does not imply control over other degradation factors – cultivation of other crops, livestock rearing, or activities such as making charcoal and collecting lumber or timber. Good environmental practices promoted by agri-chains are not sufficient if other factors and activities that degrade the environment still persist. This confirms the relevance of efforts to develop public and private regulation instruments, both at the level of agri-chains (with companies and producers) and of territories (with communities on their land).

24.3 A Proposed Integrative Approach Between Territorial and Agri-Chain Approaches for Designing PES

We propose an integrative approach for designing payments for national environmental services that addresses the limitations of current designs and developments presented in this chapter. This approach is currently proposed for Côte d'Ivoire, which is committed to implementing a REDD+ strategy (Reducing Emissions from

Deforestation and Forest Degradation). The Ivorian strategy is based on zero-deforestation agriculture supported by a mandatory national sustainability standard and a system of payments for environmental services, much like the existing instruments in Costa Rica and Mexico.[18] The goal is to delink agricultural production from deforestation for the main agri-chains by relying on private sector initiatives that favour supply chains that cause no deforestation. It is a matter of promoting agroforestry and other intensive agricultural practices with reduced environmental impacts, and to put a forest monitoring system in place.

This integrative approach seeks to benefit from synergies between different instruments and initiatives by combining a 'top-down' approach at the national level, and a 'bottom up' approach in territories that can be considered 'coherent' from the perspective of social, economic, and environmental (soil, supply basin, etc.) dynamics. The first is controlled by the State (through the national system for payment for environmental services). It encompasses the initiatives of private firms and NGOs that are part of agri-chain based rationales and those behind the improvement of producer practices (zero deforestation programme, management of supply chains, improvement in the environmental quality of territories) through common operational rules and coordination of different sources of funding (State development assistance, funding from agri-chains, etc.). The second relies on engaging the communities that use these territories and their ability to undertake common territorial projects. These two approaches intersect through the definition of these territorial projects, which translate the goals and initiatives of interest to entities at the national (and international) levels.

As far as the design of specific instruments is concerned, the first principle of the proposed approach is a rational distinction between – and combination of – payments for environmental services that are aimed at restricting land use and payments for environmental services for asset building.[19] In the first type of payment, which is based on the agreed suspension of actual rights or those deemed locally legitimate in exchange for recurring remuneration, the opportunity cost for giving up certain practices forms the basis for negotiations to determine compensation. The second type, on the other hand, concerns remuneration for the adoption and implementation of new practices.[20] The second principle involves combining individual and collective commitments by linking individual and collective conditionalities, creating a mandatory solidarity to ensure conservation and thus limiting the risk of opportunistic behaviour. Finally, the implementation of these contractual mechanisms requires the recognition of individual and collective rights over different spaces and resources, and, consequently, the securing of exclusive land rights.

[18] For a more detailed presentation, see Karsenty (2015).

[19] For details on the relevance of this distinction, see Karsenty and Ezzine de Blas (2014).

[20] The remuneration can be based on the cost of the work put in (e.g. minimum agricultural wage) or it can be variable depending on the practices concerned and the locations of implementation.

24.3.1 Payments for Environmental Services to Households, Oriented Towards Investment and Funded by Enterprises

Payments for environmental services for investments negotiated at the level of individual producers can be implemented by a specialized operator and funded by enterprises. Conditionalities of practices and technical itineraries can be based on different existing recommendations (including approaches proposed in the framework of sustainability labels and standards adopted by different agri-chains). The investments also include financial incentives for land users to plant trees or hedges, or restore degraded areas of land they own or control directly.

24.3.2 Collective Payments for Environmental Services, Mainly Oriented Towards the Conservation and Sustainability of Territories

Complementing these mechanisms by actions at the scale of territorial projects involves defining, when the social contexts permit it, collective payments negotiated at the community level, in compensation for the restriction of usage rights. Thus, this territorial approach to payments for environmental services aims to create a collective dynamic to underpin commitments regarding land use (agreement on a land use plan within clearly defined territories and, if necessary, agreement to discontinue certain practices or techniques). In an implicit manner, this reflects the idea of creating a territorial project that is sustainable or causes zero deforestation. Thus, in the approach planned by Côte d'Ivoire for payments for environmental services, the concepts of High Conservation Value (HCV)[21] and High Carbon Stock will be useful in dividing the territories into zones where new cocoa or oil palm plantations can be established. High Conservation Value (HCV) distinguishes ecosystems of varying importance depending on different criteria (e.g., biodiversity and socio-cultural ones). And High Carbon Stock distinguishes forests which, even when disturbed, provide important environmental services (carbon sequestration, of course, but also biodiversity and specific social contributions) from highly degraded forests which can be converted into agricultural plantations. Participatory zoning based on these principles could form the basis for implementation of collective payments for environmental services. The contractual commitment of a community using a piece of land could lead to the development of environmental quality indicators for the territory through discussion with the local populations. Payments for environmental services will help finance improvements in environmental quality, which depend on collective

[21] See also Chap. 22.

actions, and to provide collective benefits (land tenure security through demarcation and/or the registration of individual plots, drinking water supply, warehouse facilities, rural roads, schools, clinics, etc.), by making them conditional on the maintenance or improvement of environmental quality that is measured and recorded jointly.

24.3.3 *Public Investments Needed to Clarify Land Ownership and Secure Contracts*

Since payments for environmental services are destined for actors who have rights over the land and the resources they use, it is essential to clarify and recognize land rights (at the very least, exclusion rights) to be able to establish the rights and responsibilities without which no contractual agreement can be entered into. These collective and individual contracts will have to be in written form and be verified at regular intervals to justify the continuity of benefits or payments. To map individual plots on which farmers must fulfil their contractual obligations, the approach could be that of rural land-tenure plans (cadastral map with identification of plots and right holders, indication of the exact nature of individual and collective rights) (Lavigne-Delville 2010). While it is possible that large corporate entities may agree to contribute to the funding of such systems in their supply basins, large public investments will certainly be necessary, mainly to finance payments for collective environmental services targeting the territories. Funds earmarked for climate-related issues, in the context of the fight against deforestation, can be mobilized for this purpose.

24.4 Conclusion

Payments for environmental services are instruments that are still evolving and whose limits are still not clearly defined, to the extent that this term includes – often retrospectively – a set of practices and initiatives with certain common features (mainly the voluntary, contractual, and conditional aspects), but also with local design and implementation specificities. Although certain payments for environmental services, particularly with regard to water, adequately reflect the 'polluted pays' principle, which is often associated with the instrument, it is not so for payments for environmental services for biodiversity, carbon sequestration, and other global public goods (often associated in the objectives). In addition, although payments for environmental services have often been implemented for the suspension of usage rights, a growing proportion of initiatives aim to combine compensation for conservation and investment in sustainable agro-silvo-pastoral practices (Karsenty 2011).

Payments for environmental services are becoming increasingly similar, in their implementation, to transfers used by States to fund landowners and communities, either to reward their environmental commitment (stewardship) or to offset opportunity costs of the restrictions on land use, with impacts that are not only environmental but also social. They thus constitute public policy instruments to promote ecological transitions in rural areas.

Instead of compartmentalizing into distinct services and managed areas, which could lead, in particular, to leakage effects, this chapter shows that, to be effective and fair, payments for environmental services must reflect different territorial levels in an integrated design of environmental and developmental efficiency, by establishing coherency between instruments of sectoral (agricultural, social, environmental) public policies and those of business policies. We believe that payments for environmental services can contribute to changes in practices by building capacity of producers and other actors to achieve sustainable management of agriculture, livestock rearing, forestry, etc. We have thus proposed an integrative approach combining a territorial and collective approach with a 'top-down' approach that incorporates sustainability strategies of agri-chains.

References

Alix-Garcia J, Shapiro E, Sims K (2012) Forest conservation and slippage: evidence from Mexico's national payments for ecosystem services program. Land Econ 88(4):613–638

Aznar O, Perrier-Cornet P (2003) Les services environnementaux dans les espaces ruraux : une approche par l'économie des services. Économie rurale 273(1):153–168

Bertrand E, Destais C (2002) Le « théorème de Coase », une réflexion sur les fondements microéconomiques de l'intervention publique. Reflets et perspectives de la vie économique 41(2):111–124

Biénabe E (2013) Towards biodiverse agricultural systems: reflecting on the technological, social and institutional changes at stake. In: Hainzelin E (ed) Cultivating biodiversity to transform agriculture. Springer, London, pp 221–261

Boisvert V (2015) La compensation écologique : marché ou marchandage ? Revue internationale de droit économique 29(2):183–209

Boucher DH (2015) The REDD/carbon market offsets debate: big argument, small potatoes. J Sustain For 34(6–7):547–558

Costanza R, d'Arge R, de Groot R, Farber S, Grasso M, Hannon B, Naeem S, Limburg K, Paruelo J, O'Neill RV, Raskin R, Sutton P, van den Belt M (1997) The value of the world's ecosystem services and natural capital. Nature 387:253–260

Ezzine de Blas D, Dutilly C, Lara JA, Le Velly G, Guevara-Sanginés A (2015) Payment for environmental services in a policymix: spatial and temporal articulation in Mexico. PloS One (à paraître)

Farley J, Costanza R (2010) Payments for ecosystem services: from local to global. Ecol Econ 69 (11):2060–2068

Gómez-Baggethun E, Muradian R (2015) In markets we trust? Setting the boundaries of market-based instruments in ecosystem services governance. Ecol Econ 117:217–224

Karsenty A., 2011. Coupler incitation à la conservation et investissement. Perspective n°7. Montpellier : Cirad

Karsenty A (2013) De la nature des « paiements pour services environnementaux ». Revue du Mauss 42:261–270

Karsenty A (2015) Mettre les PSE au service de l'agriculture « zéro déforestation ». Perspective 36
Karsenty A, Ezzine de Blas D (2014) Du mésusage des métaphores. Les paiements pour services environnementaux sont-ils des instruments de marchandisation de la nature ? In: Halpern C, Lascoumes P, Galès P (eds) L'instrumentation de l'action publique : controverse, résistances, effets. Presses de Sciences Po, Paris, pp 161–189
Karsenty A, Guingand A, Langlais A, Polge MC (2014) Du Sud au Nord : regards croisés sur les paiements pour services environnementaux. Compte rendu de l'atelier international Pesmix. Les cahiers de Biodiv' 50(2), CDC-Biodiversité
Lavigne-Delville P (2010) Conceptions des droits fonciers, récits de politique publiques et controverses. Les Plans fonciers ruraux en Afrique de l'Ouest. In: Colin J-P, Le Meur P-Y, Léonard E (eds) Les politiques d'enregistrement des droits fonciers. Du cadre légal aux pratiques locales. Karthala, Paris, pp 69–103
Le Coq J-F, Soto G, González HC (2011) PES and eco-label. A comparative analysis of their limits and opportunities to foster environmental services provision. In: Rapidel B, DeClerk F, Le Coq J-F, Beer J (eds) Ecosystem services from agriculture and agroforestry: measurement and payment. Earthscan Publications, London, pp 237–264
Le Roy E (1996) La théorie des maîtrises foncières. In : Le Roy E, Karsenty A, Bertrand A, dir (eds) La sécurisation foncière en Afrique, pour une gestion viable des ressources renouvelables. Karthala, Paris
Le Velly G, Dutilly C, de Blas E, Fernandez C (2015a) PES as compensation? Redistribution of payments for forest conservation in Mexican common forests. Études et documents 28, Cerdi. Retrieved 26 June 2016, http://cerdi.org/production/show/id/1758/type_production_id/1
Le Velly G, Sauquet A, Cortina-Villar S (2015b) PES impact and leakages over several cohorts: The case of PSA-H in Yucatan, Mexico. Études et documents 29, Cerdi. Retrieved 26 June 2016, http://cerdi.org/production/show/id/1759/type_production_id/1
Legrand T, Froger G, Le Coq J-F (2013) Institutional performance of payments for environmental services: an analysis of the Costa Rican program. Forest Policy Econ 37:115–123
McAfee K, Shapiro E (2010) Payments for ecosystem services in Mexico: nature, neoliberalism, social movements, and the state. Ann Assoc Am Geogr 1000(3):579–599
Millenium Ecosystem Assessment (2005) Ecosystems and human well-being: synthesis. Island Press, Washington, DC, 155 p
Muradian R, Corbera E, Pascual U, Kosoy N, May PH (2010) Reconciling theory and practice: an alternative conceptual framework for understanding payments for environmental services. Ecol Econ 69(6):1202–1208
Norgaard R (2010) Ecosystem services: from eye-opening metaphor to complexity blinder. Ecol Econ 69(6):1219–1227
North DC (1977) Markets and other allocation systems in history: the challenge of Karl Polanyi. J Eur Econ Hist 6(3):703–716
Quispe J (2007) Caracterización del impacto ambiental y productivo de las diferentes normas de certificación de café en Costa Rica. Mémoire de master. Catie, Turrialba. 137 p
Schlager E, Ostrom E (1992) Property-rights regimes and natural resources: a conceptual analysis. Land Econ 68(3):249–262
Shapiro-Garza E (2013) Contesting the market-based nature of Mexico's national payments for ecosystem services programs: four sites of articulation and hybridization. Geoforum 46:5–15
Soto G, Le Coq J-F (2011) Certification process in the coffee value chain: achievements and limits in the coffee value chain. In: Rapidel B, DeClerk F, Le Coq J-F, Beer J (eds) Ecosystem services from agriculture and agroforestry: measurement and payment. London, Earthscan Publications, pp 319–345
Vatn A (2014) Markets in environmental governance: from theory to practice. Ecol Econ 105:97–105
Wunder S (2005) Payment for environmental services: Some nuts and bolts. Cifor occasional paper, no. 42. Cifor, Jakarta. 24 p

Wunder S (2015) Revisiting the concept of payments for environmental services. Ecol Econ 117:234–243
Wunder S, Vargas MT (2005) Beyond "markets". Why terminology matters. Retrieved 26 June 2016, http://sanrem.cals.vt.edu/1010/Wunder2005_Beyond Markets terminology matters.pdf
Zbinden S, Lee D (2005) Paying for environmental services: an analysis of participation in Costa Rica's PSA program. World Dev 33:155–272

Chapter 25
Global Strategies of Firms and the Financialization of Agriculture

Ward Anseeuw, Jean-Marc Roda, Antoine Ducastel, and Norfaryanti Kamaruddin

A sustained growth of investment into agriculture is being observed the world over. For example, in 2008, foreign direct investments (FDI) into the African continent reached US$ 87.6 billion (27 % higher than in the previous year), of which a third (i.e. US$ 27 billion) was directed towards the mining and agricultural industries (UNCTAD 2009). Even if their nature varies significantly (Boche and Anseeuw 2013), a major characteristic of these investment flows is the engagement of new actors, often global, who originate from diverse sectors and are proponents of new strategies. Even though some actors and strategies have been described and analyzed extensively (Anseeuw et al. 2012), others are less understood. These less explored strategies, which rely on financial mechanisms, such as shareholding structures (Roda et al. 2015) or private equity (Ducastel and Anseeuw 2013), often involve the financialization of agriculture. On the one hand, the global strategy of these actors and the financialization of agriculture respond to the need for capitalization and risk management inside agri-chains. On the other hand, they in turn reshape the sector and its value chains. In this way, they lead to the advanced integration and verticalization of agri-chains, and to an increasing corporatization of agriculture as a whole. These dynamics form the

W. Anseeuw (✉)
CIRAD – UMR Art-dev – International Land Coalition, Rome, Italy
e-mail: ward.anseeuw@cirad.fr

J.-M. Roda
CIRAD – UPR BioWooEB – Universiti Putra Malaisia, Serdang, Malaysia
e-mail: jean-marc.roda@cirad.fr

A. Ducastel
CIRAD – UMR Art-dev, Montpellier, France
e-mail: antoine.ducastel@cirad.fr

N. Kamaruddin
Universiti Putra Malaysia, INTROP, Serdang, Malaysia
e-mail: faryanti@gmail.com

E. Biénabe et al. (eds.), *Sustainable Development and Tropical Agri-chains*,
DOI 10.1007/978-94-024-1016-7_25

basis for new questions concerning the structuring of the sector and its capacity to act towards sustainable development.

This chapter aims to explore these dynamics. It first presents the drivers of financialization and how it is expanding into the agricultural sector. Then, two specific case studies, one from Africa and the other from Asia – although the strategies concerned in both cases are global –, are described. Finally, the chapter discusses the financialization process in light of global agriculture, the restructuring of agri-chains, and the implications in terms of sustainable development.

25.1 The Financialization of Agriculture: Definitions, Drivers, and Trends

Financialization is a fuzzy concept, which takes on different meanings according to the economic scales it refers to or to the authors who write about it. At the macro-economic scale, financialization refers to the huge increase in the share of income from financial services or derivative markets in the GDP of countries. In the last three decades, technological improvements, the expansion of electronic trading, and a new generation of investment instruments and financial products, with considerable leverage and multiplier effects, have opened up a substantial space for financial investors to make huge profits from price and currency volatility. This has logically led to a growing dominance of capital markets over traditional value chains and production systems. The 2009 financial crisis and subsequent global recession demonstrated the increased interdependency between national economies and international finance (Freeman 2010).

A few normative definitions can be derived from these observations. For example, financialization might be defined as accumulation of capital arising more from financial channels than from physical trade and traditional production. Or it might be defined as an increasing economic and political power in the hands of the new financial elites. It can then be associated with concepts of inequality and the capture of financial and economic rents (Krippner 2005). Other definitions of financialization focus instead at the scale of production by firms and actors, and then refer to the increased involvement of financial tools and investment funds in the process of production itself (Murphy et al. 2012). For companies, the end of the Bretton Woods international agreement and the beginning of fluctuation in currency values in the 1980s brought with it a new era of price and currency volatilities, and associated economic risks (Newman 2009).

25.1.1 The Drivers of Financialization

Risk management is the main driver of financialization. New financial tools allow an investor to hedge credit risk, country and political risks, price volatility and currency volatility, natural and physical risks, etc.

Through the use of different investment tools, financialization directly helps manage four categories of risk. First, it helps to disconnect a part of the revenues from the possible ups and downs of the physical cycles and unpredictable events. Transferable securities, futures and other financial derivative tools are perfect for this purpose, since they follow completely different cycles, of a more 'virtual' nature. Second, it helps to mitigate potential financial losses from exchange variations, credit defaults, and price volatility. Anchoring a part of the assets and a part of the revenues to different cycles is a rational management strategy to protect the firm concerned from risks of volatility. Third, financialization helps insulate revenues from supply disruptions, country risks, wars, political crises, etc. Finally, the risks and volatilities apply in opposite ways to physical actors, such as producers or commodity traders, and to investment funds. Revenues and assets of physical actors are subject to physical cycles specific to each commodity or country. Revenues and assets of investment funds, on the other hand, are subject to global self-reinforcing cycles of bubbles and bursts, largely independent from the physical world (Heumesser and Staritz 2013), in a process called 'reflexivity' by George Soros (Soros 2003). Thus, investment funds need to hedge their own risks by anchoring a part of their assets and revenues to the physical world, which does not react like the 'virtual' or 'informational' world from which they derive their revenues. The interests of these various groups of actors converge. This has the logical consequence of increasing the financialization of some sectors, including the agricultural sector.

Another driver of financialization is the context of accessing capital and bank loans in countries 'at risk'. In many developing countries, the country risk compels banks to be extremely cautious in sanctioning loans to local companies. It makes accessing capital more difficult or, at best, more expensive. Most small and medium enterprises internalize these difficulties or costs by becoming less competitive, or by orienting their strategies towards less investment-intensive processes. Or they tend to favour raw exports or upstream activities with higher margins and relatively lower physical investments. This tendency indirectly favours the multinational corporations, because they can more easily move their capital to lower-risk countries (Roda 2006). In particular, it is easier for these multinationals to set up their own internal financial services, which can provide cheaper loans to their subsidiaries in a risky country, while consolidating the supporting capital in Western countries or in financial centres such as Hong Kong or Singapore. In the agricultural sector, all the four leading multinational corporations, Archer Daniels Midland, Bunge, Cargill, and Louis Dreyfus, known collectively as 'ABCD', follow this model (Murphy et al. 2012).

25.1.2 History and Trends of World Financialization, in Particular as Concerns Agriculture

At a macro-economic scale, an analysis of international investments shows how countries and world regions have witnessed extremely different patterns of financialization over the last 30 years.

Advanced economies, as well as those in Latin America and Asia, followed, on the whole, a similar pattern of financialization in the 1990s, while Africa lagged behind. In the more advanced economies, investments in portfolio assets shot up to as much as four times direct investments in certain countries. The associated high volatility had contrasting effects: the large trading corporations from advanced economies with their own financial service units quickly established themselves, whereas in developing economies, agribusinesses, which are on average much smaller and do not have their own financial units and other resources, had to cut down their financial investments (Heumesser and Staritz 2013). The extremely low financial investments in developing economies during the period are probably also the reflection of the large South-South investments in infrastructure, which saw a boom during this period. The last decade seems to show a global convergence, with emerging economies slowly increasing their financial investments while developed economies drastically reduced theirs (Fig. 25.1).

At a micro-economic scale, and in particular for the agricultural sector, empirical data from the equity and liability structures (Oestreicher et al. 2013) of the world's publicly listed firms involved primarily in agriculture presents a more detailed perspective of the processes at work. The ratio of shareholder assets over total assets is a proxy of the financialization of a firm's assets (Fig. 25.2; Bureau Van Dijk 2015). Up to 2001, a common process of corporatization of the agriculture sector was observed worldwide: an increase in the number of large corporations, competition, and consolidations. Paradoxically, these consolidations increased the average of total assets, showing an apparent decrease in the financialization index of Fig. 25.2 for the Western world as well as for Latin America and developing Asia. This process ended around 2000 for both the developed world and Latin America, while it continued in Asia up to the 2008–2009 financial crisis. It appears that Latin America is no longer undergoing a process of change in the level of its financialization, with a stable equity and liabilities structure of its agriculture-based corporations, in line with the slow re-financialization of the few giant corporations from the Western world. The most dynamic trajectories are exhibited by corporations from developing Asia, which act as intermediaries for the demands from China and India, and which actively seek to grow, with a boom in commodity trading and the development of new financial units and tools (Gibbon 2014).

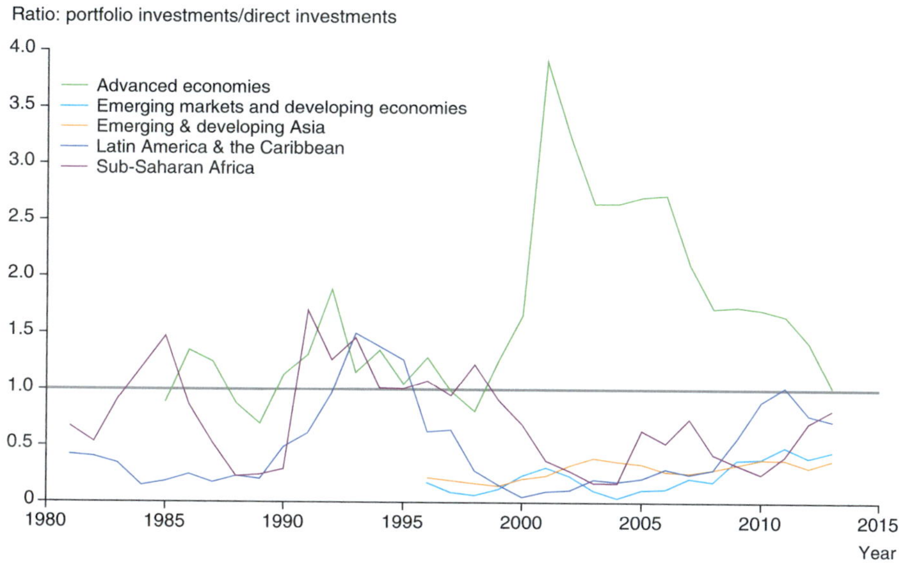

Fig. 25.1 Changes in financialization of investments for major world regions (ratio of portfolio investment assets/direct investment assets, 3 years moving average) (Source: Anonymous (2015), based on IMF data)

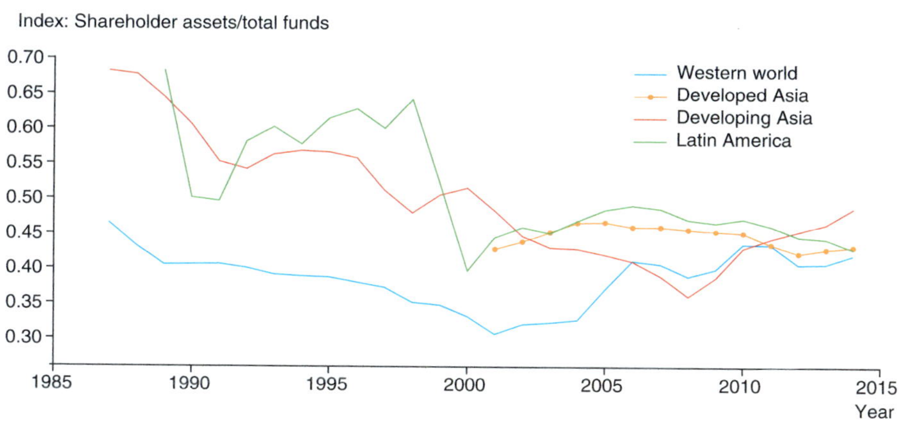

Fig. 25.2 Index of financialization of assets at the firm level in the world agriculture sector, from 1987 to 2014 (Source: Bureau Van Dijk 2015)

25.2 The Process of Financialization in Practice

In order to describe the effective process of financialization, i.e., how financialization of agriculture actually takes place, this section will present two cases. First is the presentation of two investment funds that are becoming involved in agriculture in South Africa. This work is based on active participation and participatory observations: two separate periods of three months each were spent with these funds in South Africa. The second case study presents an Asian multinational, Olam, which includes agricultural activities in its wide-ranging portfolio. This research is based on interviews with key stakeholders and Olam managers, on information provided by the company, and on the attendance of several of their board and working group meetings for developing their industrial strategy, which were held in Singapore (where Olam's head offices are located).

25.2.1 Investment Funds in South African Agriculture and Agro-Industries: The Establishment of New Financial Channels

Each of the investment funds analyzed (called 'Fund A' and 'Fund B′ for the sake of anonymity) raises capital on the financial markets and deploys it towards investment opportunities that it has identified, thus progressively building an 'asset portfolio'. To arrive at an improved understanding of these financial vehicles, their similarities as well as their differences, it is important to examine in detail the character and trajectory of both the funds' investors and their asset managers and to understand the relationship and interactions between these two sets of actors. The profile of the investors is important for understanding these funds' capital allocation and governance models. Indeed, the sources of capital, mainly related to their liability structures (Aglietta and Rigot 2009), significantly influence their investment policies, and thus their choices and expectations regarding agriculture and farmland.

Fund A was set up in 2008 with only one investor, an American endowment fund investing in this vehicle as part of its 'natural resources' portfolio, which includes agricultural investments in other countries, as well as in the domains of logging and renewable energy. In 2013, this natural-resource asset class accounted for 13 % of a large and diversified portfolio that included equities, bonds, real estate and shares in hedge funds.

Fund B was launched in 2010 with two major investors, both South African – a diversified insurance company and a public pension fund. Other minor individual and institutional investors joined the fund later through a mutual fund registered in Luxembourg. Pension funds, endowment funds, and insurance companies are all long-term investors looking for investments that generate stable returns so that they can reward their subscribers (Aglietta and Rigot 2009). To this end, they implement diversification strategies through investments in asset classes that are not correlated with one another (Campbell 2011). In such a framework, South African farms become an attractive investment.

These funds entrust their capital to a managing company, which implements the projects. A South African company manages Fund A on behalf of its investors. Initially dedicated to commodity trading, this company progressively became

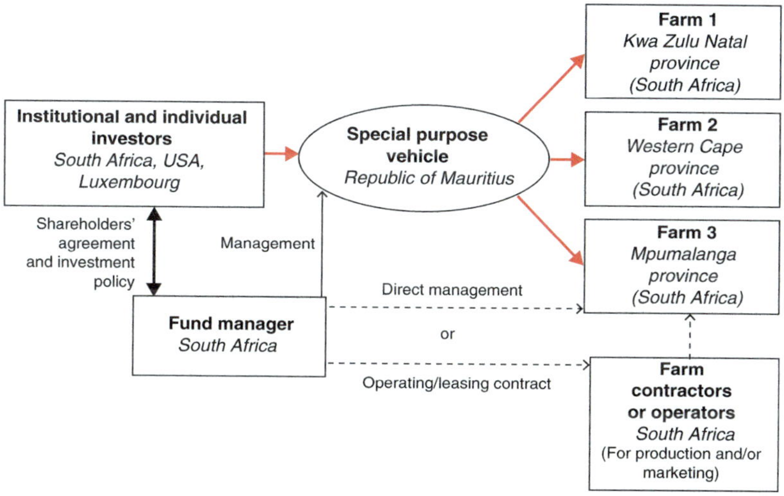

Fig. 25.3 Institutional set-up of investment funds investing in agriculture in South Africa (Source: Authors)

involved in management of farms and other agro-industrial assets on behalf of foreign investors. Fund B is managed by a South African company specialized in farmland investments in Africa. It also manages, at the same time, another fund investing in farms and agricultural production in Swaziland and is currently developing an African fund with a similar mandate. Figure 25.3 illustrates the institutional set-up of such funds.

These companies claim field experience and an extensive network in South Africa and its agricultural sector. They affirm their essential role as gateways to the country and its agricultural value chains by leveraging this 'indigenous capital'. Therefore, they are at the boundary between this indigenous capital and 'financial capital', occupying a strategic position between 'the bush' and the 'financial industry'. As they deal with non-formalized and non-benchmarked assets, these diverse managing entities compete with each other, and with others in and outside South Africa to sell their specific products and capture investment flows. Depending on their trajectories and their staff's profiles, they implement different practices and strategies, mobilizing different instruments or registers. Indeed, these backgrounds largely determine their paths to farmland investment, i.e., the integration and perception of this specific product in the portfolio of their activities. They also involve themselves with various financial sub-fields and asset classes, mobilizing different networks of investors and different persuasive registers to convince potential investors. For example, while one managing company may emphasize the contemporary consumption boom in Africa, another may focus on land as a scarce resource and highlight the returns generated from similar investments in the United States.

Analyzing the functioning of these two funds leads to the identification of a set of relevant information that illustrates both the similarity and diversity of such financial vehicles. They include the following:

The Status of the Financial Vehicle Both vehicles are registered as special purpose vehicles (SPV) with a limited life span of 10–12 years, depending on the private equity fund model. These SPVs are *ad hoc* structures owned by investors according to their contribution and governed by an investment committee, in which the major investors are represented. Both SPVs that are the subject of our research are registered in Mauritius.

The Fund Raising Process Because of its specific set-up as it pertains to its investor's exclusive position, the manager of Fund A does not have to raise funds from additional investors. The fund manager submits a proposal to the investment committee which decides whether and how much to invest in specific projects. In contrast, Fund B is a closed fund; during the fund raising period, the fund manager raised slightly more than 500 million Rand (US$ 41.5 million) from its investors.

Financial Returns Benchmarks As the South African farm is a new asset class, the target return of these funds is not standardized. Therefore, they adopt different objectives, based on different metrics. Fund B's target is the South African Consumer Price Index (CPI) + 10 %. For Fund A, on the other hand, the target has been defined based on the historical performance of farmland investments in the United States and is set at an annual return of 8 %.

The Mandate of the Fund and Its Boundaries Both financial entities are specialized investment vehicles dedicated only to South African farms and agricultural production, although we have observed other financial vehicles investing in one region (southern Africa), on the entire continent, or in a vaguely defined 'frontier market'. The investors specify the type of assets and portfolio they want through investment limits and ratios. For example, the second fund cannot invest less than 50 million Rand per farm and its final portfolio should include at least four farms. Such capping and threshold requirements tend to favour 'mature' farms and businesses, rather than 'green projects'. Indeed, each of the two funds acquired only consolidated farms resulting from previous mergers of family farms.

Type of Crops and Production The two funds have different mandates. Fund B is focused on 'permanent crops' such as fruits and vegetables. In 2014, this fund had four farms, representing 1490 ha under irrigation and producing mainly lemons and table grapes, with a smaller proportion of pears, peaches, and apples. In contrast, Fund A focuses on 'cash crops', i.e. grains, such as maize and wheat, and soya beans, on 16,342 ha on nine farms, of which 3071 ha are irrigated and 1739 ha are unirrigated, with the remaining being grasslands or used as pastures. In addition, even if it was not part of its investor's initial mandate, Fund A acquired farms with both cash crops and cattle, and decided to continue with the livestock farming on these farms.

Land Ownership and Operational Activities While other studies mention cases of investors leasing farmland, especially in other African countries (Chu 2013), in our case studies, both funds directly purchase the farms and incorporate expectations of increases in land prices into their expected cash flow models. So far, Fund A has only acquired farms at 100 % equity, while Fund B has set up two joint ventures with the former owners, but has ensured a large majority control for itself. Both funds initiated expansion or improvement programmes on their farms in order to increase property values and to obtain higher gains at the time of their exits. Such programmes focus on the productive capacities of the farms: for instance, on one of its farms, Fund B is currently increasing the production area from 12 to 73 ha.

Regarding agricultural production, Fund B has entered into contracts with operators, selecting only experienced farming companies. Such contracts are signed on a 10-year basis with a remuneration corresponding to 8 % of the annual farm value, adjusted for inflation. Fund A directly manages the agricultural operations on its farms, through farm managers hired by the company. However, it is increasingly entering into contract with external operators, often former managers, who it sponsored to set up their own companies, while still retaining control over the coordination of the entire process. These funds therefore rely on different types of returns: while Fund A relies on rents and the appreciation in value of its farms, Fund B bets on both property appreciation and income from agricultural operations.

In practice, on the farms, both funds hire field managers[1] either directly through the fund manager or through the operator. These field managers are often white Afrikaner farmers and the previous owners, or the sons of family farm owners. The field managers are mainly in charge of reconciling the economic and social reality of the farm with the financial reality of the fund manager. Indeed, the fund managers centralize the control of value adding activities, and sometimes of farm management, across the different dispersed farms. This method of control at a distance relies on a set of specific techniques and skills. The funds have thus established detailed monitoring and evaluation procedures which field managers must follow. Thus, a sequence of reporting all along the value chain is mandated, involving everyone from investors to field managers; it helps maintain control from a distance and the coordination between physically dispersed actors. In addition, the fund managers rely on advanced technologies such as Google Earth Pro, which allows for a detailed plot-by-plot monitoring of the farm and a chronological archiving of performance.

---

[1]In addition to the field managers, agricultural engineers exercise a specific function in the control of the supply chain and in ensuring value addition and the development of the farm. Each of the two funds employs an agricultural engineer, either as a consultant or an employee. He is in charge of assessing ecological, pedological, and climatic parameters prior to acquisition of a farm by the fund. He then actively contributes to the creation of a plan to develop the farm at the agronomic level (soil improvement plan, selection of the best varieties, introduction of no-tillage cropping systems, etc.). Finally, he monitors activities on the farm on behalf of the fund through regular visits. In this way, through his training and experience, the agricultural engineer knows how to juggle agricultural references and standards with the requirements of the business world.

25.2.2 Olam, A Transnational Firm Specialized in South-South Trade

Olan forms part of a wider type of transnational firms specializing in South-South trade of resources (wood, palm oil, cereals, water, ore, etc.), organized in networks, blending family control and modern management methods, and backed by ethnic affiliations. These firms have established complex parent-subsidiary company systems.[2] Olam belongs to a network of firms that is present on all continents, the Kewalram Chanrai Group, established in the 1860s. Cross investments between the different territories it operates in impart flexibility and spread risk to different sectors, either complementary or part of different economic and financial cycles. Olam was created as a sub-group of the network in 1989, as an indirect subsidiary, and launched based on an individual loan to the current chief executive officer of the company. Figures 25.4 and 25.5 help locate Olam within the group (the graph is simplified and does not show all the subsidiaries or cross investments).

Olam's strategy, inherited from the Kewalram Chanrai Group, consists of specializing in South-South trade, initially of agricultural plantations, along three axes:

- Choice of 'difficult geographies', an euphemism for 'high-risk countries', to avoid competition from agricultural firms from developed countries;
- Betting on or forecasting future demand (for agricultural products) in the large emerging markets and positioning itself in them in advance. Olam analysts develop scenarios to help the firm make choices (shifting of demand towards urban zones of emerging countries, land availability, logistical or technological disruptions, etc.);
- Investing in the control of agricultural land and plantations to respond to the growing demand for some agricultural products in India, China, and other emerging countries. Olam bets on the multiplication of demand for some products in the years to come with the growing urbanization of these countries.[3]

The global organization of Olam includes the production, the trading, the processing, and the distribution of agricultural products. The group owns complementary units and subsidiaries, positioned all along integrated value chains. Plantations, farms, warehouses, ground and maritime transport, marketing and agents, processing units, and even the distribution of consumer goods (from appliances to cars) allow an optimization of transport across different countries of the South. For example, agricultural products are exported from Africa while cars, manufactured products, and appliances are imported into Africa. Whenever necessary, local sale and resale of agricultural products help cushion differences between international

---

[2]For more about the functioning and operational aspects of this type of firm, see Roda (2005, 2010).

[3]See http://olamonline.com

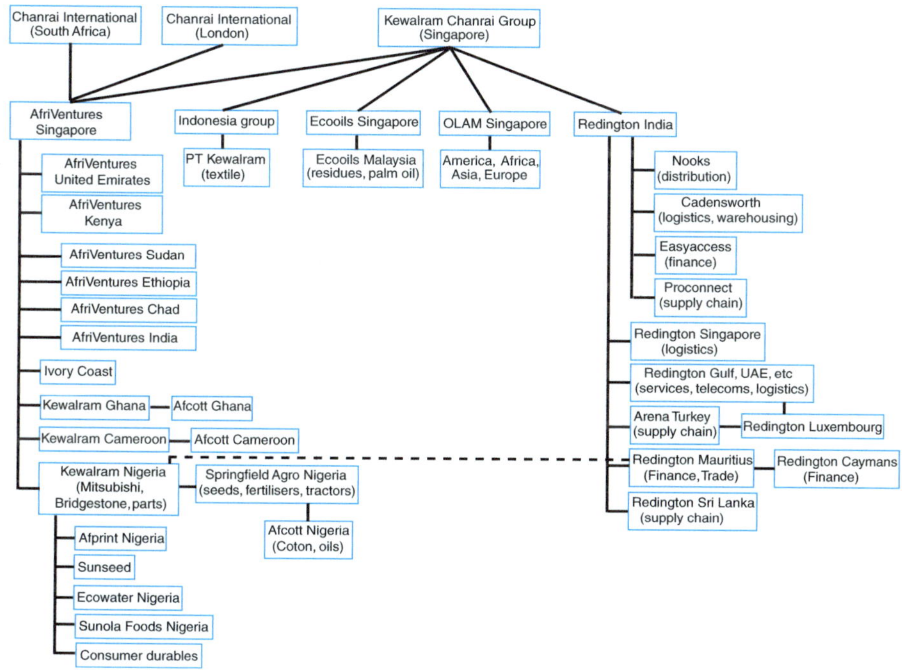

Fig. 25.4 Structure of the Kewalram Chanrai Group (major entities only) (Source: Authors, based on Olam 2013)

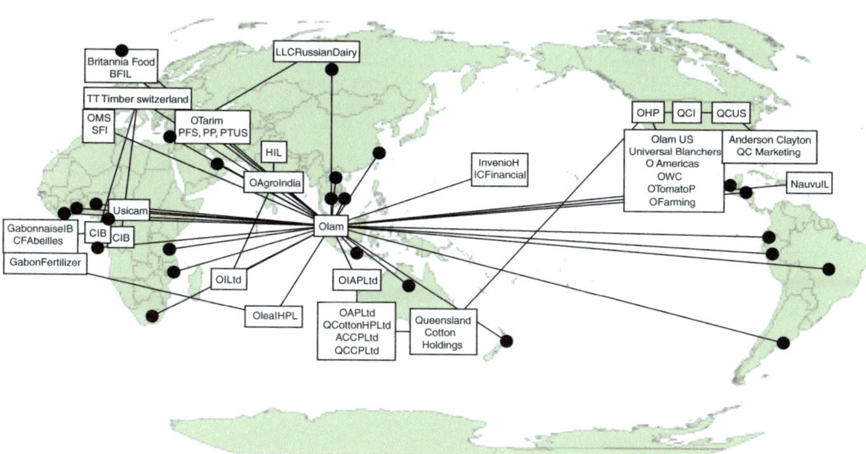

Fig. 25.5 Olam subsidiaries (direct = *black points*, indirect = *boxed texts*) (Source: Authors, based on Olam 2013)

	Upstream	Supply Chain	Midstream and Downstream
Edible Nuts, Spices and Beans	• Almond orchards (Australia, USA) • Peanut farming (Argentina) • Tomato, Garlic, Onion and other,Vegetables (California)	• Argentina shelling and blanching expansion	• Spices (SK Food, Gilroy) ; Sesame huiling in Nigeria • Cashew mechanization • Pepper and Garlic processing Vietnam, China
			• Hazelnuts procurement (Turkey, Progida) • Spices (India, VKL) • Cashew processing (Expansion) • Cashew ingredients
Confectionery and Beverage Ingredients	• Coffee plantations (Laos, Tanzania, Ethiopia)	• Coffee sourcing origins (Mexico, Guatemala)	• Speciality fats (UK, Britannia) • Cocoa grinding (IVC) • Soluble Coffee manufacturing (VN)
			• Cocoa ingredients (Spain, Macao)
Food Staples and Packaged Foods	• Dairy farm (Uruguay, NZFSU) • Palm plantations (WAF/SIFCA)	• Grains sourcing and elevation, (Australia, Russia)	• Sugar refinery (Indonesia, PT DUS, Nigeria) • Sugar miling in India • Wheat miling (Nigeria, CFM, Ghana)
	• Dairy and Grains (Russia, Rusmolco) • Rice farming (East Africa)	• Grains trading (Canada, Lansing) • Grains terminal (Australia, NAT)	• Sugar miling (India, Hemarus) • Dairy, Juice (WAF, Kayass) • Biscuit (WAF, Titan) • Sugar miling (Brazil, UAP) • Wheat miling (WAF, CFM) • Dairy processing (IVC)
Industrial Raw Materials	• Timber forestry concession (Gabon, ROC, Mozambique) • Ruber plantations (WAF) • Fertiliser manufacturing (Gabon) • Rubber(Gabon)	• Cotton (Australia, US, Africa) • Wood sourcing (Panama, Costa Rica)	
Commodity Financial Services			

Fig. 25.6 Distribution of the main activities of Olam per sector, value chain segment, and country (Source: Olam 2013)

and local prices. Figure 25.6 presents an overview of the network of activities undertaken by Olam at the international level.

25.3 Reflections on Agriculture, Structuring, and Trends

The description hereinabove of cases of financialization of agriculture highlights several trends and raises several issues.

25.3.1 Far-Reaching Verticalization of Agri-Chains and Concentration Within the Agrifood Sector, and Asymmetries in Relationships

As these two cases have shown, the organization of agricultural production is progressively approaching a highly integrated structure, comparable to those found in industrial chains. The strategies concerned are highly reliant on advanced financial rationalization, implementations of economies of scale, and the integration of upstream and/or downstream activities. The case of Olam, which owns complementary units and subsidiaries all along integrated value chains, is an apt illustration of this. But it is also evident in the case of investment funds, with the development of value chains that include a multiplicity of segments and actors, originating from diverse social spaces (institutional investors, professional asset managers, farmers, etc.) for creating or developing a specific product. Thus we propose the notion of the agro-financial chain. Rather than separating sectors, in particular finance and agriculture, the agro-financial chain straddles them. Its concept is very close to that of the 'AG space' proposed by Williams (2014), i.e., a hybrid space at the crossroads of agriculture, finance, and other sectors. This obviously raises several issues as regards sustainable development.

Whether they result from direct acquisition or from contractualization, these situations lead to the control of agri-chains by a single or a few dominant actors and to asymmetrical relationships between actors. This asymmetry occurs between big (often Western) corporations, such as the ABCD traders, and small and medium agricultural companies, traders, and even farmers from tropical countries. There also exists a certain asymmetry in the nature and extent of financialization between corporations of the South and those of the North, mainly because they did not begin operations at the same time. While the ABCD traders initiated and established financialization a few decades ago, corporations from the South, such as Olam, began their financialization processes quite recently. In fact, Olam found itself in the danger zone during its process of financialization between 2012 and 2014, when it came under attack by competing larger size corporations.[4]

Hedging against risks with financial tools such as futures and derivatives is a very flexible and convenient method, theoretically cheaper than conventional insurance, and even capable of generating profits. However, not every firm or economic actor can afford these practices. Below a certain size and economic scale, the financial information, technology, data analytics, and the human skills that are necessary become too complicated or expensive to acquire (Ahmed et al. 2014). In other words, managing a certain part of the financialization in a firm requires a certain economy of scale so that the necessary team of specialists can be maintained without it becoming disproportionately expensive as compared to its expected returns.

As such, the financialization, and consequential corporatization (see below), of the agricultural sector is leading to the domination by a few large (international) agrifood business groups (Huggins 2011) and could lead to the marginalization of the majority of African farmers. While the macro-actors of the food processing industry see their dominant positions strengthened even further, entire sections of (the rural) society are excluded from these dynamics. On the one hand, the smaller and medium size farms are being swallowed up by companies and, on the other, family farming and in particular the small-scale farmers are stagnating in languishing sub-sectors. Both categories of actors have diverse financial, social, and cultural resources that lead to biased relationships (Borras and Franco 2012), which seem to extend beyond the traditional divides within the agricultural sector. This is particularly true in South Africa, where these evolutions tend to exacerbate the dualism in the already highly divided agricultural sector. While, at the end of apartheid in 1994, South Africa had a total of 60,000 commercial farms, today it has only about 34,000 left, of which a large majority (or at least a large proportion of its production) is directly controlled by large companies (Ducastel and Anseeuw 2011). Similar tendencies of concentration have been observed in other African countries, such as Mozambique (Boche and Anseeuw 2013) or in Asia, as described in the Olam case.

[4]This attack was spearheaded by a short-selling 'research' company with the ominous name of 'Muddy Waters', which had already earned a reputation as a killer of mid-sized Asian corporations. This financial company bet against the value of Olam shares and published reports criticizing the corporation's management, recommending that investors dump Olam shares. As a consequence, Olam's market value collapsed. Olam was headed towards bankruptcy despite mounting a vigorous defence but emergency recapitalization with the help of Temasek, the Singaporean sovereign fund, saved the corporation. Today Olam is once again considered a safe investment.

25.3.2 Corporatization of Agriculture and Takeover of the Traditional Agricultural Sector

The two studied cases show that new types of actors are becoming involved in agriculture. Originating from industrial, agribusiness or financial sectors and intervening as entrepreneurs, investors, or even as pure speculators, different types of suppliers of capital are increasingly making their presence felt in the agricultural sector. In addition to their financing role, these actors bring with them business rationales, modes of actions, and regulations that originate from other sectors. The agricultural sector thus follows a process of industrialization, or, more accurately, one of 'corporatization'. Financialization is therefore difficult to separate from the rapid growth of the firms. When firms perform well and grow, they create or acquire subsidiaries, and cover their need for new capital through different financial vehicles and investment strategies. These developments tend to dominate the traditional institutional structurings and socio-economic principles of agriculture (Palley 2007). Indeed, this process is redefining the very boundaries of the agricultural sector, with significant implications for sustainable development.

An important aspect of these changes concerns speculation. The last couple of years have seen an unprecedented increase in agricultural speculation. While it has in the past been limited to an internal and short-term phenomenon, it has now rapidly evolved towards long-term strategies, led by actors internal and external to the sector. This trend, similar to speculative mechanisms in other sectors, such as real estate in particular, can have significant consequences.

In addition, corporatization goes beyond mere industrialization and pertains to a transformation of the production structures and their interactions, which tend to be structured by and integrated within corporate entities. As shown by the two case studies, the farms become territorial enclaves that seem less and less governed by the agricultural sector, to the benefit of these agro-financial and global spaces (Ducastel and Anseeuw 2013). Corporatization takes the form of specific practices and organizational forms to attain objectives of financial profitability (payroll management, remote monitoring of agricultural production, etc), thus transforming the forms of the productive organization to the benefit of institutional investors. Financialization and corporatization also call into question the model of the independent farmer, which is pivotal to the agricultural sectoral architecture. Indeed, the farmer becomes an employee of these firms, thus losing his farmer status.

25.4 Conclusion

The global agricultural sector is presently undergoing a profound restructuring. New actors, bringing with them external references, practices, and experiences have entered the sector. Their interactions and inputs impact on the sector's 'traditional' modes of action, investment, and production. The 'agricultural exception', as debated upon during negotiations for the inclusion of the sector within the ambit of the WTO, can probably be considered buried. A new agricultural development paradigm is emerging (de Janvry 2010), manifesting itself both at the national and international levels.

The issue is all the more relevant since a large majority of farmers in Africa and Asia remain part of small-scale family farms. Financialization and corporatization have to do with large-scale enterprises, agribusinesses, and projects. Hence financialization concerns a significant proportion of agricultural investment and will therefore have major repercussions on the entire sector. Its consequences are directly linked to the transformation of the agrarian society of the countries concerned, of the agricultural sector and its value chains, through the corporatization and the concentration of the sector. Even though the resulting agrarian transformations are as yet not broad-based, their control by a few has far-ranging implications for agriculture. The on-going restructurings are designed for the long term as they are based on a development paradigm being openly promoted by dominant actors, a paradigm that relies on corporate agriculture. Being also supported by the different governments in the regions concerned, this paradigm is being spread by the public development agencies in the form of foreign direct investment in agriculture and support for large investment projects and corporate land acquisitions (Anseeuw et al. 2012).

References

Aglietta M, Rigot S (2009) Crise et rénovation de la finance. Odile Jacob, Paris, 362 p

Ahmed G, Hamrick D, Gereffi G (2014) Shifting governance structures in the wheat value chain: implications for food security in the Middle East and North Africa. Duke Working Papers. Durham, 28 p

Anonymous, 2015. World economic outlook database: balance of payments statistics. World Economic Outlook, Washington, DC. Retrieved August 25, 2016, http://www.imf.org/external/datamapper/index.php

Anseeuw W, Boche M, Breu T, Giger M, Lay J, Messerli P, Nolte K (2012) Transnational land deals for agriculture in the Global South. Analytical report based on the land matrix database. Research Report, Bern/Montpellier/Hamburg, CDE/Cirad/Giga, 64 p

Boche M, Anseeuw W (2013) Unraveling land grabbing: different models of large-scale land acquisition in Southern Africa. LDPI working paper, 46, 32 p

Borras SJM, Franco JC (2012) Global land grabbing and trajectories of agrarian change: a preliminary analysis. J Agrar Chang 12(1):34–59

Bureau Van Dijk (2015) Osiris database. Bureau Van Dijk, Amsterdam

Campbell J (2011) Investing and spending: the Twin Challenges of University Endowment Management. Conference Paper. Boston, Harvard, Forum for the Future of Higher Education 2011, 8 June, 32 p

Chu JM (2013) Creating a Zambian Breadbasket: "Land Grabs" and Foreign Investments in Agriculture in Mkushi District. LDPI working paper, 33, 27 p

de Janvry A (2010) Agriculture for development: new paradigm and options for success. Agric Econ 41:17–36

Ducastel A, Anseeuw W (2011) La libéralisation agricole post-apartheid en Afrique du Sud: nouveaux modèles de production et d'investissement. Afr Confid 237:57–70

Ducastel A, Anseeuw W (2013) Situating investment funds in agriculture. Farm Policy Journal 10 (3):35–44

Freeman RB (2010) It's financialization! Int Labour Rev 149(2):163–183. doi:10.1111/j.1564-913x.2010.00082.x Retrieved 26 August 2016

Gibbon P (2014) Trading houses during and since the great commodity boom: financialization, productivization or...? DIIS working paper 2014 12, Copenhagen, DIIS, 34 p

Heumesser C, Staritz C (2013) Financialization and the microstructure of commodity markets: a qualitative investigation of trading strategies of financial investors and commercial traders, Working Paper 4. OFSE, Vienne, 40 p

Huggin C (2011) Commercial pressure on land in its historical perspective, Research Report, Commercial Pressures on Land. ILC, Rome 66 p

Krippner GR (2005) The financialization of the American economy. Soc Econ Rev 3(2):173–208

Murphy S, Burch D, Clapp J (2012) Cereal secrets: the world's largest grain traders and global agriculture. Oxfam International, Hoboken 80 p

Newman SA (2009) Financialization and changes in the social relations along commodity chains: the case of coffee. Rev Radic Political Econ 41(4):539–559

Oestreicher A, Koch R, Vorndamme D, Hohls S (2013) ASSERT-assessing the effects of reforms in taxation. A micro-simulation approach. Göttingen (Germany), Georg-August-Universität Göttingen Working Papers, 72 p

Olam (2013) Building relevant scale: delivering the Olam strategy. Powerpoint presentation. Available on Olam's website: www.olamnet.com

Palley T (2007) Financialization: what it is and why it matters, Working Paper No. 525. The Levy Economics Institute, Annandale-on-Hudson 31 p

Roda JM (2005) Réseaux d'entreprises et stratégies industrielles dans les forêts tropicales/Inter-firm networks and industrial strategies in tropical rainforests, Doctoral thesis in geography. Paris Diderot Univeristy, Paris, 393 p

Roda JM (2006) Diagnostic Économique Et Fiscal De La Filière Bois Au Cameroun. Cirad, Research Report, Yaoundé

Roda JM (2010) On the nature of intergenerational and social networks in the African forest sector: the case of Chinese, Lebanese, Indian and Italian business networks. In: German LA, Karsenty A, Tiani AM (eds) Governing Africa's forests in a globalized world. Routledge, London, pp 335–353

Roda JM, Kamaruddin N, Palhiarim TR (2015) Deciphering corporate governance and environmental commitments among southeast Asian transnationals: uptake of sustainability certification. Forests 6(5):1454–1475

Soros G (2003) The alchemy of finance. Wiley

UNCTAD (2009) World investment report: transnational corporations, agricultural production and development. UNCTAD, New York/Geneva. Retrieved August 23 2016, http://unctad.org/en/docs/wir2009_en.pdf

Williams JW (2014) Feeding finance: a critical account of the shifting relationships between finance, food and farming. Econ Soc 43(3):401–431

Chapter 26
General Conclusion and New Research Perspectives

Estelle Biénabe, Patrick Caron, Alain Rival, and Denis Loeillet

The renewed interest shown in agriculture by the international community and national governments since the 2008 crisis has firmly placed agri-chain approaches at the heart of development issues and debates. *Filières* or agri-chains constitute a framework for regulating relationships between various stakeholders and help generate private investment (Chap. 4). Downstream industries involved in processing, international trading, and distribution have adopted a strategy of investing in tropical agri-chains to secure and diversify their supplies and, at the same time, control the quality of products and the risks to their reputations. Standards, as well as certification and traceability mechanisms, are the tools used in this strategy. Since the limitations and shortcomings of the policies of liberalization are now clear and given the repeated failures of markets (of credit, inputs, etc.), the actors of development believe that supporting agri-chains and promoting public-private institutional arrangements to guarantee the interests of the various stakeholders should once again become priorities. The challenge is to improve the integration of producers into markets and to ensure a better distribution of added value among all the actors on the basis of a common development strategy. The

E. Biénabe (✉)
CIRAD – UMR Innovation, F-34398 Montpellier, France
e-mail: estelle.bienabe@cirad.fr

P. Caron
CIRAD – DG-DRS, Montpellier, France
e-mail: patrick.caron@cirad.fr

A. Rival
CIRAD, DGD-RS, Jakarta, Indonesia
e-mail: alain.rival@cirad.fr

D. Loeillet
CIRAD – UPR Geco, F-34398 Montpellier, France
e-mail: denis.loeillet@cirad.fr

E. Biénabe et al. (eds.), *Sustainable Development and Tropical Agri-chains*,
DOI 10.1007/978-94-024-1016-7_26

strengthening of collective action is one of the cornerstones of this objective (Chap. 6).

The challenges of sustainable development compel us to define new arenas and new forms of action – which are now multiplying through standards and other regulatory instruments – and to come up with innovative practices and novel ways of deriving value. Exploring the links between transformations of tropical agri-chains and sustainable development allows us to conclude this book by listing the lessons learnt about the ability of these agri-chains to act as vectors of development, spaces of regulation, frameworks for innovation, and objects of assessment. A number of new fundamental transformations have begun to take place, such as the emergence of forms of capitalism in the South, new investment strategies in agriculture, and the financialization of this sector. The proliferation of these dynamics calls for further exploration and the identification of research perspectives that can help in this endeavour.

26.1 Agri-Chains as Spaces of Regulation of Sustainable Development?

26.1.1 Agri-Chains and Development Models: Highly Topical Issues

The historical review of a particular period and a specific type of region in Chap. 3 helps explain how tropical agri-chains were formed and structured, and places the discussion on the role of agri-chains as spaces of regulation for sustainable development in a long historical trajectory. The historical processes of specializing on individual commodities structured the production and marketing by distinct agri-chains. Consequently, agricultural research in the twentieth century too was organized for the most part according to separate commodities. This historical background reflects the political vision of accompanying increases in production in a context of a technological revolution and of the structuring of markets and international trade that went together with the increase in the planet's population from one to seven billion in just two centuries. It shows how agri-chains – a term that encompasses the different concepts of value chains, supply chains, and *filière* – are a particular form of development based on specializations and the division of labour between different operators.

However, agricultural specialization and a sole focus on improving producer incomes are not always sufficient or even desirable given the necessity of a diverse diet and the importance of the sanitary environment for vulnerable populations (Chap. 6).

The current interest in sustainability once again raises the question of the trade-offs between the efficiency of specialization and the ability to manage risk through diversification. These reflections remind us of the importance of considering – including while undertaking research – alternative forms of support for development and of analyzing how conducive to sustainable development are the different forms of organization. The recent interest that actors of development have shown in reframing agri-chain approaches by incorporating them in new public-private partnerships is proof of the relevance of the issues involved. Furthermore, the growing diversity of strategies for investing in agri-chains and the financialization of agriculture are new factors to consider (Chap. 25). This diversity reflects the changes in international aid and forms of capitalism designed to combine financial returns with social and environmental impacts: philanthropic capitalism, social entrepreneurship, responsible capitalism, corporate foundations, impact investment strategies, etc.

26.1.2 New Forms of Regulation Associated with Agri-Chains

The proliferation of standards and the mechanisms associated with them is a key factor in the evolving links between agri-chains and sustainable development. As concerns about sustainability have grown, so have the criticisms of the growth of tropical agri-chains because of their negative environmental and social impacts. In response, standards have been used to strengthen the ability of agri-chains to become arenas for action and for regulating behaviour. These standards are defined by different types of actors (NGOs, private actors, States, etc.) and differ widely, both in purpose and in scope: fair trade, organic or 'reasoned' agriculture, zero deforestation, sustainability standards, and other commitments made by agro-industries. A number of them are associated with labels that are promoted to and valued by consumers, allowing producers to be better paid for their efforts in some cases. Others are part of social and environmental responsibility (CSR) strategies of individual companies or formalize collective voluntary commitments (zero deforestation approaches, sustainability standards). Their rise is linked to the growing role of civil society in the emergence of new forms of regulation, especially that of NGOs, which are both whistleblowers and partners of multinationals and coalition of actors striving to reorient activities at the agri-chain level and change markets. International organizations such as the World Bank, OECD, and FAO also contribute to this standardization by producing, beyond health and food safety standards, lists of core principles, and guides for adopting responsible practices.[1]

The standards and the way of assessing and ensuring compliance (certification, indicators, thresholds, etc.) with them are established through negotiation by a growing number of categories of stakeholders: companies, NGOs, scientists, governments, and producers. These private voluntary standards are evolving in tandem

[1] For example, 'OECD-FAO Guidance for Responsible Agricultural Supply Chains' (http://www.oecd.org/daf/inv/investment-policy/rbc-agriculture-supply-chains.htm, retrieved 9 June 2016).

with public regulations. Some of them complement the action of public authorities by mandating compliance with regulations such as on deforestation (Chap. 22). Sometimes these voluntary standards are made official and compulsory by national governments.[2] In other cases, government standards are less strict. One of the proposals advanced in the book is to combine the implementation of these standards with national mechanisms for payment for environmental services controlled by the States. In all cases, their scope and their consequences remain the subject of much current debate.

26.1.3 The Limits of Regulating Sustainable Development Through Agri-Chains

This book has explored the effectiveness of the standardization mechanisms put in place in agri-chains in producing regulations for a more sustainable development and has highlighted their limitations. These mechanisms are being promoted as an effective way of including effects of activities – called 'externalities' – not previously considered by agri-chain operators in their transactions. In doing so, they themselves generate various unregulated and indirect environmental and social effects. They also displace some problems outside the scope of action of the agri-chains: population migration and increased risk of conflict due to the local attraction of the activities of agro-industries, shifting of deforestation outside monitored areas, etc. Moreover, the very nature of mechanisms used to develop sustainability standards excludes a number of concerns of the local actors from consideration: doubts about some production models, inequality in access to resources, etc. Despite the promotion of participatory and inclusive approaches, asymmetries in the ability to act remain key concerns and complicate the issue of control of agri-chains through standards.

In addition to analyzing mechanisms and proposing improvements, this book examines the limitations of acting via the agri-chains for a more sustainable development. The expansion of markets through long-distance trade and the growth of tropical agri-chains lead to an externalization of the social and environmental effects and costs, which are obscured due to distance between locations of production and consumption. This distance can be geographical, of course, but also strategic and cultural (Princen 1997). By examining the strategies implemented by multinational corporations for the management of sustainable development, Godard and Hommel (2005) show, like Princen, how market competition puts pressure on operators not to internalize all the costs of their activities. These reflections show the limitations of mechanisms intended to expose the environmental and social impacts within agri-chains involved in long-distance trade, whose

[2] For example, French companies with more than 500 employees are required by law to submit an annual report on their corporate social and environmental responsibility (CSR) activities.

competitiveness is based partly on the fact that some costs are kept hidden. One proposal is to reduce the distance between production and consumption, by reducing asymmetries between producers and the downstream sections of the agri-chains, or through a relocation of marketing channels (Chap. 23). These issues also refer to the debate on the reliance on the market or the State for internalizing externalities and managing public goods, and the necessary linkages between private and public regulation.

26.2 Agri-Chains and Territories

Current initiatives for increased social and environmental sustainability in agri-chains will not be sufficient to manage and preserve the commons and environmental public goods and to reduce inequalities between actors. Given the limitations of these initiatives in taking certain social and environmental issues into account, territorial approaches have emerged as a complementary way forward.

26.2.1 *Agri-Chains in the Territories and the Management of Local Resources*

The need to ensure not only the sustainability of activities in the agri-chains but also the sustainability of territories that these agri-chains are transforming is being widely debated in this book. Hence, it addresses, first, the role of agri-chains in producing value from territorial resources and as a driver for territorial development (Chaps. 4, 5, and 22) and, second, the strain that agri-chains put on resources (water and land in particular). The territories sometimes become arenas of competition between different agri-chains. Several authors warn against managing a territory's resources exclusively in a manner to benefit the economic interests of agri-chains that are present. On the other hand, Fusillier and Lejars (Box 5.1) show, in a context of sustainable water management and of pronounced scarcity, the importance of involving the actors of agri-chains downstream of production because they play a key role in the health of the economic activities in the territories.

The growth of the circular economy also calls for an examination of the links between the consumption of resources as part of the agri-chains' activities and the management of these resources at the territorial level. Thus, the traditional segmentation into agri-chains and the varying capacity of actors to implement collective regulations and to manage a territory's resources lead to three types of processes (Chap. 16):

- the creation or strengthening of systems of local intra- or inter-farm exchanges of products;

- the emergence of new agri-chains through the creation of new economic relationships between actors;
- the territorial anchoring of agri-chains, with connections between agri-chains taking place through dialogue between actors at the level of a territory.

These distinct processes rely on different tools and methods of analysis, foresight, and support. They differ in their technical dimensions as well as in terms of organization.

26.2.2 The Territory as a Regulatory Space Complementary to the Agri-Chain

The need to add a territorial dimension to the goal of sustainable development has led to a search for and reflections on the complementarities between territories and agri-chains. As Godard points out (2005, p. 23), it is a matter of 'articulating common action at the highest possible territorial level to align expectations, set directions and ground rules, and the decentralization of individualizable actions in order to harness the innovation capacities and detailed knowledge of situations that characterize local actors.' More broadly, the issue concerns the link between the local and the global, between social constructions and local policies, and global regimes. Both territories and agri-chains connect, each in its own way, actors and processes acting at these different scales. To avoid, on the one hand, the reliance on the very local and, on the other, the exclusive control by international mechanisms and a centralizing rationale, this interaction between agri-chain and territory constitutes a crucial space for action. Combining agri-chain approaches with territorial ones helps in taking environmental and social issues into account at a much broader scope.

Given the manifest limitations of market regulation within agri-chains, various authors discuss the role of the State. For example, local professional networks, structured around distinct agri-chains, and national public systems complement each other to ensure a proper monitoring and control of diseases and anticipation of health risks (Box 6.1). The territory, as a link between a framework of public intervention and collective action (Caron 2011), appears as a key regulatory space to manage rural and agricultural land and access to resources, complementary to that of agri-chains, as emphasized in Gaël Giraud's Foreword. The capacity for regulation and control by local actors, including by local authorities, of trajectories of development depend on how deeply anchored are the agri-chain's actors (producers and agro-industries) to the territory (Chap. 5). One proposal is to promote territorial projects in order to encourage this anchoring. These projects are based on dialogue and agreements between local communities, agro-industries, and public authorities.

Various territorial development approaches are proposed in this book. They bring together public stakeholders and private-sector actors of the agri-chains in territorial projects that embed the local into the global. These projects have to be created on the basis of compromises arrived at through a local dialogue and by taking external orientations and regulations into account (Chap. 22). As discussed in Chap. 24 in relation to payments for environmental services, these orientations can be formalized through mechanisms implemented by the State at the national level and by taking the strategies and commitments made at the international level into account. Research should also be undertaken on systems of indicators suitable for monitoring projects in these territories.

26.3 Agri-Chains as Spaces of Innovation for Sustainable Development

26.3.1 The Bioeconomy and New Ways of Producing Value and Using Resources: New Innovation Perspectives

As presented in the third part of this book, another major source of transformations of agri-chains that pertain to sustainability issues is the explosion of new ways of producing value and uses from biomass, promoted as part of the bioeconomy and a necessary energy transition. A large number of innovations have been adopted for improved material and energy efficiency in tropical agri-chains. Bright new prospects are opening up, due mainly to the circular economy, which uses the concepts of – and knowledge produced by – industrial ecology to propose intersectoral approaches (Chap. 16). Relationships between different agri-chains or with other activities within a territory are multiplying in order to transform and recycle waste and agro-industrial (or urban) effluents into resources for agricultural systems and, in this way, limit the recourse to external inputs. These innovations sometimes even lead to the birth of new – and economically viable – agri-chains for trade in these resources. Even when recycling and the use of by-products of an activity have been going on for a long time, such as between crop cultivation and animal husbandry, gains of efficiency are still possible. Thus, the increasing scarcity of resources is leading today to innovative inter agri-chain connections that mobilize and leverage technical solutions and economic and institutional regulations.

Moreover, bioenergy is now a major component of many agricultural and forestry chains in the South. Energy is no longer a by-product; it has instead become a co-product, whose importance is likely to increase over the next few years. It may even become a product in its own right. Changes and radical breaks with the past in tropical agri-chains thus lead to a remodelling of cropping systems depending on the purpose of production: food or non-food. Green chemistry also helps produce value from multiple products and to propose options for the future, in line with policy adopted to ensure energy and food transitions. Goebel et al. (Chap. 15) show the need to conduct research to better integrate these new uses in the objectives of cropping systems, thus allowing the design of suitable itineraries. However, current research investment remains low in the field of bioenergy. In the North, as in the South, public policy on bioenergy is still struggling to define clear frameworks for intervention and is thus currently encouraging modes of production which confine bioenergy to a secondary production (Chap. 14).

26.3.2 Embedded and Interlinked Technical, Political, and Organizational Aspects

Technical choices and innovation cannot be considered in isolation of organizational and institutional factors. It will be a delusion to believe otherwise. Technical and policy choices cannot be made independently; they have to be coordinated. The desire for sustainable development reinforces the need to make technical choices in view of policy objectives that are themselves currently undergoing profound changes and to make new trade-offs and compromises. The exclusive focus on a logic of supply through technical and economic support to agri-chain operators has become outdated. While uninterrupted and reliable supply remains a key objective, even more so in a changing and uncertain context, the methods of achieving this goal are no longer those of the past. Zakhia-Rozis et al. (Chap. 13) propose a global perspective for a research agenda in the field of the biotechnical sciences to confront the challenges of food availability. These authors emphasize the combinations between qualitative and quantitative improvement in production, new relationships between actors within the agri-chains and territories, and a reduction in losses and waste.

The strategic choices to be made in the domain of bioenergy are apt illustrations of the complexity of trade-offs and linkages between technical and political factors, as revealed in the report by the High Level Panel of Experts on Food Security and Nutrition of the Committee on World Food Security (HLPE 2013). Indeed, two equations require resolving:

- that of the energy transition and the opportunity of substituting biomass for fossil fuels;
- that of food security and the possible competition between food and non-food uses of agricultural production, land, and co-products.

This second equation itself involves two underlying trade-offs concerning technical options, explored in Chap. 15:

- between the use of dedicated crops and the use of by-products;
- between possible alternative uses of the land in terms of their agronomic characteristics.

It is interesting to note that innovative technical solutions can reduce the importance of the trade-offs to be made between objectives and between uses, such as processes that enable a more efficient and complete utilization of the plant. Indeed, this is a key aspect of the bioeconomy.

Throughout the book, we have noted the unavoidable embedding and interlinking of technical and organizational aspects at all stages of the innovation process. The growing concern for sustainability has enriched technical and economic criteria that orient technological trajectories. Thus, Chap. 12 illustrates how the design of innovative processing methods and support for different forms of organization of artisanal or industrial processing require us to consider technical and organizational aspects together, regardless of the objectives (environmental efficiency, job creation, food security, etc.).

26.4 New Challenges for Evaluation

Given the urgent issues that have to be resolved and the intellectual and institutional renewal associated with evaluation requirements, assessment mechanisms are increasingly being mobilized to address issues of sustainability. For proof, one has only to look at the emphasis placed on the indicators and monitoring framework associated with the 169 targets that constitute the 17 Sustainable Development Goals of the UN and the global partnership to achieve them formed through a resolution of the UN General Assembly in September 2015. The objectives of sustainable development require a fundamental re-examination of how the performance of agricultural activities is measured. It is no longer enough – as the concept of multifunctionality already pointed out 20 years ago – to evaluate these performances solely in terms of production volumes, an objective that assumed so much importance in the nineteenth and twentieth centuries that it overshadowed all other metrics of performance. Even though job creation; the fight against poverty and food insecurity; social cohesion; resilience of social groups; political stability; mitigation of the effects of climate change; cultivation of biodiversity; prevention of pollution and desertification; etc. are now accepted as necessary and desirable outcomes of agriculture, we are far from being able to measure them reliably. To complicate matters further, the evaluation of these objectives is interlinked. Furthermore, even if they concern a measurable process or have an impact on a global scale (for example, the mitigation of the effects of climate change), these various objectives retain their specificity. Indeed, the structural characteristics of the agricultural sector and the policy objectives assigned to it still differ from one local production area to the next, from one country to its neighbour. The process of evaluation is therefore especially complex and assessments only acquire meaning through specific institutional mechanisms. It requires the linking of criteria, indicators, and standards resulting from, on the one hand, political trade-offs required by the local situation, and, on the other hand, a globally negotiated framework. The challenges of adaptation to and mitigation of the effects of climate change and the measurement of agricultural performance illustrate this tension perfectly.

The evaluation of the performance of agri-chains and of all that happens within them is no exception to these intellectual and operational exigencies. This is especially so since agri-chains are extremely fertile spaces of innovation, and because, with rare exceptions, they connect actors to global and local processes. They are an essential vehicle for impacts at scale, necessary given the challenges faced and the importance policymakers and donor agencies assign to them. The aim is not the widespread dissemination of a locally proven solution – which appears most often doomed to failure given the non-reproducible nature of the context and the resources invested to ensure success – but to design mechanisms for experimentation, learning, and supporting innovation that enable transformations by acting in a complementary manner at several organizational levels.

Different methods of evaluation are presented in this book's fourth part for measuring a wide variety of environmental and social impacts of production systems and agri-chains. They are used to estimate *ex ante* the impacts of different scenarios, to assess *ex post* the outcomes of actions already taken, or to control and orient trajectories in real time through *in itinere* monitoring of systems of indicators (Chap. 20). The choice of criteria and indicators is crucial to being able to make the correct trade-offs. Scientific methods and knowledge provide us with the tools and information to inform and guide these choices, and establish sustainability thresholds. In this regard, life cycle assessments, designed to analyze the impacts all along a supply chain, play a special role. The illustrations presented show how important it is to use new time steps and spatial scales in order to reassess performance, reconsider the design of innovations, and take criteria that could reflect discrepancies into account. To be able to predict the long-term consequences of decisions and actions and take global environmental problems such as climate change into account, we have to consider unusual and unexpected temporal and spatial scales as well as interdependencies within production systems and supply chains. This issue of the scale of evaluation goes hand in hand with that of representation of the system under evaluation and the ability to take inter-system transfers – for example, of pollution – into account. In this sense, the use of methods that combine life cycle assessments undertaken at the scale of the supply chain with more territorial approaches appear to be a promising option. Similarly, to ensure that ecological imperatives do not overshadow the social dimension during the construction of sustainable development indicators at the supply chain level (Feschet and Garrabé 2013), it is important to include this dimension explicitly in assessment methods.

Furthermore, the objectification of the assessment methods remains open to debate (Chap. 19). The criteria and indicators used are based ultimately on political choices, sometimes negotiated, sometimes not. Their choice depends on decision-making processes, the nature of the choices to be made, and assessment mechanisms. It is a matter of finding a balance between normative sustainability (externally defined indicators and thresholds) and systemic or procedural sustainability (negotiations, choices, trade-offs, compromises, negotiated thresholds) (Box 18.1). This raises the questions of how the actors organize themselves to negotiate and revise these criteria and of problems of asymmetry in the ability to implement original mechanisms and produce and use indicators.

The development of tools and indicators should lead to a better assessment of the environmental and social functions fulfilled by agricultural systems and chains, and to identify others and link them to specific practices. Characterizing practices and their effects in this manner opens the way for the necessary regulations by relying on new mechanisms that implement novel types of coordination between actors. Recognition of the services that agriculture provides is an essential step in their promotion and, consequently, the transformations of the agri-chains concerned. It opens up new perspectives, whether to provide remuneration through payments for environmental services, produce or develop standards, or support the organization of new agri-chains (Chap. 16).

26.5 Challenges of Knowledge Creation: Research Approaches and Agenda

Giving thought to the links between transformations of agri-chains and sustainable development strengthens the relevance of research to act at political and technical levels for supporting:

- local, national, and international mechanisms and vehicles at the level of agri-chains and territories in defining and making trade-offs, and the policy frameworks and instruments necessary to implement them;
- the different operators of agri-chains, individually or collectively, in making appropriate technical choices and broadening the range of possibilities.

As this book repeatedly notes, the challenges of sustainable development require us to revisit the design of innovation systems and processes associated with agri-chains. This has major implications for research, which is evolving to meet the diversity of the criteria, actors, and situations to account for, with a growing awareness that research activities in themselves are rarely neutral. It is a matter of changing attitudes and practices, and of reconsidering objects and conceptual categories that form the basis for action. The second and third parts of the book discuss the transformation of practices, research and partnerships mechanisms for innovation, and of research themes. The lessons learnt and the feedback obtained from the different aspects that constitute the development of products presented in this book (varietal improvement, cropping systems, artisanal and industrial processing) inform the reflections on the positioning and modalities for agricultural research and its future agenda.

26.5.1 *Contextualizing Knowledge In Order to Innovate Sustainably*

The linkages between the technical and organizational aspects have to be addressed differently depending on particular economic and social contexts, with each agri-chain raising specific issues of sustainable development. For example, the issues of energy transitions and food security are addressed in different ways in different countries, regions, and agri-chains, even if they also make sense at a global scale (Chap. 14).

Developing and evaluating technical options requires an accurate diagnosis of the context. For example, the recourse to selected genetic resources depending on the end uses (Chaps. 11 and 15) helps in designing cropping systems and processing methods differentiated according to political contexts and trade-offs. The relationship between the genotype and the environment must be formed in an integrated manner, not only in the design and choice of crop itineraries and processing methods, but also by linking these choices between them and between different agri-chains present in a same territory, as shown by the seasonal complementarity between sorghum and sugarcane (Chap. 15).

26.5.2 *Positioning of and Approaches to Research*

Even though partnerships between the research community and the various actors of agri-chains have existed for a long time, the desire to incorporate sustainability issues has led to a renewal of the manner in which innovation is engineered and partnerships are formed. The emphasis is thus on defining research topics and processes in multi-stakeholder platforms in order to ensure greater relevance of knowledge production and sustainable capacity building (Chaps. 8 and 9). It is no longer a matter of proposing turnkey solutions but rather of supporting agroecological, climatic, energy, health, and urban transitions while taking into account the diversity of production and processing systems (Chap. 12). Participatory approaches rely on several different types of tools: modelling, production of scenarios, etc.

The need to take environmental and social issues into account underscores the importance of designing innovations in conjunction with an assessment of their expected effects and impacts. Trajectories of adaptation of systems are thus constructed and fine-tuned in a series of design and evaluation cycles by embracing a learning perspective (Chap. 10). As Godard and Hubert (2002) point out, these two dimensions often pertain to different scientific skills, thus constituting a significant challenge for the research community.

26.5.3 Integration of Knowledge

Meeting the ambitions of sustainable development is a major challenge of integration for the research community. Revising our notion of performance, reconciling different scales, understanding better the links between the technical and political aspects, etc. require us to go beyond disciplinary divisions. Given the narrow specializations that dominate the sciences, the integration of knowledge produced by different disciplines on distinct objects remains a major challenge. In such a context, agri-chain approaches are a veritable boon. Designing techniques and their contributions to innovation to accompany transformations of agri-chains requires cross-disciplinary contributions from the biological, technical, and social sciences. Considering the agri-chain in its entirety means combining genetics, agronomy, process engineering, analysis of actor systems and of instruments of regulation, evaluation, etc. As emphasized in this book's Introduction, the agri-chain or *filière* has long been a major arena and object for the integration for targeted research. This agri-chain approach, whose developments this book has presented, has proven its worth from the standpoint of integration of research into innovation processes.

From the perspective of the social sciences, different types of agri-chain analysis have also shown themselves to be important tools for understanding the insertion of actors into economic processes: insertion of producers into markets, quality management, governance of standards and power relationships, distribution of added value, etc. However, other dimensions at the household level – such as women's labour, access to a varied diet, the health environment, exposure to risks, etc. – are not sufficiently taken into consideration, especially from the perspective of food and nutrition security (Chap. 7). Conversely, 'sustainable livelihoods' approaches, which focus on the diversity of household strategies to ensure the livelihoods of families and their ability to access and combine different resources, do deal with these dimensions. But as Scoones (2009) emphasizes, these latter approaches, rooted in the local environment and context, fail to sufficiently take global transformations into account. Combining them with agri-chain analyses, which connect the local to the global, is an interesting way of understanding the processes of change between scales.

Similarly, links are yet to be constructed from the viewpoint of environmental issues between agri-chain analyses, which historically have given short shrift to them, and the very many approaches which focus solely on them: whether from the viewpoint of sustainable management of resources, which has resulted in an extensive literature but with little reference to agri-chains (Bolwig et al. 2010); or that of global environmental changes, with the emergence of sustainability studies (Steffen et al. 2015; Ericksen 2008).

The transformations taking place in production and processing systems in order to ensure a more sustainable development may need to bridge sectoral divides, and therefore divisions by agri-chains specialized each on a single product. The renewed interest in ecological intensification in pursuit of more integrated production systems (agroforestry, integration between crop cultivation and livestock rearing, etc.) lead us to rethink the connections with downstream segments and markets. The circular economy is also shaking up the modes of organization by agri-chain. Notions of sustainability can also differ and thus call into question our development models depending on particular perspectives (Chaps. 17 and 23).

To combine the above-mentioned approaches even more closely, different avenues are being investigated or need investigating: enrichment of life cycle assessments by incorporating territorial and social considerations; approaches focused on sustainable food systems, whether at the level of territories – localized agrifood systems (LAS) (Muchnik et al. 2007, 2008) and foodsheds (Kloppenburg et al. 1996; Peters et al. 2009) – or at wider scales where global environmental changes and risks become relevant (Ericksen et al. 2010; International Panel of Experts on Sustainable Food Systems 2015). In all these different perspectives, the linkages of processes structured within agri-chains, on the one hand, and territories, on the other, is a major challenge. It is from this that our notions and visions of the role of agriculture in sustainable development will be redefined. It is on this basis and through the interfacing of the communities of scientists, policymakers, and agricultural practitioners that it will be possible to reconcile global and local forms of management of commons and public goods.

References

Bolwig S, Ponte S, du Toit A, Riisgaard L, Halberg N (2010) Integrating poverty and environmental concerns into value-chain analysis: a conceptual framework. Dev Policy Rev 28 (2):173–194

Caron P (2011) Réconcilier agricultures et sociétés : dévoiler les territoires et repenser les limites, 3 vols, habilitation à diriger des recherches. Université Paris Ouest Nanterre La Défense, Nanterre. 39 p., 254 p., 242 p

Ericksen P (2008) Conceptualizing food systems for global environmental change research. Glob Environ Chang 18:234–245

Ericksen P, Stewat B, Dixon J, Barling D, Loring P, Anderson M, Ingram J (2010) The value of a food system approach. In: Ingram J, Ericksen P, Liverman D (eds) Food security and global environmental change. New York, Earthscan/Taylor and Francis, pp 25–45

Feschet P, Garrabé M (2013) ACV sociale et développement durable. In: Macombe C (ed) ACV sociales. Effets socio-économiques dans les chaînes de valeurs. Cirad, Paris, pp 69–83

Godard O (2005) Le développement-durable, une chimère, une mystification ? Mouvements 41:14–23

Godard O, Hommel T (2005) Les multinationales, un enjeu stratégique pour l'environnement et le développement durable ? Revue internationale et stratégique 60(4):101–112

Godard O, Hubert B (2002) Le développement durable et la recherche scientifique à l'Inra. Rapport intermédiaire de mission pour la directrice générale de l'Inra. décembre 2002. Versailles, Inra, 58 p

HLPE (2013) Biofuels and food security. A report by the High Level Panel of Experts on Food Security and Nutrition of the Committee on World Food Security. Committee on World Food Security, Rome 131 p

International Panel of Experts on Sustainable Food Systems (2015) The new science of sustainable food systems. Ipes Food, 22 p

Kloppenburg JJ, Hendrickson J, Stevenson GW (1996) Coming in to the foodshed. Agric Hum Values 13(3):33–42
Muchnik J, Requier-Desjardins D, Sautier D, Touzard JM (2007) Les systèmes agroalimentaires localisés (SYAL): introduction. Econ Soc 29:1465–1484
Muchnik J, Sanz Canada J, Torres Salcido G (2008) Systèmes agroalimentaires localisés: état des recherches et perspectives. Cahiers agricultures 17(6):513–519
Peters CJ, Bills NL, Wilkins JL, Fick GW (2009) Foodshed analysis and its relevance to sustainability. Renewable Agric Food Syst 24(1):1–7
Princen T (1997) The shading and distancing of commerce: when internalization is not enough. Ecol Econ 20(3):235–253
Scoones I (2009) Livelihoods perspectives and rural development. J Peasant Stud 36(1):171–196
Steffen W, Richardson K, Rockström J, Cornell SE, Fetzer I, Bennett EM, Biggs R, Carpenter SR, de Vries W, de Wit C, Folke C, Gerten D, Heinke J, Mace GM, Persson LM, Ramanathan B, Reyers B, Sörlin S (2015) Planetary boundaries: guiding human development on a changing planet. Science 347(6223)

MIX
Papier aus verantwortungsvollen Quellen
Paper from responsible sources
FSC® C105338

If you have any concerns about our products,
you can contact us on
ProductSafety@springernature.com

In case Publisher is established outside the EU,
the EU authorized representative is:
Springer Nature Customer Service Center GmbH
Europaplatz 3, 69115 Heidelberg, Germany

Printed by Libri Plureos GmbH
in Hamburg, Germany